Alexander Wilson

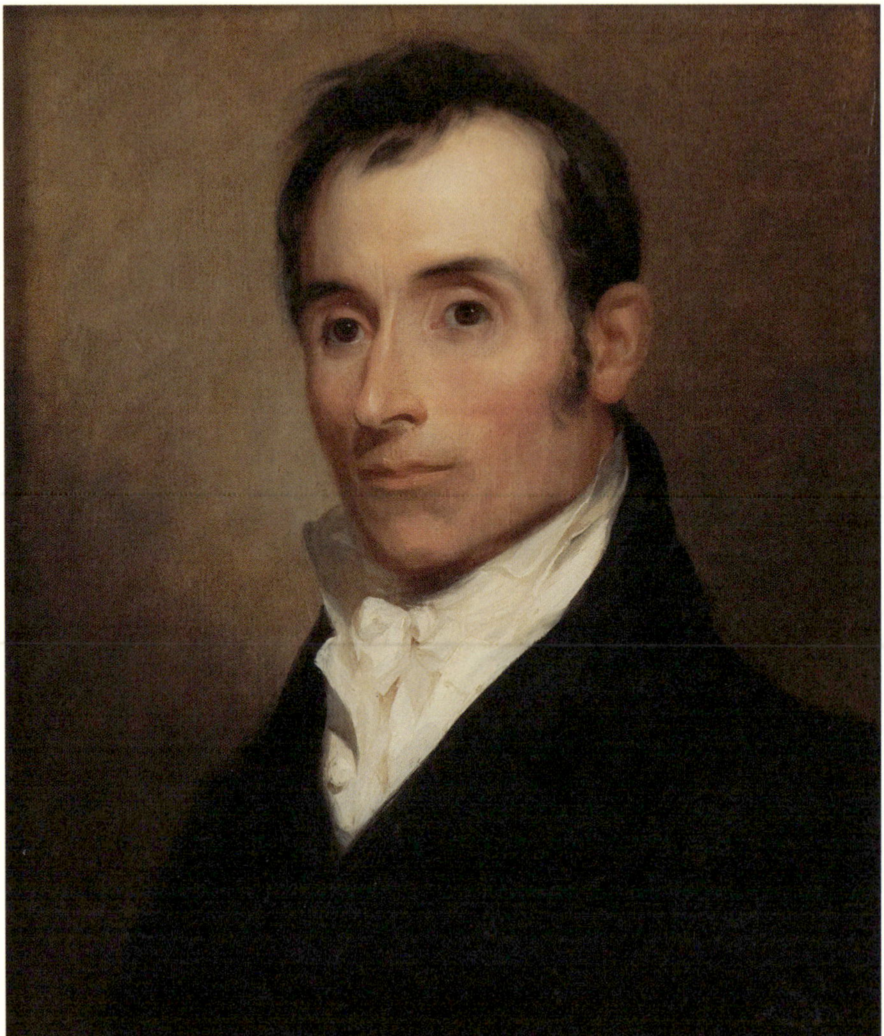

ALEXANDER WILSON

The Scot Who Founded American Ornithology

EDWARD H. BURTT, JR.

WILLIAM E. DAVIS, JR.

The Belknap Press of
Harvard University Press
CAMBRIDGE, MASSACHUSETTS
LONDON, ENGLAND
2013

Frontispiece: Portrait of Alexander Wilson attributed to Thomas Sully, most likely painted around the time of Wilson's election to the American Philosophical Society in 1812. (From the collection of the American Philosophical Society, gift of Dr. Nathaniel Chapman, 1822.)

Printed in Canada
Book Design by Dean Bornstein

Library of Congress Cataloging-in-Publication Data

Burtt, Edward H., Jr., 1948–
Alexander Wilson : the Scot who founded American ornithology /
Edward H. Burtt, Jr., William E. Davis, Jr.
pages cm
Includes bibliographical references and index.
ISBN 978-0-674-07255-8 (alk. paper)
1. Wilson, Alexander, 1766–1813. 2. Ornithologists—United States—Biography.
3. Ornithology—United States—History. 4. Wilson, Alexander, 1766–1813. American ornithology—Illustrations. 5. Birds—United States—Pictorial works. I. Davis, William E., Jr., 1936– II. Title.
QL31.W7B87 2013
598.092—dc23
[B] 2012036742

To my best friend, Thelma, with love—Jed

To my wife, Betsy Davis, with love—Ted

Contents

Preface · ix

1. Themes in Wilson's Life and Writings · 1

2. A Varied Life · 13

3. Illustrating *American Ornithology* · 63

4. Pioneer Ornithologist · 281

5. Wilson's Legacy · 331

Appendix A. On the Shoulders of Giants: Wilson's Predecessors · 355

Appendix B. Wilson's Contemporaries and Correspondents · 385

Notes · 413

Selected Bibliography · 429

Acknowledgments · 433

Index · 437

Preface

Over the last ten years our colleagues, students, friends, and families occasionally asked, "What are you doing?" to which we responded enthusiastically, "Writing a book about Alexander Wilson: naturalist, illustrator, and Father of American Ornithology." That response usually elicited a noncommittal "how interesting" or occasionally, bewilderment followed by "Wasn't that Audubon?" Rarely genuine interest led to the question, "Was Wilson's Warbler named after him?" The species in question might be Wilson's Storm-Petrel, Wilson's Snipe, Wilson's Plover, or Wilson's Phalarope. Wilson has more species named after him than any other American ornithologist, one measure of his stature among American ornithologists.

To provide context for understanding Wilson's achievement, the first chapter explores themes in Wilson's life, writings, art, and science. The second chapter is a brief biography. Chapter 3 explains the publication process of the early nineteenth century. But at its core this chapter and the book are about Wilson's observations of birds and the artwork he created for the plates in his nine-volume *American Ornithology.* The drawings vary from a few lines and notes penciled in the field to remind him of shape and color—for example, the simple sketch of a Canvasback—to detailed pen, ink, and watercolor paintings, such as his perched Red-tailed Hawk, a stunning work of art. Most of Wilson's drawings and draft plates are reproduced here for the first time. Collectively they are witness to his artistic achievement, his development as an artist, his method of working, his observational skill, and his scientific insight. The detail and elegance of his portraits are astonishing, especially when you realize that the knowledge portrayed in these illustrations was based on Wilson's own discoveries, that many of these species were first described by him, and that the ecology and behavior evident in the artwork were unknown until he observed and illustrated them.

Beyond his artistic achievements, Wilson wrote at length about the ecology and behavior of birds that he observed. Unlike his European contemporaries and those European naturalists who preceded him, he spent countless hours observing birds in the habitats in which they lived. He also raised and kept many species in order to observe their behavior and developmental changes. These activities led to a text that is filled with details unavailable in the biological literature prior to 1808, when the first volume of *American Ornithology* was published. In Chapter 4, we explore how these and other written contributions led Wilson to profoundly affect the development of American ornithology and American science in general. Chapter 5 is devoted to how Wilson's legacy shaped the styles, choice of topic, and accomplishments of subsequent generations of ornithologists, artists, and writers, whether or not they acknowledged this influence. In addition, because few of Wilson's predecessors or contemporaries are known to those living in the twenty-first century, we have included two appendices that describe those whose prior work influenced Wilson or who shared their knowledge with him as he developed his *American Ornithology*.

Throughout the book, we have tried to capture Wilson's conviction that American birds symbolized the wilderness, which defined a uniquely American culture, as well as his passion for discovering, naming, and learning about these remarkable creatures. As you turn the pages and absorb Wilson's life and work, as you contemplate the character expressed in the few penciled lines of some sketches or the detail evident in a feather or foot of a pen-and-ink drawing, as you read about Wilson's observations of birds, often in his own words, we hope that you will develop a rich appreciation not only of American birds, but also of the Father of American Ornithology, Alexander Wilson.

In descending the Ohio by myself, in the month of February, I often rested on my oars to contemplate their [Passenger Pigeons'] aerial manœuvres. A column, eight or ten miles in length, would appear from Kentucky, high in the air, steering across to Indiana. The leaders of the great body would sometimes gradually vary their course, until it formed a large bend, of more than a mile in diameter, those behind tracing the exact route of their predecessors. This would continue sometimes long after both extremities were beyond the reach of sight; so that the whole, with its glittery undulations, marked a space on the face of the heavens resembling the windings of a vast and majestic river.

—*Alexander Wilson,* American Ornithology, *vol. 5, p. 108*

Alexander Wilson

CHAPTER ONE

Themes in Wilson's Life and Writings

Alexander Wilson, Scottish poet and Father of American Ornithology, was a child of the Enlightenment. Like Rousseau he saw in wilderness a natural order to be enjoyed and observed, not an alien world to be feared and conquered. He was an observant teenager whose poetry portrayed Scotland and its people with remarkable insight. Jefferson's words and principles as stated in the Declaration of Independence, written when Wilson was ten, inspired him. Thomas Paine's *The Rights of Man* spoke to his soul. A journeyman weaver, Wilson rebelled against the industrialization of weaving and the dishonesty of mill owners. Ultimately, he left Scotland for the United States, where he renounced his British citizenship and embraced the experiment in democracy.

Wilson's nine-volume *American Ornithology* was a pioneering achievement in science. It established the United States as an equal partner with Europe in the study of natural history. The *Ornithology* was widely admired among European scientists for the excellence of its information, its stunning illustrations, and the superb quality of its printing and binding. The admiration voiced by Europeans may have been in part a reaction of surprise at the magnitude and excellence of a work about which they had no prior knowledge, written and illustrated by a man of whom they had never heard. Indeed *American Ornithology* was Alexander Wilson's first and only venture into ornithology, his first and only venture into science. Nonetheless, one can find in his life and writing a number of themes that coalesced into his vision of the United States as a vibrant, young democracy on the threshold of greatness, whose vast wilderness of resources—symbolized by its diverse, colorful birds—made it different from the old, oppressive monarchies of Europe.

A Commitment to Individual Rights

Wilson spent his teenage years living in the country outside Paisley, Scotland, with his father who distilled and smuggled whiskey in violation of English law. Such activity was regarded with tolerance, even admiration, by the Scots who were restless under English rule. The sense of distrust of authority and the importance of the individual that young Sandy, as the boy was called, may have seen in his father found expression later when he completed his weaving apprenticeship and became active in the early effort to unionize the weavers. Regardless of his success in this endeavor, and regardless of the legality of his actions, about which there is some question, there can be little doubt that he was strongly committed to the concept of individual rights and the worth of the individual who worked at the loom. His sense of individualism is further expressed in his dislike of the emerging factory system, where he and fellow journeymen wove in crowded conditions, and his preference for roaming the countryside selling his cloth door to door.

Wilson's commitment to the individual and the commoner was also evident in his poetry. In 1788 Robert Burns's poetry about daily life, written in the Scottish dialect, captured the heart of Scottish culture. Wilson, too, decided to abandon his earlier English style and use his native Scottish dialect to write about people and situations he knew. His was a language that protested English dominance, and, like Burns, he focused on the trials and successes of daily Scottish life.

Wilson was ten when the American colonies declared their independence, twenty-one when the U.S. Constitution was signed, and twenty-three when George Washington was sworn in as the first president of the United States of America. Smuggled goods, especially whiskey and, later, gunpowder, flowed readily between western Scotland and the American colonies, where many local families had relatives. Undoubtedly sailors would have been eager to tell and local residents eager to hear news from America. During his formative years, Wilson would thus have heard and read about the rebellion in the colonies and experienced the increased vigilance of the English government as first the American colonies rebelled and then the French monarchy collapsed. As his own troubles with the local authorities grew, he came to the conclusion that his future lay in the United States, and he left Scotland forever.

Upon arriving in the United States, Wilson encountered the writings of Thomas Jefferson and became an advocate of Jefferson's ideals. The difference between Jefferson's ideal of the small landowner/farmer as the essential strength of America and Wilson's commitment to the individual is impossible to measure because Wilson's early views can only be inferred from his actions, poetry, and his few early letters. But Wilson's focus on the rights of workers is certainly similar to Jefferson's focus on the rights of small landowners, even though the backgrounds of the two men were dissimilar. Wilson grew up in one of Europe's earliest industrial centers, Jefferson's youth was spent in an agrarian setting. Wilson, like Jefferson, saw the North American fauna and flora as very different from those of Europe and reacted with passionate disbelief to the assertion of European scientists, most prominently the Comte de Buffon, that the animals and plants of the New World were degenerate versions of European species.

One of Wilson's earliest experiences in the United States was related in his first letter home after landing near Wilmington, Delaware: "As we passed through the woods on our way to Philadelphia, I did not observe one bird such as those in Scotland, but all much richer in colour. We saw great numbers of squirrels, snakes about a yard long, and some red birds."[1] Over the next few months, Wilson held several jobs before taking a job as a schoolteacher, which he held for nine years. During this period, Wilson's interest in natural history grew, as we can see from his enthusiastic accounts of students who eagerly brought him all sorts of small animals, from insects to birds and mammals. His commitment to the United States grew as well, as demonstrated by his declaration of citizenship and speech in honor of Jefferson's first presidential inauguration. From these changes emerged Wilson's plan to celebrate the wonders of democracy's expanding across a wondrously rich continent by documenting all its birds.

Wilson's commitment to the individual, the common man, also provided the impetus to sell *American Ornithology* as a subscription to individuals throughout the United States. His experience as a peddler selling cloth in Scotland stood him in good stead in this endeavor, but it was his commitment to the Jeffersonian citizen that made this exhausting task so attractive to him. Prior to Wilson's project, works on natural history had been commissioned by wealthy Europeans who collected specimens or artworks depicting them. Such individuals financed expeditions and naturalists, and retained for their personal display or publication all that was found. By contrast, through his subscription plan, Wilson's exploration was funded by contributions from hundreds of persons, many of whom were not wealthy and all of whom were Americans. He also established the first network of what today we would call citizen scientists, who corresponded with him concerning their observations and were recognized generously by Wilson for their contributions. Finally, Wilson's commitment to the common man was underscored by donation of his specimens to the Peale Museum in Philadelphia, which was organized and curated by Charles Wilson Peale and open to the public for a small entrance fee.

Related to the democratization of his financial plan, Wilson used his species descriptions, and the prefaces to each volume of his collection, to stress the excellence of the American workmanship involved in *American Ornithology*'s publication: the paper made of American rags, the type designed and manufactured by Philadelphia printers, the inks and paints compounded from native materials by Charles Wilson Peale and sons of Philadelphia. He also emphasizes in these writings that European accounts of American birds are based on specimens, whereas his species accounts are based on observation of live birds in their native habitat. Whether his assertions of the excellence of *American Ornithology* are an effort to reassure customers of the value of their purchase, a reaction to Buffon's and most Europeans' stated belief that American birds were degenerate forms of the more perfect European species, or a response to his own experiences in Scotland prior to his flight, they each signify the prominent theme of American excellence.

Understanding God through Nature

Another prominent theme is Wilson's concept of deism. Wilson saw in the distribution of birds and their interactions with the habitat and each other the pattern of life as it was before the arrival of Europeans. He also believed that these natural patterns reflected the presence of some superior being. As he noted in the opening paragraphs of volume 1: "A study thus tending . . . to lead us, by such pleasing gradations, to the contemplation and worship of the *Great First Cause,* the Father and Preserver of all, can neither be idle nor useless, but is worthy of rational beings, and doubtless agreeable to the Deity."[2]

Wilson's belief in a deity seems to have been a generalized one that probably emerged during his early childhood, perhaps as a result of tutoring by a theological student his mother had hired to supplement his education. As a religious woman she may have intended to prepare him for the Presbyterian ministry, the Protestant denomination into which Wilson was baptized and the prevailing Scottish sect at the time. Yet despite his frequent reference to a generalized deity, Wilson does not attribute the creation of species to a supernatural being. Similarly, when he marvels at how well adapted a species may be, he may make a generalized comment about how an adaptation reflects the wisdom of a supernatural presence, but he does not specifically attribute the adaptation to God's manufacture. Wilson's belief in God can be found throughout his text, but he believed in a remote presence that established the system, one that takes satisfaction in the pattern of natural interaction. Nowhere does he describe a God who tinkers with the daily details of adaptation.

Nature: The Love of His Life

Wilson began his literary career as a poet in Scotland, where he could draw on a rich literary tradition and participate in an active literary community that included Robert Burns, Robert Tannahill, several minor poets, and poetry contests. The infant United States had no such tradition, no such community. He chose to become a citizen in a nation without a history, without an oppressive class system to protest, without legends or ruins to contemplate.

What the United States did have in abundance, however, was nature. Not surprisingly then, while Wilson continued to write poetry and prose after his immigration to the United States, he no longer wrote about working conditions or interpersonal relationships, and he no longer composed protest poetry: instead, he wrote about nature. Yet unlike the naturalist writers who preceded him, such as Cotton Mather, John Smith, Mark Catesby, even William Bartram, he did not write about nature as something to be feared, conquered, and tamed. Instead Wilson wrote about the wonder of wilderness. He wrote as an observer of nature, an observer who was fascinated by what he saw. Even William Bartram, who would figure prominently in Wilson's achievement, even Bartram who described nature in loving detail, always had an eye on what an excellent farm a stretch of land would be or how navigable the river was. Wilson wrote about the wilderness itself, about its aesthetic and spiritual value, about the wonder of birds and their lives. He established this theme in a wonderful anecdote in the preface to volume 1:

> In one of my late visits to a friend in the country, I found their youngest son, a fine boy of eight or nine years of age, who usually resides in town for his education, just returning from a ramble thro the neighbouring woods and fields, where he had collected a large and very handsome bunch of wild flowers, of a great many different colors; and presenting them to his mother, said, with much animation in his countenance, "Look my dear 'ma, what beautiful flowers I have found growing on our place! Why all the woods are full of them! Red, orange, blue, and 'most every color. O I can gather you a whole parcel of them, much handsomer than these, all growing our own woods! Shall I 'ma? Shall I go and bring you more?" The good woman received the bunch of flowers with a smile of affectionate complacency; and after admiring for some time the beautiful simplicity of nature, gave her willing consent; and the little fellow went off, on the wings of ecstasy, to execute his delightful commission.
>
> The similitude of this little boy's enthusiasm to my own, struck me; and the reader will need no explanations of mine to make the

> application. Should my country receive with the same gracious indulgence the specimens which I here humbly present her; should she express a desire for me *to go and bring her more,* the highest wishes of my ambition will be gratified; for, in the language of my little friend, *our whole woods are full of them!* And I can collect hundreds more, *much handsomer than these.*[3]

Wilson clearly relates to the innocent pleasure of the little boy, to the delight the boy expresses in what he has found in the woods. Indeed, he uses the little boy's own words to describe his collection of birds and its display to the reader. There is no hint of the terror of the woods so often conveyed by earlier authors. The theme is nature as a source of inspiration and joy, as it clearly is for the mother, her son, and Alexander Wilson. Such a theme as applied to the American wilderness was new, but Wilson carried it throughout his own journeys in search of birds and throughout his writing. This deep appreciation of wilderness, a perspective of wonder at the natural world, was picked up by those like Emerson and Thoreau who came after him.

Preservation amid Abundance?

Wilson's sense that the wonder of nature revealed the presence of a deity and his intense pleasure in the American wilderness led to his concern for its preservation. Time and again he refers to the habitat preferences of birds, the natural patterns of distribution and migration that result from these preferences, and frequently, how European settlers' alterations to landscapes used as habitats had changed the birds' distribution. At times he even speculates briefly on what a future without wilderness will be like. Both William Bartram, Wilson's teacher and mentor, himself a traveler in the wilderness, and Thomas Jefferson also wrote about the disappearance of the wilderness, but with only a tinge of regret; they seemed to feel that the loss was more than compensated for by the possibility of farms and towns dotting the landscape and commerce traveling the rivers. Alexander Wilson shares their vision of the future, but for him the loss of wilderness causes genuine, poignant regret. One cannot read his text or look at his drawings without realizing that part of what drove him to write about America's birds was a desire to make every American realize the value of our birds and thereby want to ensure their survival for all time.

A More Meaningful Excellence

Throughout his literary career Wilson was driven to excel. While not specifically stated in his poetry or prose, he mentions his need to excel and his desire for recognition in letters to friends. As a young poet he was concerned with his public reputation and his standing among the poets of his generation. He recognized that Robert Burns was on a different literary plateau and was sufficiently realistic to regard such heights as beyond him, but that did not dissuade him from entering poetry competitions in which he recited his own work and in which he finished as high as second, a considerable achievement given the large number of young poets in Scotland in the 1780s and 1790s. His narrative poem "Watty and Meg," about domestic strife, sold tens of thousands of copies. His books of poems sold less well and earned little, although Mrs. Dunlop, a patron and friend of Robert Burns, praised the book in a letter to Burns.

Wilson's failure to achieve widespread recognition for his poetry may have contributed to his decision to leave Scotland for the United States. After his arrival, he continued to write poetry and his work was regularly published in *Port Folio,* a literary magazine published in Philadelphia. The audience for poetry in the United States around the turn of the nineteenth century, however, was a mere shadow of the audience in Scotland. Wilson may have gained satisfaction from the frequent publication of his work, but his success in terms of recognition as a poet was limited at best. Did his failure to achieve recognition as a poet contribute to his idea to describe all the birds of the United States? The attempt had never been made. If he succeeded, excellence would be his.

There can be little doubt that Wilson wished his *American Ornithology* to be outstanding in every detail. He states as much in the introduction:

> It is intended to comprehend description, and representation of every species of our native birds . . . engraved in a style superior to any thing of the kind hitherto published and colored from nature, with the most scrupulous adherence to the true tints of the original.
>
> . . . As time may prey on the best of colors, what is necessary in this respect will by no means be omitted, that the figures and descriptions may mutually corroborate each other. It is also my design to enter more largely than usual into the manners and disposition of each respective species; to become, as it were, their faithful biographer, and to delineate their various peculiarities, in character, song, building, economy &c. as far as my own observations have extended, or the kindness of others may furnish me with materials.[4]

Wilson's physical pursuit of his stated goal underscores his commitment to excellence. He traveled far more widely than any naturalist before him, often journeying alone through areas that offered little more than animal trails for guidance. Later, when his funds were nearly exhausted, he undertook to color all the plates in volumes 7 and 8 himself and worked eighteen to twenty hours daily to complete the work to his satisfaction.

By this point in his endeavor, the excellence he sought was not the public recognition he craved as a young poet. He was older now and his drive was for a more personal satisfaction.

> Biassed, almost from infancy, by a fondness for birds, and little less than an enthusiast in my researches after them, I feel happy to communicate my observations to others, probably from the mere principle of self-gratification, that source of so many even of our most virtuous actions; . . . I ask only support equal to my merits, and to the laudability of my intentions. I expect no more; I am not altogether certain even of this. But leaving the issue of these matters to futurity, I shall, in the meantime, comfort myself with the good old adage, "Happy are they who expect *nothing*, for they shall not be disappointed."[5]

The audience that would judge the excellence of *American Ornithology* was Alexander Wilson. From 1803 until his death in 1813 he was the writer, artist, and final judge. As so often happens when one works solely to meet one's own standards of excellence, Wilson's *American Ornithology* brought him the national and international recognition he had so longed for as a young Scottish poet.

Order Out of Chaos

Wilson also sought to bring some semblance of order to the taxonomic chaos that characterized description of the North American avifauna at the turn of the nineteenth century. Exactly fifty years before publication of the first volume of *American Ornithology,* Linnaeus published the tenth edition of his *Systema Naturae,* in which he standardized the genus and species names of those North American birds that were known to him. Prior to that time writers on North American birds, Mark Catesby for example, provided names without regard to standardization and often with little knowledge of names already in use. After Linnaeus's tenth edition, the situation was largely unchanged. Thomas Jefferson followed Catesby's names rather than adopting those of Linnaeus, and William Bartram published a list of names that were uniquely his own. Meanwhile European taxonomists such as Latham were adopting Linnaeus's system and applying it to American birds, but working only with specimens preserved in brandy, which introduced many mistaken identifications (and not for the reason you may think, although sailors often drank the brandy during the long voyage back to Europe from North America, much to the detriment of the specimens and the consternation of the European taxonomists). Thus Wilson faced a chaotic array of names for the birds he proposed to describe and a set of ordinal and family names that was to prove inadequate for discussion of the many new species he would discover.

In his first volume, Wilson lamented the naming problem: "One principal cause of the great diversity of classifications, appears to be owing to the neglect, or want of opportunity, in these writers, of observing the manners of the living birds, in their unconfined state, and in their native countries."[6]

He then proposed the solution that was to characterize *American Ornithology* throughout. "It is only by personal intimacy, that we can truly ascertain the character of . . . the feathered race; noting their particular haunts, modes of constructing their nests, manner of flight, seasons of migration, favourite food, and numberless other minutiae, which can only be obtained by frequent excursions in the woods and fields, along lakes, shores, and rivers, and requires a degree of patience and perseverance, which nothing but an enthusiastic fondness for the pursuit can inspire."[7]

Wilson clarified his use of "personal intimacy" a few pages later, explaining: "The greatest number of the descriptions, particularly those of the nests, eggs, and plumage, have been written in the woods, with the subjects in view, leaving as little as possible to the lapse of recollection: as to what relates to the manners, habits, &c of the birds, the particulars on these heads are the result of personal observation, from memorandums, taken on the spot."[8]

From the perspective of a modern ornithologist, ecologist, animal behaviorist, or student in introductory biology this would seem to be the obvious way to learn about unknown birds, but with the possible exception of Mark Catesby, who described a few species in this manner, none of the previous writers on birds of North America had studied live birds. Astonishing as this fact may be to modern readers, it makes sense given how difficult observation of live birds was at the time: most of the species were undescribed so that watchers would not know which birds they were hearing and seeing, and there were no binoculars with which to view a vireo or warbler foraging in the top of a tree. Wilson's assertion that he took notes while he observed the bird represented a novel approach to the study of birds and to natural history, and we know that he did so because of his sketches, the notes made on those sketches, and his detailed accounts of many species. His intent to do so is attested by his frequent apologies for incomplete accounts of birds that he saw only rarely. Direct observation of birds in the wild was an important theme throughout the books and may be Wilson's greatest contribution to ornithological science. He also introduced Linnaean taxonomy to American ornithology and resolved many misconceptions concerning North American species. The success of *American Ornithology* was due to its achievement as an American work of art, as a text that promoted a new perspective toward wilderness, and as a work of innovative science, in both its methodology and the interpretation of the results.

CHAPTER TWO

A Varied Life

Alexander Wilson was born on 6 July 1766. His father, also Alexander Wilson, was a silk weaver, a former soldier, and a smuggler of fine whiskey, which was a tacitly approved expression of Scottish nationalism in the face of British rule. His mother was the beautiful Mary McNab from Rhu. The family was respected with a comfortable income and a modest house at the edge of the White Cart River where it passed through Paisley, Scotland (Figure 2.1). When "Sandy" was four days old, he was christened at the nearby Presbyterian church by Dr. John Witherspoon, who ten years later would leave Scotland to become president of the College of New Jersey (later Princeton) and a signatory to the Declaration of Independence.

Childhood and Education

As a boy, Sandy Wilson did not stand out from the other boys running about the row of houses built along the riverbank and known as the Seedhills. He was a dark-haired, dark-eyed boy with a fair complexion, wiry build, and above-average height.[1] He was also intelligent and mild mannered.[2] In the winter he and the other boys played in the snow. In the summer they walked upstream, stripped in the woods, and plunged into the river to splash and swim the afternoon away. A favorite playground was a patch of rough water around some emergent rocks known as the Hammills.

Sandy attended the Latin-Grammar School in Paisley where he studied reading, writing, arithmetic (his least favorite and poorest subject), church music, and Latin.[3] His mother wanted Sandy to be educated for the Presbyterian ministry and to that end his parents hired a divinity student, Mr. Barlas, to tutor the boy in theology.[4] Sandy and his teacher probably read from the Bible, undoubtedly the King James translation. (James VI of Scot-

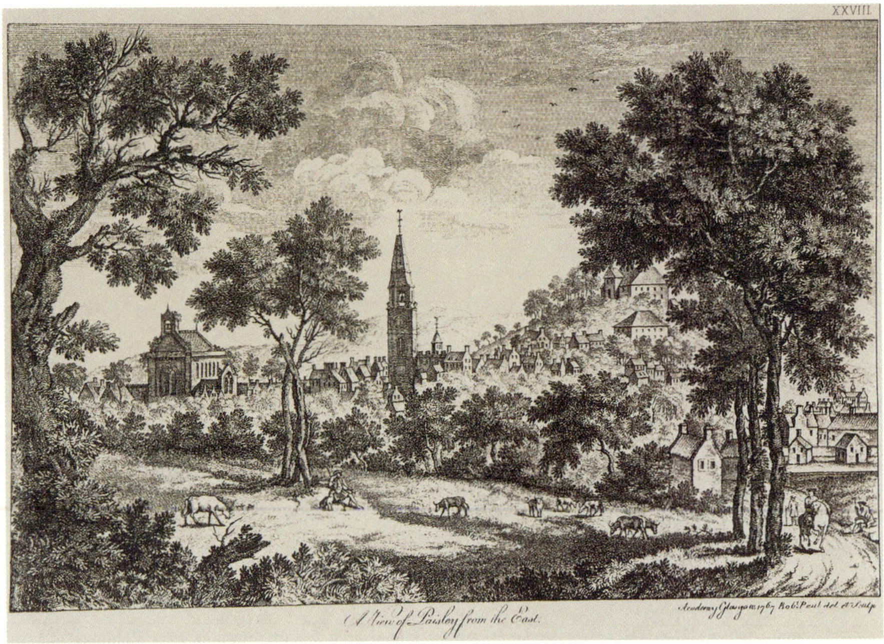

Figure 2.1. Paisley, Scotland, a village of forty-five families and sixty-six looms located west of Glasgow, about 1770. (From the archives of the Paisley Museum, Renfrewshire Council.)

land became King James I of England and Scotland when his cousin Queen Elizabeth I died.) The lilt and beauty of its poetic text may have influenced Wilson's lifelong appreciation of language.

How long Mr. Barlas taught Sandy Wilson is unknown, but in 1776 Sandy's mother died and his formal education ended. Within the year his father had married a widow with three children of her own and the family, now with six children, moved to Auchinbathie Tower, near Lochwinnoch, a long walk from Paisley. To supplement the family income Sandy went to work herding cows for Mr. Stevenson of Threepwood.[5] At the time, land had no fences, walls, or hedgerows, and livestock roamed freely. The only way to keep them in designated grazing land and out of the neighbor's crops was to have them tended by someone who could keep them from wandering. Sandy Wilson thus spent long solitary days walking the nearby hills tending cows. When his charges slowed to graze or settled down to chew the cud, he watched birds or threw himself in the grass and invented verses. He undoubtedly recited these to himself and probably to a small, but admiring, audience of cows. Eventually he began to write down his musings. His earliest surviving poem dates from the period 1776–1779, probably 1778–1779:

> Castle Semple stands sae sweet,
> The parks around are bonnie, O;
> The ewes and lambs ye'll hear them bleat,
> And the herd's name is Johnnie, O,[6]

The poem continues for six more verses. For a boy of twelve or thirteen these few lines show Sandy Wilson's considerable facility with language and his delightful sense of rhythm. No other extant poems can be dated from these years, but he very likely wrote others.

In 1779 Sandy returned to Paisley where, against his wishes, he was apprenticed as a weaver to his brother-in-law, William Duncan.[7] During his three-year indenture he spent his free time reading and exploring the countryside outside Paisley. As his brother-in-law descended into drunkenness, the young Wilson played with his nieces and nephews, read to them, and tutored them. The eldest son, William, came to view his uncle as a friend and father, a close relationship that lasted throughout Wilson's life.[8]

While an apprentice, Wilson mastered the flute and learned to dance, social skills that were highly valued by the teenage apprentices. During these years he also acquired a gun. In the long evenings of late spring and summer, Wilson walked the ten miles to visit his family at Auchinbathie. He also spent Saturdays with his dog on the moorlands below Misty Law, where he poached game for the family table.[9]

In addition to providing food for the family table, he observed birds and wrote about the waterfowl of Lochwinnoch, spent time watching Northern Gannets (*Morus bassanus*) and other seabirds that frequented the Firth of Forth.[10] These days are probably responsible for his intimate knowledge of European birds, which was the foundation for his later study of American birds. They also support his assertion that he had had a fondness for birds since his youth and refute the widespread opinion that Wilson's interest in birds developed rather suddenly in his mid- to late thirties.[11] There can be little doubt that these days were the source of his marksmanship, which was considered remarkable by his peers. Imagine sighting a migrating warbler or kinglet high in an oak with its emergent foliage not through a pair of lightweight binoculars or a telescopic gunsight, but along the barrel of a seven-pound shotgun. Marksmanship and a personal knowledge of European birds were among many skills that Wilson would need as an ornithologist.

Wilson fulfilled the terms of his apprenticeship and in 1782 celebrated his graduation to journeyman in his first surviving poem in the Scottish dialect, a verse that suggests considerable relief at his liberation. He also moved home for the summer to find a beautiful, smart, and ambitious fifteen-year-old girl, Meg Duncan, serving as the family maid. Meg fell in love with Alexander and doted on him, though Wilson spent most of his daylight hours hunting on the moors with his dog. She also attracted the attention of other youths as well as John Craig, the family's seventy-five-year-old landlord and the richest man in the neighborhood. This was not a grandfatherly relationship on either side. Katherine, Alexander's stepmother, soon stepped in to rescue young Alexander from Meg's romantic entanglements without offending their elderly landlord. She found her stepson a position in Matthew Barr's shop at Lochwinnoch. But Wilson often walked home in the evenings, whether to see his family or Meg is not clear. Meg's numerous suitors made life in the tower chaotic and finally the exasperated Katherine dismissed her.

Meg was not to be so easily dismissed. Within weeks she had married John Craig. His sons were furious and he threatened to disinherit them. Although Meg's position was tenuous, she had a plan. Nine months later, in February 1785, she produced a son, Andrew Craig. Was he John's?[12] She pointed to the decorous nine months, but a note in the *Cairn of Lochwinnoch* states, "Auld Fauldheads [said] this son was not his begetting frequently, altho' he had to father it."[13] The community considered young Wilson a much more likely father. Whatever the truth, John Craig lived another ten years. He separated from Meg, but she received an annuity from the Craig family, raised her son quietly in Lochwinnoch, and passed out of Wilson's life.

At Barr's shop in Lochwinnoch, Wilson wove with fellow journeymen John Orr and John Allan, teenagers like himself. His peers noted his fondness for books, his facility with words, and the ease with which he composed verses. They especially appreciated his ability to cut to the heart of the matter, often to the amusement and delight of his small audience of journeymen weavers and bobbin girls.[14] This characteristic is well illustrated by his poetic response to John Allan's repeated pleas for Sandy to write him an epitaph. Young Wilson put him off again and again, arguing that Allan was a good sort, but lacked a distinctive characteristic on which to base an epitaph. One day, however, the crescendo of entreaties became unbearable and young Wilson responded with an amusing verse based on Allan's fruitless searching for birds' nests on Sundays:

Below this stane John Allan rests
An honest soul, though plain,
He sought hail Sabbath day for nests,
But always sought in vain.[15]

So John had his epitaph. One wonders how many threads broke in the merriment of the moment. Unfortunately we do not know if John had this epitaph inscribed on his gravestone, but he certainly kept it: many years later the scrap on which the epitaph was written was given to the Paisley public library by his widow.[16]

When the shop at Lochwinnoch closed, Sandy returned to Paisley where he entered a two-loom shop. Almost three weeks later Wilson ventured a conversation with his shopmate, David Brodie, only to discover that they shared an interest in poetry. Brodie was a couple years older than Wilson and better educated. He brought copies of Salust, Virgil, and other Latin poets to the shop; also Shakespeare, and Goldsmith, Pope, and Gray. When the weavers paused they would read selections aloud and when they resumed their weaving, they would discuss what they had read. Brodie's and Wilson's association blossomed into a lifelong friendship. Many years later Brodie recalled with pleasure their days together. He had enjoyed their many discussions and appreciated Wilson's thoughtful reflections on the poetry. He also marveled at the young Wilson's composing aloud while seated at his loom, his shuttle flying back and forth as he wove.[17] The months shared with Brodie mark Wilson's transition from studying the poetry of others to struggling to express his own poetic thoughts.

Poet and Peddler

Wilson now spent much of his free time writing poetry. He dreamed of fame as a poet, but his poetry was largely imitative of the poetry he and Brodie read: he had not yet found his own voice. That all changed in 1786 when Robert Burns published his first book, *Poems, Chiefly in the Scottish Dialect.* Burns, a genius and farmer from Ayrshire, wrote in the dialect in which he, his friends, neighbors, and Wilson spoke and thought. His work had wide appeal because of its insight and imagery and because its subjects were those familiar to every Scotsman. The success of *Poems* changed Alexander Wilson's life. Wilson had used dialect and idiom sparingly in his early poems, and he had written mostly in the standard English of his models Goldsmith, Pope, and Gray. After Burns's *Poems,* however, he became intensely observant of the life and people around him and wrote about them in his native dialect. Poems began to accumulate in his room.

Wilson left the shop he shared with David Brodie sometime in 1788 to join his brother-in-law, William Duncan, whose weaving business was in Queensferry on the banks of the Forth. He remained there for several months and then accompanied Duncan on a peddling trip through eastern Scotland. The experience convinced him that peddling better suited his temperament than throwing a shuttle all day at a loom. With the financial help of friends he assembled a pack of goods and set off. As he traveled, he visited historical sites, former homes of his favorite authors, and places of natural beauty. In his free time he wrote.

On one of Wilson's peddling trips he read a notice for an upcoming village dance. He loved to dance and he was a handsome, single, young man. He had the requisite breeches, shirt, tie, waistcoat, and coat, but his silk stockings were stained and threadbare. To make matters worse, he had no gaiters. Loath to miss the dance, he procured some chalk to whiten the upper parts of his stockings and painted black gaiters on the lower parts. Now "properly" attired, he attended the dance. In the soft candlelight of the dance he spent a wonderful evening.[18]

After several frustrating months of peddling Wilson returned to Paisley unable to meet his expenses, but with a collection of new poems that included "The Disconsolate Wren," the best of his poems in the Scottish dialect and his first poem devoted to birds. He resumed weaving to replenish his savings and spent his evenings editing his poems for publication.

At the suggestion of Tom Witherspoon, a lifelong friend, Wilson sought out Thomas Crichton, governor of the Town Hospital, a well-educated, widely respected man, for advice about his poems. Crichton later described their first contact: "Wilson introduced himself, and, with a great deal of modesty, expressed his wish for a little conversation. He told me his name, and informed me that he had a volume of poems in manuscript, which he intended for the press, and requested that I would look over them at my leisure. He put the small volume, which was neatly written, into my hand, and left it with me saying that he would soon call again to hear my opinion."[19]

Crichton was impressed with the poems and introduced Wilson to John Neilson, Paisley's leading printer, who agreed to publish the book if Wilson could enlist about six hundred subscribers. He suggested that Wilson find a patron whose name on the subscription list would encourage others to subscribe. Wilson thought of his acquaintance William McDowall of Castle Semple, who had been elected five times to Parliament from the county of Renfrew. Wilson's request for a meeting was accepted and he was received in the library at Castle Semple, where McDowall read through Wilson's poems, asked many questions, and then asked for two subscription forms with the promise that he would enlist some subscribers. Wilson was ecstatic.

With the financial backing of James Kennedy, a boyhood friend and successful businessman whose love of poetry exceeded his own talent as an author, Wilson purchased the goods to stock his peddler's pack for another trip. This time he would sell not only cloth, but also subscriptions to his book. Two months later he returned to Lochwinnoch from eastern Scotland. His assets were four hundred subscriptions, twelve shillings, a pocket watch, and a first-ever journal of his trip, which his friends had encouraged him to keep. The journal was his first foray into prose and is notable for its detailed observations of human character and the depth and elegance of its descriptions of the places he visited.

Despite the shortfall in subscriptions, Neilson agreed to proceed with publication and the small volume, titled simply *Poems,* was released in 1790.[20] The "Journal as a Pedlar" was printed as the introduction. In order to collect on the subscriptions, Wilson returned to eastern Scotland, his pack loaded with books and cloth goods, but he received less money than he had planned: only two hundred subscribers honored their commitments. Back at Lockwinnoch he returned to weaving and to contemplating how to repay his debts to James Kennedy and John Neilson.

Early in 1791 Wilson published an essay in *The Bee,* a prominent literary magazine of the time. About the same time Kennedy alerted Wilson to a debate that was scheduled for the Edinburgh Pantheon on the question "Have the exertions of Allan Ramsay or Robert Fergusson done most honour to Scottish poetry?"[21] Kennedy planned to compete and urged Wilson to do the same. After consulting with David Brodie, Wilson decided to speak on behalf of Robert Fergusson and travel to Edinburgh to present his perspective in person. During the next week he wove forty ells (fifty yards) of silk gauze, enough to pay for his trip to Edinburgh; read and studied Brodie's copy of Fergusson's poetry; wrote "The Laurel Disputed," his poetic argument on behalf of Fergusson; and committed the poem to memory.

He walked to Edinburgh. Keep in mind that Paisley is on the west coast of Scotland and Edinburgh is about fifty-six miles to the east. Needless to say Wilson was not equipped with the accoutrements of modern distance walkers. When he arrived he found six other competitors, all advocating on behalf of Allan Ramsay. When the five hundred members of the audience voted, Wilson finished second, but by only seventeen votes. The winner, a Mr. Cumming, had paid six shillings a ticket for forty tickets, which he presented to ladies planning to attend on the condition that they vote for him. Wilson's friends thought that he, Wilson, should have won first place. Be that as it may,

Wilson had scored something of a moral victory and was greatly buoyed by his success.

Upon his return to Paisley, Wilson prepared a second edition of poems to be published by Peter Hill, a prominent bookseller in Edinburgh. He deleted several of his weakest poems from the first edition and added new ones, including "The Laurel Disputed," which he thought would attract members of the fashionable set in Edinburgh who had made up most of the audience at the Pantheon. The second edition was released in 1791 under the expanded title *Poems: Humorous, Satirical and Serious.* The second collection is generally better than the first, reflecting Wilson's increasing skill not only as an author, but also as an editor. Although a greater economic success than the first, his second edition did not sell enough to offset his debts. But Wilson was now recognized as a young Scottish poet with a promising future (Figure 2.2).

Although Wilson resumed his seat at the loom, he continued to observe the people and places around Lochwinnoch and to write about both. He wrote several new poems that were published in *The Bee* and was invited to become its assistant editor. Also in 1791 he wrote verses for a book of hymns by Robert Gilmour.

Publication in March 1791 of "Tam o' Shanter," Burns's masterpiece, combined with Wilson's recent successes, encouraged him to think about writing his own ballad. The result was "Watty and Meg," Wilson's adaptation of *Taming of the Shrew.* Eighty years later Grosart concluded that the poem was unique in the Scottish literature for its rough, coarse, and accurate word painting, and that its language gave the tale a vigor and realism that was thoroughly Scottish.[22] Wilson had Neilson publish it anonymously, which meant that Neilson received all the profits. The poem, which was wildly popular, sold more than 100,000 copies and cleared Wilson's debt to Neilson.[23] Its anonymity and popularity led to widespread speculation that it was the work of Robert Burns, a claim that Burns denied with regret.[24]

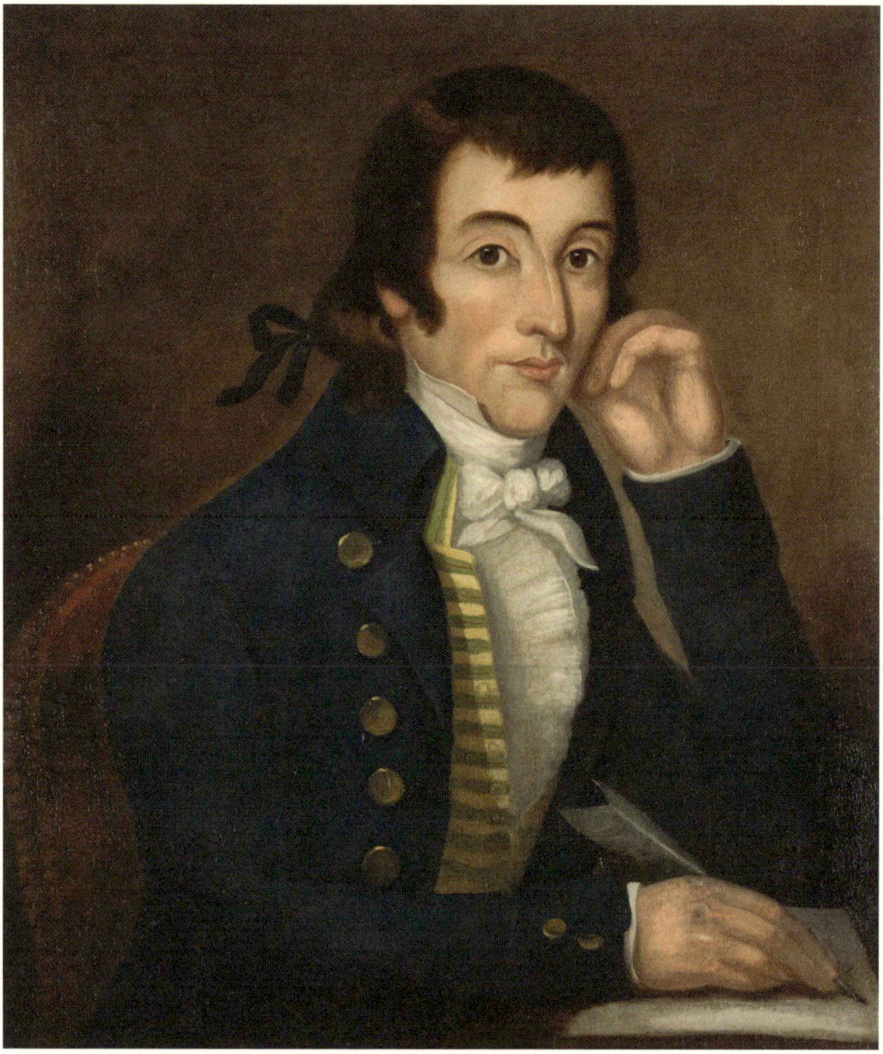

Figure 2.2. Alexander Wilson, age twenty-two, as originally painted by James Craw in 1788 and later copied by Cairns, a Paisley amateur. The original Craw portrait, now darkened with age, hangs in the Wood Library of Ornithology, McGill University, in Montreal and the Cairns portrait shown here is in the archives of the Paisley Museum, Renfrewshire Council.

Politics, Poetry, and Emigration

The weavers of Paisley were politically liberal and interested in political philosophy, natural history, and literature. Informal associations for the pursuit of these interests were common, and in this environment the young Wilson found friends with whom to share his love of natural history, his interest in poetry, and his commitment to individual rights. Early in 1790 Wilson published "The Hollander, or Light Weight," a satirical poem in which he described the grim working conditions of the weavers and the dishonest practices of the mill owners. The poem made him an immediate hero among the weavers and is one of the first poems to comment on the social dislocations of the early Industrial Revolution. In fact, although Wilson probably never appreciated what he had done, "The Hollander" and the poems that followed it are arguably the first protest literature of the Industrial Revolution, the first account of the grievances of workers against owners and management, and the first expression of the case for unionization.

William Henry, the mill owner who is described, although not named, in "The Hollander," filed charges of criminal libel and incitement to unrest against Wilson in the summer of 1790. These charges and the financial failure of his book contributed to Wilson's physical collapse in the late summer, and he returned to his family in the Tower of Auchinbathie to recover. By November 1790, however, the legal proceedings against Wilson had been allowed to "go to sleep" and no trial ever occurred.

"The Hollander" established Wilson as a poet in the eyes of the local population, and his depiction of workers and working conditions brought him to the attention of reformers active in the late eighteenth century in Paisley and throughout Scotland. In particular, Wilson, an accomplished flutist, made the acquaintance of James and William Mitchell, musicians and leaders of the radical faction in Paisley. Wilson sympathized with the view of individual rights and responsibilities advocated by the Mitchell brothers, who were quick to recognize and employ the poet's writing skills. The watchful eye of the British government necessitated that reports and resolutions proposing parliamentary reform be written and circulated anonymously, but

Wilson's authorship of such reports and resolutions originating in the Paisley region is suggested by the skill with which such circulars were written; by the expression of the principles of Paine's *The Rights of Man,* which Wilson had read and greatly admired; and by Wilson's close association with James Mitchell, who by 1791 was secretary to the Friends of Reform and who circulated the resolutions and reports in the Paisley region during that year. Wilson's advocacy for the rights of weavers, his involvement in the reform movement, and his poetic commentaries were about to converge and redirect his life.

On Tuesday, 22 May 1792, William Sharp, owner of Long Mills and a prominent manufacturer in Paisley, received an anonymous letter demanding £5 to suppress publication of the poem "The Shark; or Lang Mills Detected," which was enclosed. The poem describes the wretched conditions in which the weavers worked in the mills of Willy Shark, who short-measured their production of cloth in order to cheat the weavers, who were paid by the yard. The poem's description of working conditions and its advocacy of the workers' cause are characteristic of Wilson, as evidenced in his earlier poem "The Hollander" and his association with the Friends of Reform. The letter, which was in Wilson's handwriting, was not characteristic of him. Within twenty-four hours William Sharp had petitioned for Wilson's arrest, stating that Wilson was known for his "highly libelous, incendiary and dangerous publications . . . some of which are at this moment the subject of enquiry."[25] Recall that the libel case against Wilson for writing "The Hollander" had been allowed to rest, but was not resolved. The sheriff issued a warrant for Wilson's arrest, and he was taken into police custody as he left Neilson's print shop, where he had been checking proofs of "Watty and Meg." He was taken to the Tollbooth, as the Paisley prison was known, and imprisoned for a month. On 27 June 1792 the court granted £50 damages, fined Wilson £10 and court costs, and banned publication of the poem. Wilson was again imprisoned pending payment of the fine and costs, but this time William Duncan, his brother-in-law, posted bond and Wilson was freed. On 22 January 1793 Wilson was jailed for fourteen days following publication of the poem. The fines were increased, but his friend Tom Witherspoon paid a deposit on the fines and Wilson was again released—after he promised to "keep the public peace for the space of two years."[26] He was also sentenced to burn the offending poems "The Hollander; or Light Weight"; "Hab's Door; or the Temple of Terror," and "The Shark; or Lang Mills Detected" in the public square. This he did on 6 February 1793, on the steps of the Tollbooth, while shielded from public view by officials and friends. Despite Wilson's desire to reform the factory system and his genuine commitment to individual rights and responsibilities, his troubles throughout 1792 and 1793 were the result of blackmail and disdain for the court's rulings.

Time has done little to clarify these troubling events. In 1961 Robert Cantwell published the court records, but they shed little light on the core mystery.[27] Wilson admitted in court that the poem and the letter were in his handwriting, but denied sending them to William Sharp. When questioned about who sent them, he refused to answer. The attempted blackmail was foolish in the extreme. Not only did it result in Wilson's serving three separate jail sentences, but it also allowed William Sharp to avoid prosecuting Wilson for libel, a proceeding in which the truth of the charges that he had lengthened the cloth-measuring device so as to cheat his weavers could have been confirmed. The Rev. Alexander Grosart, who published Wilson's Scottish letters and poetry, concluded that Wilson's charges were correct in every detail.[28] Furthermore, he agrees with Sir William Jardine that Wilson was used by the Friends of Reform, namely James and William Mitchell, who may have conceived of the blackmail scheme and convinced Wilson to write the letter.[29] Regardless of who hatched the plan, the poem is in Wilson's style. The content is in keeping with "The Hollander" and "Hab's Door" and with Wilson's outrage at the working conditions in the mills. The blackmail letter is unlike Wilson, but the writing is his and it was sent. That is as close to the truth as we are likely to come.

Throughout 1792 and 1793 Wilson's sense of humor and his friends sustained him, but on 4 January 1794 Wilson was once again in jail for writing and "industriously circulating an advertisement addressed to 'The Friends of Liberty and Reform.' Calling a General Meeting of the friends of Reform to have been held this night in Falconer's Land, Stories Street at five o'clock."[30]

For the third time a friend paid bail and Wilson was released. Later the charges were dropped for lack of evidence.

France was in the midst of the Reign of Terror and the sentences passed on Scottish reformers found guilty of treason and sedition included transportation and death. Perhaps because Wilson was known widely for his poems about working conditions, the court chose not to create a martyr for the reformers' cause by prosecuting him. The libel and blackmail charges were not dropped, however, and Wilson faced the constant threat of renewed prosecution and imprisonment. His situation had become intolerable. In addition, the tension between those friends who shared his commitment to the cause of reform and those friends who were satisfied with the status quo and who had paid his bail may have contributed to Wilson's decision to leave Scotland for the United States. Not only was the United States the land of equal opportunity for all, but it also represented a new start, free from the mistakes that threatened his future in Scotland.

Wilson's decision to leave was made in January soon after his release from prison. He immediately returned to weaving and over the next four months saved all but sixteen shillings of his pay. By mid-May he had saved enough money to buy passage on a ship to America. He visited his family one last time to say goodbye. During that parting his young nephew, William Duncan, pleaded to accompany him and so it came to pass that in mid-May 1794 Alexander Wilson, twenty-eight, and William Duncan, sixteen, walked to Portpatrick where they caught a ship to Belfast and then bought deck space on the *Swift,* which set sail for Philadelphia on 23 May 1794. Wilson never returned to Scotland. His time as a Scottish poet and labor activist was over.

Wilson's career as a poet contributed to his development as an ornithologist in several important respects. By the time he left Scotland he was an accomplished writer who wrote easily and clearly. He could paint clear, often stunning word pictures, and had become a remarkable observer of people—an ability that he would use with even greater success as an observer of birds. He had experience with editing and proofreading. He also knew how to sell subscriptions to his books, a technique he would use to finance *American Ornithology.*

America: Land of Struggle and Mystery

During his first two years in the United States, Alexander Wilson worked as an engraver, weaver, and peddler; settled briefly in Sheppardstown on the Virginia frontier; and, in 1796, settled about twenty miles northwest of Philadelphia in Milestown where he taught school until July 1801. Because the school was small and the schoolmaster's income depended on the number of students, Wilson studied mathematics, taught himself how to survey, and supplemented his income by surveying land for the local farmers.

In 1798 Wilson wrote to his father that he and his nephew had decided to buy "a piece of land in some healthy and fertile part of the country, convenient to a market for the disposal of produce. In the State of New York there is a tract of perhaps the richest land in the United States, situated about 270 miles to the north, or north and by west of Philadelphia, lying between the Senica [*sic*] and Cayuga lakes."[31] After writing this letter, William visited the tract "to see the country and learn further particulars."[32] Like many recent immigrants, Alexander Wilson and William Duncan sent money home, in this case for William's sister, Isabel, to come to America and join him. After Wilson borrowed money and bought the land, and shortly after Isabel arrived from Scotland, she and William left Philadelphia and settled on the newly purchased land to carve a farm out of the wilderness. Having settled his nephew, Wilson was free to devote himself to teaching and long rambles in the neighborhood of Milestown and beyond. One of his trips was an eight-hundred-mile, twenty-eight-day ramble to the family farm in Ovid, Cayuga County, New York, to visit William and Isabel.

Wilson's interest in poetry and politics was irrepressible, and in 1800 he emerged from his self-imposed silence. His first poem was an elegy to George Washington. His second poem, "Jefferson and Liberty," also published in 1800, an election year, may have represented campaign literature of the day. While they probably reflected his genuine esteem for both men, they are neither controversial nor memorable: you have probably never heard of either poem and will not find them in any compilation of American poetry. That they were widely circulated in the newspapers of the time speaks to either the poor literary taste of the editors or the lack of American poetry at the turn of the nineteenth century.

On 4 March 1801, Thomas Jefferson was inaugurated as president of the United States and Wilson was asked to celebrate the event with a speech to the citizens of Milestown, where he was teaching school. The speech, titled simply "Oration," was published in newspapers throughout the United States.[33] Its wide circulation along with the poem "Jefferson and Liberty" may have contributed to Wilson's later friendly correspondence with Jefferson and to their eventual meeting.

Wilson had gained the respect and admiration of his neighbors, but the good months were few. On 23 July 1801 Wilson was writing to his friend Charles Orr from Bloomfield, New Jersey, near Newark: "As to the reports circulated in the neighbourhood of Milestown, were I alone the subject of them they would never disturb me, but she who loved me dearer than her own soul, whose image is forever with me, heart is broken for her friendship to me, she must bear all with not one friend to whom she dare unbosom her sorrows. Of all the events of my life nothing ever gave me such inexpressible misery as this."[34]

The affair to which Wilson refers remains shrouded in mystery, but it caused him to flee Milestown and settle briefly in Bloomfield where, as described in a letter to Charles Orr, he again taught school. "I took the first school from absolute necessity that I could find. I live six miles north from Newark and 12 miles from New York, in a settlement of canting, preaching, praying and sniveling ignorant Presbyterians. They pay their minister 250 pounds a year for preaching twice a week, and their teacher 40 dollars a quarter for the most spirit-sinking laborious work, 6, I may say 12 times weekly."[35] Not surprisingly, Wilson did not remain long in Bloomfield. On 14 February 1802 Wilson wrote to Orr: "On the 25th of this month I remove to the School House beyond Gray's Ferry to succeed the present Teacher there."[36]

Unknown to Wilson this move would be the watershed event of his life. The school was the Union School at Kingsessing, where he was to receive one hundred dollars a quarter and spend the happiest, quietest four years of his life (Figure 2.3). His new residence was not important for its contentment

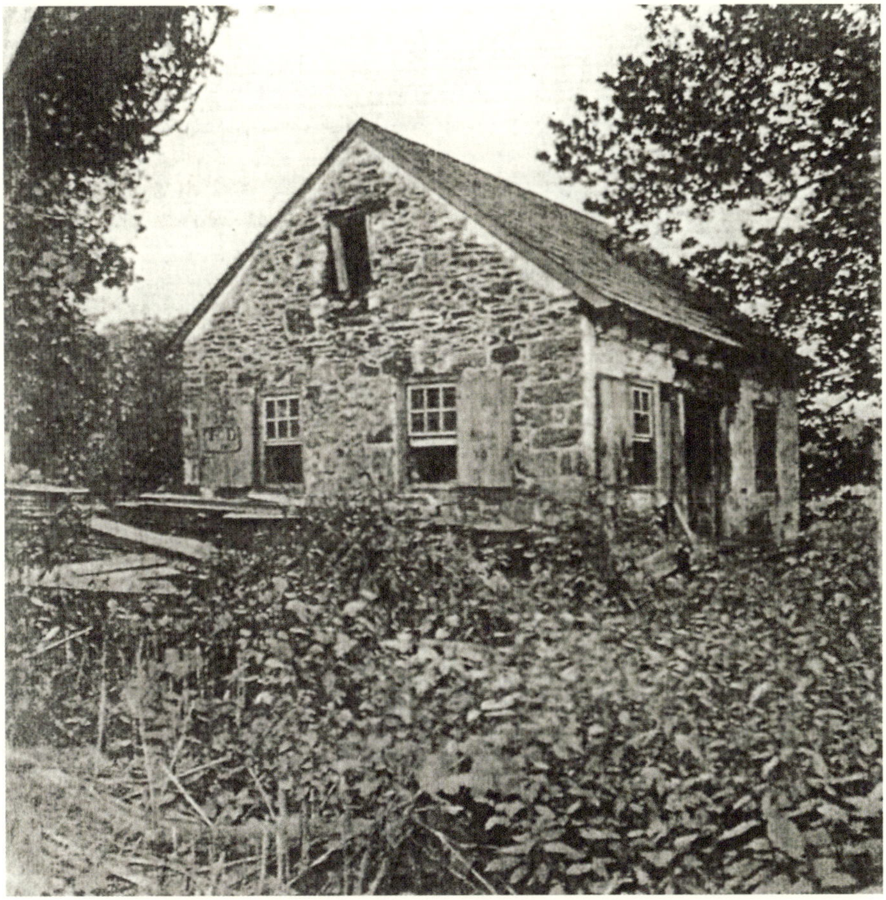

Figure 2.3. The Kingsessing School at Gray's Ferry where Alexander Wilson taught from February 1802 to April 1806. Mary and Isaac Leech were his students throughout his tenure, their younger siblings, Henry and Hannah, began their schooling under Wilson's tutelage. In 1803 William Wood, whose illustrated exercise book is shown in Figure 2.5, was another student of Wilson's at the school. The school was built in 1796 on a plot of land donated by John Bartram on the southeast side of Darby Road. In 1852 a new school was built on land sold to the school district by Isaac Leech, and the old school became a blacksmith shop run by the cousins Henry Leech and William Rively. The photograph was taken in 1869 or 1870; the building was torn down in 1872. (Courtesy of the Leech-Fetters-Webb family.)

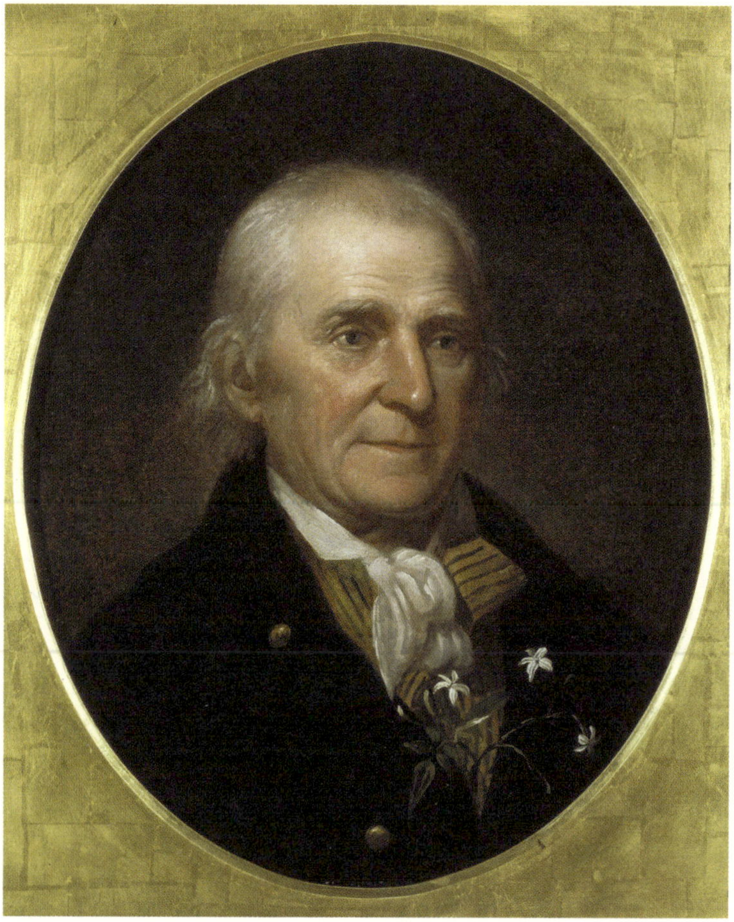

Figure 2.4. This portrait of William Bartram (1739–1823) was painted around the time of his acquaintance with Alexander Wilson, circa 1802–1813. (From the collection of the Independence National Historical Park, Philadelphia.)

or quietude, however, but because of its proximity to the garden and home of the Bartram family, with whom he soon became acquainted.

John Bartram (1699–1777) was a botanist and correspondent of Linnaeus. In 1728 he established a botanical garden on his farm. His son John inherited the farm, but his son William inherited his father's passion for natural history (Figure 2.4). William Bartram's *Travels through North and South Carolina, Georgia, East and West Florida* (1791) is a travel log that describes the flora, fauna, geography, and native Americans of the southeastern United States. The book was widely acclaimed by naturalists and by the early romantic writers of Europe. It remains in print.

In his *Travels*, William Bartram listed 215 species of North American birds. It was the most complete list of North American birds published up to that time, but neither its organization nor its use of Latin names followed the Linnaean genus-species system. Bartram followed the list with some comments on molt, song, nesting, and migration. These were based on his observation of birds and demonstrated the importance of observing the living bird. Unfortunately, Bartram's observations were of only a few species, were not organized by species or topic, and his conclusions were stated briefly and without logical development. Nonetheless his book kindled Wilson's desire to learn all he could about America's birds.

Wilson met William Bartram sometime in 1802, a few months after he took the teaching position at the Union School. There is no record of their meeting. Probably it was a quiet acknowledgment of each other's presence: two reserved men nodding to each other across a patch of plants, with the gardener straightening up to begin a quiet conversation. William Dunlop, a painter and playwright who met William Bartram in 1797, describes just such a scene at their first meeting:

> Arrived at the Botanist's Garden, we approached an old man who, with a rake in his hand, was breaking the clods of earth in a tulip bed. His hat was old and flapped over his face, his coarse shirt was seen near his neck, as he wore no cravat or kerchief; his waistcoat and breeches were both of leather, and his shoes were tied with leather strings. We approached and accosted him. He ceased his work, and entered into conversation with the ease and politeness of nature's noblemen. His countenance was expressive of benignity and happiness. This was the botanist, traveler, and philosopher we had come to see.[37]

The following letter reveals that their friendship had prospered.

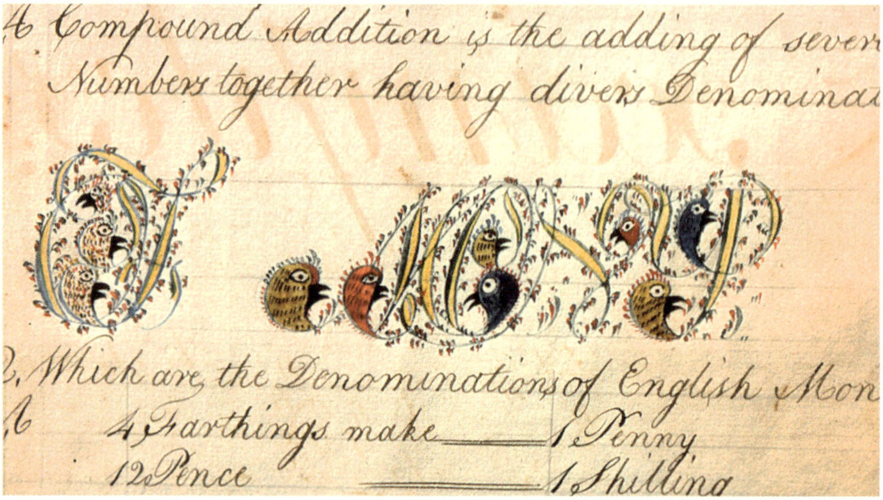

Figure 2.5. One of Wilson's stylized sketches drawn in William Wood's exercise book, presumably in recognition of the boy's outstanding work. The paisley design not only reflects Wilson's training as a weaver, but also indicates his considerable artistic ability, his creativity, his interest in birds, and his commitment to encouraging his students. (From Charles and William Wood's arithmetic notebooks collection, part of the special collections of Monroe C. Guttman Library, Harvard Graduate School of Education.)

To William Bartram

Monday noon 4 March [1803]

Dear Sir,

This Bird I take to be the female Yellow Rump. I suppos'd it on first sight to be some other. If Miss Bartram thinks it worth drawing it is at her service. I have this moment rec'd yours, which like all the letters you have honoured me with are to me as valuable as Bank Notes to a Miser.

Yours

Alex. Wilson[38]

The older man nurtured Wilson's interest in natural history, answered his questions, and encouraged the romantic, sensitive view of nature that is evident in Bartram's writing, in Wilson's poems, and later in Wilson's descriptions of birds. By 1 June 1803 Wilson's direction was clear enough in his own mind for him to write to a friend in Paisley, "I have had many pursuits since I left Scotland, Mathematics, the German Language, Music, Drawing, &c., and I am now about to make a collection of all our finest birds."[39] Undoubtedly Wilson was shooting birds. As indicated in the letter just cited, he had sent a Yellow-rumped Warbler to Bartram in March of that year. But is the collection he refers to here a collection of birds themselves or of bird illustrations?

Wilson had no formal training, yet his artistic ability is evident in the stylized birds with which he decorated his students' notebooks in 1802 (Figure 2.5).[40] Moreover, in 1803 William Bartram, already his friend, became Alexander Wilson's teacher, artistic adviser, and mentor. An undated letter from this year shows that Wilson was sketching birds and eagerly seeking William Bartram's artistic advice:

> My Dear Friend,
> I send with more diffidence than on any former occasion some further attempts. If from the rough draughts here given you can discover what Birds they are, please to give me their names. Any advice for their amendment from you will be truly welcome.
>
> With sincere esteem and affection,
> I am, Dear Sir,
> Your's
> Alex Wilson[41]

The close relationship that had developed between these two men is evident in this letter, especially when compared with the letter of 4 March 1803 presented earlier.

Bartram often loaned paintings to Wilson for him to copy (Figure 2.6). Other early "attempts at drawing" were from life.[42]

> 10 November 1803
> I have murdered your Rose. I traced the outlines with great patience but in colouring and shading I got perfectly bewildered. After I have gained a little more practice I shall make one desperate attempt more on these Roses.[43]

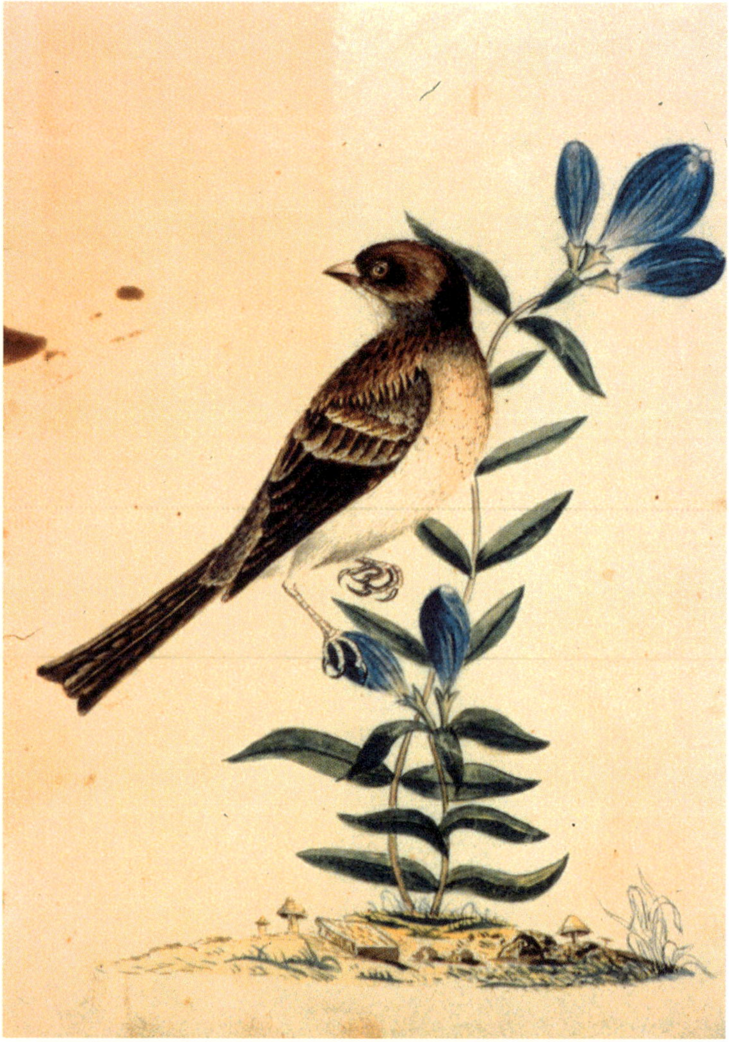

Figure 2.6. A drawing of a Field Sparrow (*Spizella pusilla*) by William Bartram from the early 1790s, found tucked among the pages of an exercise book by William Wood. Apparently, Bartram had loaned the drawing to Wilson as an example to be copied. When Wilson rose to leave the Bartram home, he placed it in the book he had been grading to protect it during his walk home—and there the drawing remained for almost two hundred years, until it was discovered by descendants of William Wood and presented to Harvard University along with William's student workbooks. (From Charles and William Wood's arithmetic notebooks collection, part of the special collections of Monroe C. Guttman Library, Harvard Graduate School of Education.)

17 November 1803

> I have taken the liberty of sending you another Specimen of attempts to imitate your beautiful Engravings, presuming on your goodness.[44]

20 November 1803

> . . . I was quite delighted with the Anemone [sent by Bartram's niece for Wilson to draw], but fear I have made but bungling work of it. Such as they are I send them for your inspection and opinion . . .
>
> The duty of my profession will not admit me to apply to this study with the assiduity and perseverance I could wish. Chief part of what I do is sketched by candle-light.[45]

Bartram taught Alexander Wilson the names of plants, as the note on the "Anemone" indicates, and of birds, but he also provided something less tangible. A few months later, Wilson wrote:

To William Bartram

Kingsessing, 31 March 1804

> . . . I take the first few moments I have had since receiving your letter, to thank you for your obliging attention to my little attempts at drawing; and for the very affectionate expressions of esteem with which you honour me.[46]

Reading this, one can easily imagine the gentle criticism softened by encouragement that Wilson received from Bartram, a quiet Quaker who in the late eighteenth and early nineteenth centuries was recognized throughout Europe and the United States as America's foremost naturalist.

Wilson's life was changing in other ways, too. On June 9, 1804, he appeared in the Court of Common Pleas in Philadelphia, renounced his allegiance to the King of England, and swore to uphold the Constitution of the United States.[47] After ten years of numerous odd jobs and teaching positions, Alexander Wilson had found professional success. Wilson was a good teacher—a strict disciplinarian, but also popular with his students. He rewarded them by drawing stylized sketches of birds in brightly colored inks on the best pages in their workbooks, and they often brought him birds and small mammals to draw.[48]

> William Bartram 31 March 1804
>
> . . . a boy, not long ago, brought me a large basket full of crows. I expect his next load will be of bull-frogs, if I don't soon issue orders to the contrary. One of my boys caught a mouse in school, a few days ago, and directly marched up to me with his prisoner. I set about drawing it that same evening, and all the while the pantings of its little heart showed it to be in the most extreme agonies of fear. I had intended to kill it, in order to fix it in the claws of a stuffed owl, but happening to spill a few drops of water near where it was tied, it lapped it up with such eagerness and looked in my face with such an eye of supplicating terror, as perfectly overcame me. I immediately untied it, and restored it to life and liberty. . . . I felt at that moment the sweet sensations that mercy leaves on the mind when she triumphs over cruelty.[49]

The incident illustrates Wilson's warm relationship with his students who sought to please him by supplying subjects for him to sketch. It also demonstrates the sensitivity that Wilson brought to his studies of drawing and birds.

Wilson's letters indicate that ornithological questions and requests for artistic advice continued to command his attention throughout 1804. A comparison of the March and August letters shows that his knowledge of birds was expanding rapidly during this period:

> 29 March 1804
>
> I send for your amusement a few attempts at some of our indigenous birds . . . Be pleased to mark the names of each with a pencil as except 3 or 4, I do not know any of them.[50]

> 16 August 1804
>
> I have been drawing Woodpeckers this sometime. Pray be so good as inform me if there is not 4 different species besides the Fliccer in these parts. The common redheaded, the speckled hairy and small

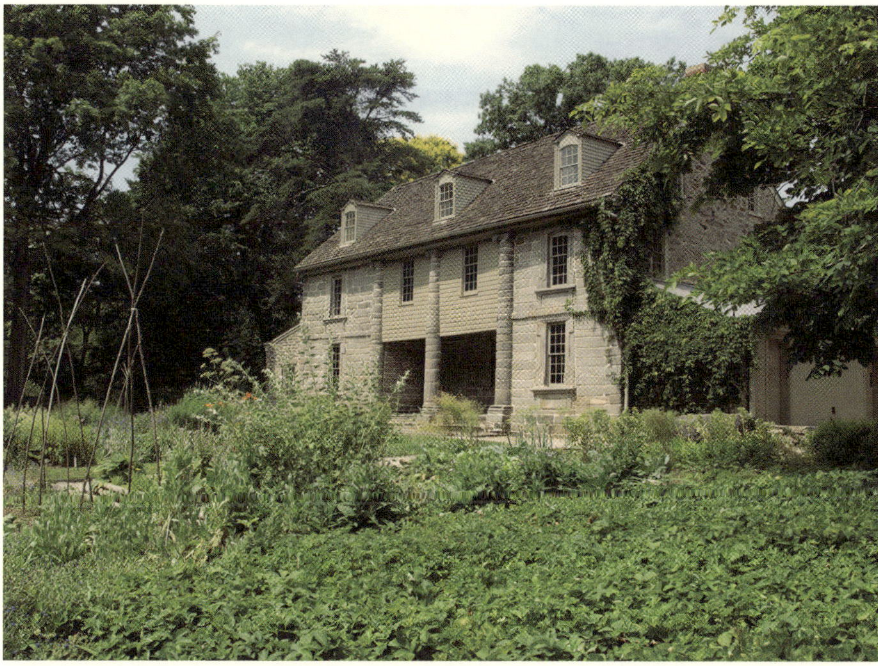

Figure 2.7. The home of William Bartram, built by his father John in 1728–1729. The house is surrounded by extensive gardens where John and William collected native plants and grew samples for shipment to wealthy, European collectors. The house was a moderate walk from Wilson's school and it was here that Alexander Wilson often retreated to write and discuss ornithology with William Bartram, his confidant and mentor. (From the collection of the John Bartram Association.)

> speckled hairy Woodpecker and another speckled one with a crimson coloured throat.[51]

Wilson was undoubtedly roaming the countryside, increasing his knowledge by observation, but he was also reading. William Bartram had become Wilson's librarian. William and his father before him were voracious readers who had amassed a substantial library that emphasized natural history, especially plants, but also contained a number of important ornithological works. Wilson, who was a frequent guest in the Bartram home, undoubtedly read a large number of these books, many of which he cited in his species accounts in the *American Ornithology* (Figure 2.7). Through Bartram's sponsorship

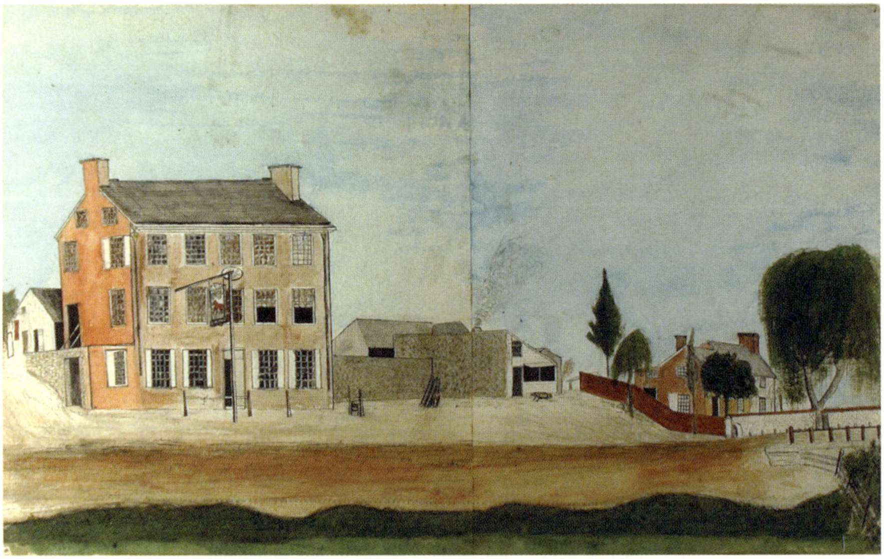

Figure 2.8. The Sorrel Horse Tavern stood at the corner of 51st Street and Woodland Avenue, Philadelphia. The original building of stone and plaster was built in 1704; the third story and brick front were added in 1860. The building to the right of the tavern is the blacksmith shop of Maximilian Leech, a close friend of Wilson's. Later Isaac Leach—Max's eldest son, one of Wilson's students, and Wilson's traveling companion on his hike to Niagara Falls—inherited the shop and it continued as a smithy throughout Isaac's life. The house to the right of the shop is the home of Max and Betsy Leech, where they lived and raised their family, which included not only Isaac, but also a daughter Mary, who became one of Wilson's colorists while still in her teens, and twins, one of whom was named Wilson, after Alexander, the family's friend and teacher to four of their children. The watercolor was probably painted by Mary Leech Fetters in 1860 shortly before she moved west to live with her son's family. (From the archives of the Ernst Mayr Library of the Museum of Comparative Zoology, Harvard University.)

Wilson also had access to the library of the American Philosophical Society and the Library Company of Philadelphia, which contained a number of books devoted wholly or in part to ornithology. (For a list of books available to Wilson at all of these libraries, see Table A.1.)

Wilson's wanderlust, his desire "to make a collection of all our finest birds," and the example of his mentor's travels through the Carolinas, Georgia, and Florida some thirty years earlier all came together in the fall of 1804 when he decided to journey to the falls at Niagara. He left in early October accompanied by Isaac Leech, a teenage pupil and eldest son of his neighbor, and walked to Ovid, New York. There he was joined by his nephew, William Duncan, for the final walk to the falls. Wilson's return trip was hurried by the approaching winter and his nearly two-month absence from the school in Kingsessing. On the last day of the trip he and Isaac walked forty-seven miles to get home. Upon his return, Wilson went straight to the home of Maximilian and Betsy Leech to congratulate them on the birth of twins, Ann and Wilson, and to give the couple six dollars as a present for his namesake (Figure 2.8). This left him enough money to buy some firewood and seventy-five cents for bread and milk.[52] His detailed notes of the journey inspired a 2,219-line poem, "The Foresters," which he published in serialized form in *Port Folio,* a prominent literary magazine of the time.

In addition to his notes, Wilson collected two birds. He illustrated these and, after seeking Bartram's advice, sent the illustrations to President Jefferson. If this seems like an odd choice of correspondent, it was not. In his *Notes on the State of Virginia* Jefferson had published a list of birds based on those described by Mark Catesby.[53] In addition to Catesby's names, Jefferson had provided for each species listed the Linnaean equivalent name, the current common name, and the page and volume of "Oiseaux" in the Comte de Buffon's *Histoire naturelle.* (In an accompanying table he had listed additional species by common name only and without documentation.) This work represents the first known state bird list and is a testament to Jefferson's interest and familiarity with birds of the region. With his illustrations Wilson sent the following letter:

> 18 March 1805
>
> On my return from a visit to the Falls of Niagara in October last I killed two Birds on the shores of the Mohawk river and conceiving them to be little known, particularly the Jay, I have taken the liberty of transmitting under favour of Mr. Bartram as faithful a sketch of them as I was capable of taking. The Jay approaches nearly to the *Corvus Canadensis* of Linnaeus and *Le Geay brun* of Buffon differing however in the colour and article of crest so much as to seem to

> be a distinct species. From several other Birds found while on the same Tour I am inclined to believe that many subjects still remain to be added to our Nomenclature in the Ornithology of the middle and northern states.[54]

Jefferson's reply from Monticello on 7 April 1805 opened graciously and then entered into a detailed discussion of the birds' features as revealed in Wilson's drawings. Jefferson compared Wilson's jay with his own observations of European birds (he had been ambassador to France from 1780 to 1789) and concluded that it was not European. He identified the second bird as a flycatcher (*Muscicapa*) similar to one killed by a neighbor a few days earlier. Jefferson also described a bird that he had been unable to identify and hoped that Wilson would be able to solve the mystery. Wilson expressed his elation at Jefferson's reply in his letters to Bartram (on 18 April 1805) and William Duncan (on 8 May 1805).[55] The letter to Duncan also requested that he collect birds and information.

Wilson's correspondence with Jefferson and Duncan signal a new dimension in his ornithological development. Not only did he continue to work hard on his own—observing, collecting specimens, drawing, and reading—but he also began to exchange information with others and enlist their help with his work. The letters to Thomas Jefferson and William Duncan were the first in which Wilson inquired beyond his immediate circle of friends for ornithological information. On 30 September 1805 Wilson wrote to Jefferson, enclosing in his letter an additional set of drawings and suggesting that the mysterious bird was a Wood Robin, now called a Wood Thrush (*Hylocichla mustelina*).[56] The exchange of letters is much like a modern conversation among birders discussing a difficult identification.

Another important development in late 1804 was the publication of William Bartram's account of the Brown Creeper (*Certhia americana*).[57] The article opened with a title, no English species name, and a six-word Latin descriptive name (he was still in violation of Linnaeus's binomial system) and continued with a detailed physical description and three short paragraphs on migratory behavior, ecology, and voice. A picture of the creeper was prepared by his niece, Anne (Nancy) Bartram. This species description, the only one Bartram ever published, presages the orderly species descriptions found throughout *American Ornithology* and may have served as a model for Wilson's work.

In early 1806 Wilson read that Jefferson planned to send an expedition to explore the Ohio and Mississippi rivers. Shortly thereafter he outlined a proposed trip in a letter to Jefferson.

> Kingsessing, 6 February 1806
>
> Sir,
>
> Having been engaged, these several years, in collecting material, and furnishing drawings from nature, with the design of publishing a new Ornithology of the United States of America, so deficient in the works of Catesby, Edwards, and other Europeans, I have traversed the greater part of our northern and eastern districts; and have collected many birds undescribed by these naturalists. Upwards of one hundred drawings are completed; and two plates in folio already engraved. But as many beautiful tribes frequent the Ohio, and the extensive country through which it passes, that probably never visit the Atlantic States; and as faithful representations of these can only be taken from living nature, or from birds newly killed; I had planned an expedition down that river, from Pittsburg to the Mississippi, thence to Neworleans, and to continue my researches by land in return to Philadelphia.[58]

Jefferson never answered Wilson's letter, but the failure of Wilson's petition to Jefferson was soon forgotten in the excitement of developments that he described in a note to his friend about two months later:

> To William Bartram 22 April 1806
>
> I take the liberty of informing you that, having been importuned to engage as assistant editor of that comprehensive and voluminous work, [Abraham] Rhees' new Cyclopaedia, now publishing here,

> and a generous salary offered me, I have now accepted of the same, and will commence my new avocation on Monday next.
>
> This engagement will, I hope, enable me in more ways than one to proceed in my intended Ornithology, to which all my leisure moments will be devoted.[59]

The Birth of American Ornithology

As the newly hired assistant editor in the publishing house of Bradford & Inskeep, the foremost publisher in Philadelphia and perhaps the United States, Alexander Wilson received nine hundred dollars a year, but more importantly could realize his dream of publishing "a collection of all our finest birds" (Figure 2.9). He wasted little time. Within weeks of assuming his position, he met with Samuel Bradford and broached the possibility of writing a work on the birds of the United States to be published by the firm. Samuel Bradford agreed to publish *American Ornithology,* as it was christened, but only on the condition that Wilson obtain two hundred subscriptions to ensure the financial success of the venture.

Throughout the remainder of 1806 and into the spring of 1807, Wilson labored over the *Cyclopaedia* and devoted his free time to preparation of his *American Ornithology.* Twenty-five hundred copies of the prospectus for *American Ornithology* were printed on 6 April 1807. Written as a letter to "Lovers of Natural History," it is the first public statement of Wilson's vision for the project. In large type he points to its important features:

> the first books on the natural history of birds published in America;
> printed on the finest vellum paper;
> printed in a typescript specially designed for this work;
> available bimonthly as separates that would contain 3 plates each and cost $2. [This plan would be modified later to 10 volumes of 10 plates/volume at a cost of $12/volume];
> would cover the natural history of all resident and migratory species of the United States as well as descriptions and illustrations of new species.[60]

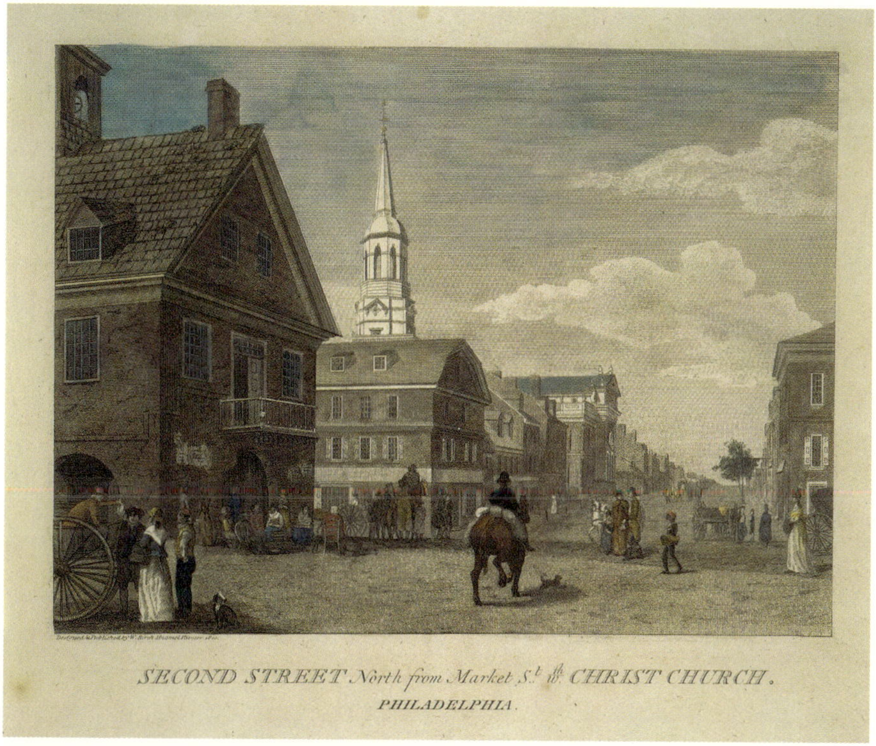

Figure 2.9. Central Philadelphia, Second Street viewed from Market Street, as it looked when Wilson was an editor working for Samuel Bradord of Bradford & Inskeep in central Philadelphia. (From the archives of the Library Company of Philadelphia.)

The letter that follows describes the intended work and touches on a number of themes that figured prominently in Wilson's life and his vision of *American Ornithology*. One of these is the potential of natural history in general and birds in particular to provide a cultural alternative to the "idle prattle of novels, . . . noisy discord of politics . . . [and] short-lived joys of scandal, ambition, avarice or debauchery."[61] Wilson saw wilderness as a common cultural experience that could unite Americans much as a shared history and ethnicity united the citizens of European nations.

In the prospectus Wilson writes of seeing "the great Author of the Universe," "Nature's God," and "Nature . . . ever open and inviting . . . replete with a divine and inexhaustible store of pleasure and instruction!"[62] Wilson's view of nature as a window on divine will, which he had already described in his letters, would be a recurrent theme in *American Ornithology*. Like Jefferson, Wilson was a deist. He believed in God, but was not an adherent of a specific religion.

In the prospectus Wilson also articulates publicly for the first time his commitment to writing a scientific treatise on American birds. He promises to specify the genus and species of every type of bird found in the United States of 1807. He also pledges to correct the mistakes of past authors who had based their work on preserved birds by drawing from live birds and by writing from hours of personal observation of many living individuals.

Wilson makes a revealing comment in the prospectus about his view of art when he promises to portray America's birds in the "simplicity of truth and nature," a style that he contrasts to the "glare of false and gaudy coloring, [and] extravagant distortion of posture" of previous illustrations.[63] He attributes such artistic excesses to the use of stuffed specimens, and clearly articulates a stylistic preference for simplicity, an opinion that would have appealed to Americans of the early nineteenth century.

Wilson's prospectus must have seemed like a promising business venture: his letter to William Bartram just two days after the prospectus was issued indicates that Samuel Bradford, his employer and publisher, had offered important assistance in the search for subscribers.

> 8 April 1807
>
> Enclosed is a proof-sheet of our prospectus; . . . We mean to bind in the prospectus at the end of the next half volume (of Ree's Cyclopaedia), for which purpose twenty-five hundred copies are to be thrown off and an agent will be appointed in every town in the Union. The prospectus will also be printed in all the newspapers, and everything done to promote the undertaking.[64]

That Bartram's advice, knowledge, and especially his encouragement continued to be important to Wilson is evident in another letter, sent later that month:

To William Bartram

Philadelphia, 29 April 1807

My Dear Sir,

The receipt of yours of the 11th inst., in which you approve of my intended publication of American Ornithology, gave me much satisfaction; and your promise of befriending me in the arduous attempt commands my unfeigned gratitude. From the opportunities I have lately had, of examining into the works of Americans, who have treated of this part of our natural history, I am satisfied that none of them have bestowed such minute attention on the subject as you yourself have done. Indeed, they have done little more than copied your nomenclature and observations, and referred to your authority. To have you, therefore, to consult with in the course of this great publication I consider a most happy and even auspicious circumstance; and I hope you will, on all occasions, be a rigid censor, and kind monitor, whenever you find me deviating from the beauties of nature, or the truth of description.[65]

Wilson continues later in the same letter to muse on a problem that is still very much with us.

> The more I read and reflect upon the subject, the more dissatisfied I am with the *specific* names which have been used by almost every writer. A name should, if possible, be expressive of some peculiarity in colour, conformation, or habit; if it will equally apply to two different species, it is certainly an improper one. Is *migratorius* an epithet peculiarly applicable to the robin? Is it not equally so to almost every species of *turdus* we have? *Europea* has been applied by Pennant to our large *sitta* or nuthatch, which is certainly a different species from the European, the latter being destitute of the black head, neck, and shoulders of ours. Latham calls it *carolinensis,* but it is as much an inhabitant of Pennsylvania and New York as Carolina. The small red-bellied *sitta* is called *canadensis* by

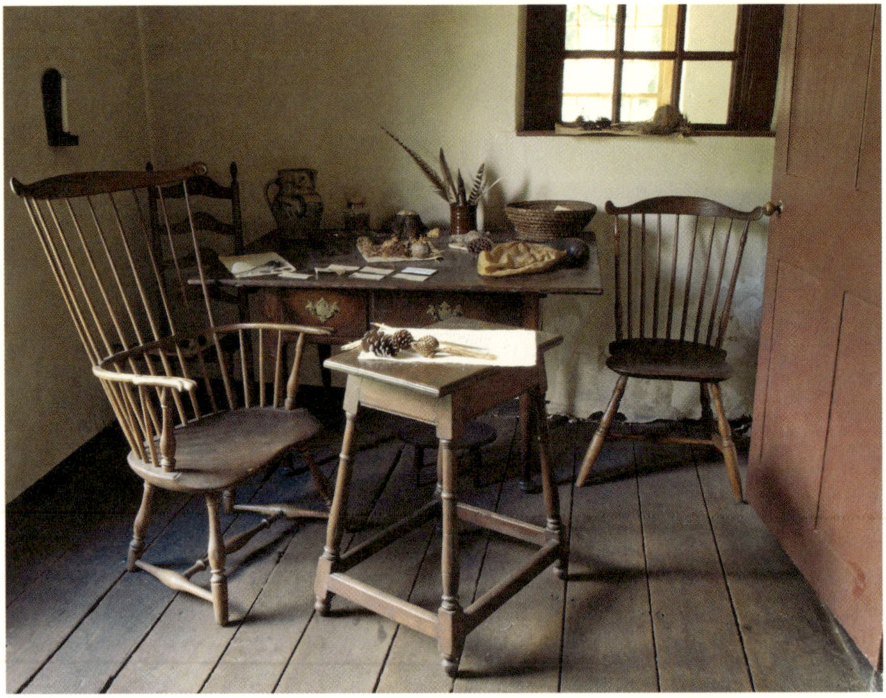

Figure 2.10. The room at Bartram House used by Wilson as his study and bedroom when he visited William Bartram. The furniture is from the eighteenth century and may have been used by Wilson. (From the collection of the John Bartram Association.)

> Latham, a name equally objectionable with the other. *Turdus minor* seems also improper; in short I consider this part of the business as peculiarly perplexing; and I beg to have your opinion on the matter, particularly with respect to the birds I have mentioned, whether I shall hazard a new nomenclature, or, by copying, sanction what I do not approve of.[66]

This letter of 1807 gives us a sense of Wilson's progress from student of ornithology to ornithologist, as well as an idea of the tenor of Wilson's and Bartram's discussions over the next few years. Unquestionably Bartram had taught Wilson not only the names of birds and plants, but also something

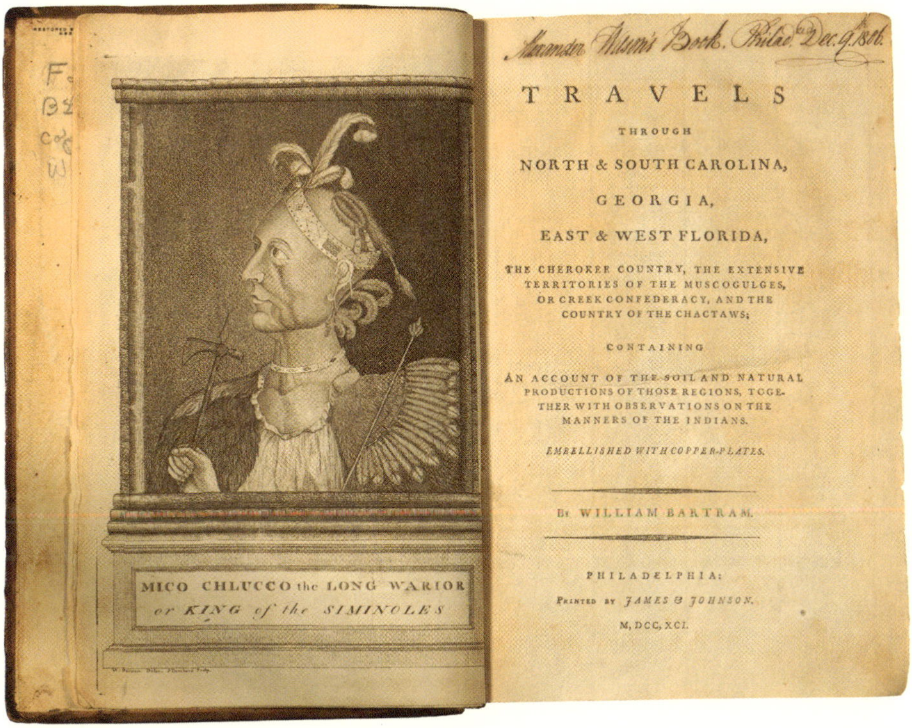

Figure 2.11. Title page of Wilson's copy of Bartram's *Travels through North and South Carolina* . . . with "Alexander Wilson's Book, Philad, Dec. 9th. 1806." at the top of the page. (From the Academy of Natural Sciences of Philadelphia Library and Archives Collection, ANSP 79 [Drexel University].)

even more valuable—the importance of observation. In his poetry Wilson had shown himself to be a keen observer of human nature and behavior. Just imagine, then, the thoughtful discussions about birds and plants that the two men must have had during Wilson's evenings at Bartram's house, where he had his own room (Figure 2.10). Certainly the value Bartram placed on observation in his own written work struck a responsive chord in Wilson (Figure 2.11).

Toward the end of September 1807, Bradford sent Alexander Wilson to New York City to obtain subscriptions for the *Cyclopaedia* and *American Ornithology*. Wilson carried with him copies of the prospectus and samples of the plates of his birds. Once again he was on the road peddling. His efforts on behalf of the *Cyclopaedia* met with success, but he managed to sell only a few subscriptions to *American Ornithology*. Moreover, Brisbane and Brannan, New York's leading booksellers, wanted to see the first volume before deciding whether or not to act as agents. Soon after his return to Philadelphia, he was greatly cheered by a letter from Thomas Jefferson in which the president requested a subscription.

From October 1807 to September 1808 Wilson edited the *Cyclopaedia* and wrote, illustrated, edited, and oversaw production of the first volume of his *American Ornithology*, 250 copies of which were printed and bound in September 1808. Volume 1 represents the first detailed scientific and literary treatment of American birds. The species accounts are often lengthy and provide descriptions of anatomy, behavior, ecology, diet, and economic value. The treatment is thorough and remarkably modern, a testament to Wilson's observational and scientific skills. The literary style, too, is light and clear, often poetic, reflecting Wilson's years of experience as a writer.

Success

With the publication of volume 1, Bradford had fulfilled his part of the bargain. If publication was to continue, Wilson had to find subscribers. In late September he left Philadelphia on a trip that would take him to New York and New England, where he visited Columbia, Dartmouth, Harvard, Princeton, and Yale. Columbia and Dartmouth not only welcomed him, but also subscribed to *American Ornithology*. Elsewhere, however, he met with compliments, but only forty-one subscriptions—far fewer than he needed. His disappointment is evident in a letter he wrote later that fall to Alexander Lawson, who had encouraged Wilson's interest in drawing and had engraved his plates.

> To Alexander Lawson Albany, 3 November 1808
>
> In short, the book . . . so far exceeds the ideas and expectations of the first literary characters in the eastern parts of the United States, as to command their admiration and respect. The only objection has been

the 120 dollars, which in innumerable instances has risen like my evil genius between me and my hopes. Yet I doubt not but when those subscribed for are delivered, and the book a little better known, the whole number will be disposed of and perhaps encouragement given to go on with the rest.[67]

Wilson was less restrained when writing in his journal:

Arrived at ________ waited on Dr. ________, Principal of the Seminary. It was near dusk before I could see him, and our conversation, which was held on the steps leading to his house occupied about five minutes. He considered the volume as too expensive for any class of reader about this town. He behaved with cold indifference—turned over a few leaves without any seeming interest, and said, that as far as he could see, (for it was nearly dark) it looked well. He returned the volume and we parted. If, as Principal of this college, this literary luminary shed no more cheerful influence over the exertions of his pupils than he did on the author of the *American Ornithology* I do not much wonder that storms and tempests should desolate this seminary, and damp the energies of the inhabitants.[68]

Despite the disappointment evident in his letter to Lawson and the bitter tone of his journal entry, both documents also convey a sense of the strength and commitment that sustained Wilson throughout the Herculean effort required to write, produce, and sell *American Ornithology*. Aside from its meager monetary return, the trip had provided an opportunity to observe and write about the birds of New England and to meet bird watchers throughout the region. To a friend he wrote:

Boston, 10 October 1808

. . . I shall not sit down with folded hands, whilst anything can be done to carry my point, since God helps them who help themselves. I am fixing correspondents in every corner of these northern regions,

> like so many piquets and outposts, so that scarcely a wren or tit shall be able to pass along, from New York to Canada, but I shall get intelligence of it.[69]

Alexander Wilson returned to Philadelphia in late November, but his lack of subscribers and his need for information meant that by early December he was planning another trip, one that would take him through Maryland, Virginia, the Carolinas, and Georgia. Almost a hundred years earlier Mark Catesby, the first person to write at length on American birds, had traveled and subsequently written about the natural history of the area.[70] Only thirty years earlier William Bartram, Wilson's friend and mentor, had also traveled and written about the Carolinas, Georgia, and Florida.[71] Their widely acclaimed books undoubtedly influenced Wilson's itinerary, though wealthy southern planters also represented a promising, untapped market for his *American Ornithology.* Furthermore, while he was keenly aware of the pioneering accomplishments of Catesby and Bartram, he also knew of inconsistencies in their work that might be resolved by new observations.

Wilson was passing through Washington on 17 December when he stopped at the White House and knocked on the door. It was answered by a young man who asked for identification. Wilson responded by introducing himself as Alexander Wilson, the ornithologist, who wished to meet with President Jefferson and present him with a copy of the *American Ornithology.* When the secretary asked if he had an appointment, Wilson acknowledged that he had none, but again asked to see Mr. Jefferson. The secretary asked him to wait. He returned a few minutes later to say that the president would be delighted to meet Mr. Wilson. After being ushered into the Oval Office, Wilson greeted the president and presented to him volume 1 of *American Ornithology,* thereby fulfilling the president's subscription. The two men then spent much of the afternoon discussing birds, science, and the importance of refuting the European assertion that American animals and plants were degenerate forms of those in Europe. At the end of their meeting, Jefferson gave Wilson letters of introduction to all members of his cabinet and many lesser officials urging them to buy subscriptions, as well as a letter of introduction to John Tyler, governor of Virginia and father of the future president, asking that he provide information and further introductions. Today it is difficult to even conceive of an informal visit such as Wilson's.

Wilson sold seventeen subscriptions in Washington before continuing to Georgetown, Alexandria, Richmond, and Charleston. On his way to Charleston he rode through the extensive pine forests characteristic of the Carolina coastal plain. Not far from Wilmington, he shot and killed two Ivory-billed Woodpeckers and slightly wounded a third. With the two specimens in his pack, he wrapped the injured bird in his coat, placed it in front of him on his saddle where he could keep hold of it, and rode into Wilmington where he planned to spend the night. As he rode through the streets and up to the hotel, the piteous cries of the woodpecker attracted a worried crowd. They joined the landlord in a relieved laugh when Wilson dismounted and unfolded his coat, revealing the furious woodpecker.

After he registered, Wilson took his pack and his woodpecker to his room, then left to tend to his horse. Upon his return he could hear chopping sounds as he ascended the stairs to his room. He opened the door and the woodpecker cried out, possibly in frustration at being discovered in its effort to escape. It had climbed the window frame and cleared a fifteen-inch square section of plaster from the wall, plaster that now lay in chunks on the bed. It had also cut a hole larger than a man's fist through the lathing and begun chiseling on the back side of the exterior weatherboards. In less than an hour it would have escaped. Wilson caught the woodpecker, tied its leg to the leg of a heavy mahogany table, and went to find it some suitable food. When he returned, the Ivorybill had cut the leg off the table, but could not separate itself from the mahogany leg.

At this point, Wilson settled down to draw the woodpecker and it is this very individual that appears in the *American Ornithology*. Despite many attempts over the next three days, he was unable to coax it to eat and it died. His devotion to collecting and nurturing this bird underlines Wilson's emphasis on working with live birds and collecting his own specimens for anatomical study. The bird and its portrait are important because they constitute the northernmost record of this magnificent and now extinct species.

From Charleston, Wilson wrote several detailed letters that describe the countryside and his travels in such beautiful prose that one wonders whether he intended them for later publication. From Charleston he continued on to Savannah and thence to St. Augustine, Florida. In mid-March 1809, he sailed home to Philadelphia.

As with his trip to New York and New England, this journey allowed Wilson to establish a network of correspondents and obtain both notes and specimens for future volumes. This trip, however, was also a financial success. He sold 250 subscriptions, thereby assuring publication of *American Ornithology*.[72] One senses his commitment and pride in the letter to his father that accompanied a copy of volume 1:

> To Alexander Wilson, Sr. Philadelphia, 15 June 1809
>
> In giving existence to this Work, I have expended all I had been saving since my arrival in America. I have also visited every town within 150 miles of the Atlantic coast, from the river St. Lawrence to St. Augustine in Florida, from whence I returned about two months ago. Whether I shall be able to realize a fortune by this publication, or recover first costs, or suffer the sacrifice of my little all is yet doubtful. I met with a most honourable reception among many of the first characters in the United States, and have collected such a mass of information on this branch of Natural history, as will entitle the work to the merit of originality at least.[73]

Western Travels

Soon after Wilson's return from St. Augustine in March 1809, he established *American Ornithology* as a separate department at the publishing house of Bradford & Inskeep and negotiated contracts with the engravers, colorists, and printing specialists required for the colored plates. During the spring and summer, he prepared fifty illustrations and wrote accounts of forty-two species, some of which he described for the first time. He also spent this period of intense work evaluating his notes on birds, outlining his interpretations, and soliciting Bartram's views.

Bartram had been a crucial influence during Wilson's development as an ornithologist, scientist, and artist. He was the person to whom Wilson turned for guidance as he developed the skills he would need to complete *American Ornithology,* and he continued to consult Bartram as his skills and knowledge matured. After publication of volume 1, however, Wilson's letters to Bartram became less frequent and the tone changed:

> *To William Bartram*
>
> 11 October 1809
>
> Dear Sir,
>
> Thanks for your bird, so neatly stuffed, that I was just about to skin it. It is the Rallus virginianus of Turton, and agrees exactly with his description. The one in company was probably the female, Turton mentions 4 species as inhabitants of the United States. I myself have seen 6. Mr. Abbot of Savannah showed me two new species. I found the Sora, as the Virginians call it, in the rice flats near Savannah, in March. Gen. Wilkinson told me that the Sora was in multitudes at Detroit. Query—don't you think they breed in the north, like the rice-birds? Are not the European naturalists mistaken in saying that the reed-bird and rice-bird pass from the Island of Cuba in September, to Carolina?[74]

Wilson's knowledge and confidence are apparent; this 1809 letter represents a discussion between equals. Wilson was an ornithologist sharing his knowledge with a valued colleague and discussing questions of interpretation based on published information known to both men.

Our modern field guides make the problems Wilson faced difficult to comprehend. Sexual dimorphism and maturational changes in the American Kestrel (*Falco sparverius*), for instance, seem to be the sources of variation described in the following letter, which also reveals the care with which Wilson noted and pondered details.

To William Bartram

Philadelphia, 25 October 1809

. . . the Small Hawk, commonly though improperly called the Sparrow Hawk, above of a bright red body beautifully crossed with black particularly the tail, the head blue. Crown red cheeks spotted with black—10½ inches—streaked laterally below with brown—vent white.

. . . described as having yellow eyes—all the specimens I have met with had the eyes dusky.

I have frequently shot a smaller Hawk than this last with very long slender legs and blue wings much like the preceding but certainly a different species—that had yellow eyes.[75]

On 1 January 1810, Wilson signed the preface to volume 2 of *American Ornithology* and three weeks later, Robert Carr, his printer, presented him with several copies of the second volume. With these and copies of the first volume, Wilson left Philadelphia on 30 January 1810 on the most ambitious expedition of his life. From Philadelphia he walked west to Lancaster (which was then the capital of Pennsylvania), crossed the Allegheny Mountains, and arrived in Pittsburgh on 15 February 1810. He obtained nineteen subscriptions; bought a rowboat on the stern of which he painted "ORNITHOLOGIST"; and on the morning of 23 February loaded his trunk, greatcoat, gun, and supplies into the boat and left Pittsburgh for Cincinnati, Ohio, and Louisville, Kentucky, which was 720 miles downriver. As he drifted along, he dodged floating ice; landed frequently to observe, describe, and sketch birds; and took notes on American frontier life.

At Big Bone Lick, Kentucky, Wilson put ashore to observe what he called "Carolina Paroquets." He also shot several for anatomical study and picked up a wounded one, for which he built a cage. Within an hour the parakeet was eating cockleburs, its favorite food, from his hand. During the overland portion of his trip, as he journeyed from Louisville to Lexington to Nashville to Natchez and New Orleans, Poll, as Wilson called her, rode in one of his pockets wrapped and tied in a handkerchief. Whenever he stopped for meals, Wilson would set her free, and she would sit on his baggage and eat the cocklebùrs he provided. Mealtimes along the Natchez Trace were often a time of socializing as the Chickasaws or Chactaws, particularly their children, collected around Wilson and watched with delight as Poll climbed out of his pocket to perch on his shoulder and take cockleburs from his mouth.[76] Poll was very much an ambassador of goodwill between Wilson and the Native Americans. In fact, the relationship promoted by Poll may have saved Wilson's life when he developed dysentery; the natives gave him advice on what to eat to cure himself and even brought him eggs and wild strawberries to help him heal.

Upon arriving in Louisville, Wilson secured a room at the Indian Queen Tavern.[77] The very next day, he sold his skiff for exactly half what it cost him; ". . . and the man who bought it wondered why I gave it such a droll Indian name (Ornithologist) 'some old chief or warrior I suppose' said he."[78]

With the first two volumes of *American Ornithology* under his arm, Wilson visited the local merchants in pursuit of subscribers and in due course entered the store of one John James Audubon. The several accounts of their meeting differ and no account appears in Wilson's letters to Lawson. Audubon did not buy a subscription to Wilson's *American Ornithology,* which may have been due to his limited means. Wilson looked at Audubon's portfolio of drawings, complimented him on his sketches of birds, and was surprised to learn that Audubon had no plans to publish them. They spent two days hunting together, but Sandhill Cranes (*Grus canadensis*) and Passenger Pigeons (*Ectopistes migratorius*) are the only species that Wilson mentions seeing with Audubon. On 23 March, Wilson left Louisville with no new species and no new subscriptions. He walked to Shelbyville, Frankfort, and Lexington, where he sold fifteen subscriptions and purchased a horse before continuing on to Nashville. In Nashville he spent eight days painting birds, among which the Tennessee (*Vermivora peregrina*), Nashville (*V. ruficapilla*), and Kentucky (*Oporornis formosus*) warblers were new to science. He sent the paintings, notes, and a long narrative letter to Lawson on 28 April 1810. The letter arrived, but the paintings and notes were lost, delaying publication of volumes 3 and 4.

Leaving Nashville on 4 May 1810, Wilson journeyed through the wilderness to Natchez. On the night of 10 October 1809, Meriwether Lewis, explorer and governor of the Louisiana Territory, had been fatally wounded at Grinder's Stand in Tennessee. Wilson had published the first description of Lewis's Woodpecker (*Melanerpes lewis*), which he had named for his friend, and he resolved to visit Grinder's Stand to investigate Lewis's mysterious death.[79] His account raises a number of questions concerning earlier reports of the circumstances surrounding the death, but he could not decide between suicide and murder.[80] Before he left Grinder's Stand, he used his own money to arrange for a marked gravesite for Lewis's remains.

Soon after arriving in Natchez on 17 May, Wilson obtained twelve subscribers. While in the area, he was the guest of William Dunbar, a prominent planter and scientist, who provided information on local birds and later sent Wilson a specimen of a Roseate Spoonbill (*Platalea ajaja*). From Natchez, Wilson journeyed to New Orleans, where he embarked for New York on 24 June 1810. When his ship was becalmed in the Gulf of Mexico, Wilson used the time to visit several nearby islands and study their birds.

During this leg of the trip, Poll, Wilson's parakeet companion of several months and many miles, met an unfortunate end: "In this short space she had learned to know her name; to answer, and come when called on; to climb up my clothes, sit on my shoulder, and eat from my mouth. I took her with me to sea, determined to persevere in her education; but, destined to another fate, poor Poll, having one morning about day break, wrought her way through the cage, while I was asleep, instantly flew overboard, and perished in the Gulf of Mexico."[81]

Wilson reached Philadelphia in early August 1810 only to discover that his drawings and notes mailed from Nashville had never arrived. For a month he worked furiously to replace the lost material, not notifying Bartram of his return until early September. He published an account of his seven-month journey through the American wilderness in the fall issue of *Port Folio*. Throughout the fall and winter he wrote about and painted birds; conferred with Lawson, his friend and engraver; and supervised the colorists. On 12 February 1811 he dated the preface to the third volume of *American Ornithology*. He had already completed several plates for volume 4, but maintained his furious pace throughout the spring.

Devoted to American Ornithology

In the summer of 1811 Wilson resigned the editorship of Rhees's *Cyclopaedia*, a move made possible by the financial success of his western trip. The number of subscribers was now 450.[82] To meet the increased demand, the first volume had had two additional press runs, and the second volume had had one additional run. Because he had exceeded Bradford's original goal of two hundred subscriptions, Wilson had a small cash reserve. He could afford to forgo his salary and devote all his time to *American Ornithology*. His resignation also enabled Wilson to spend more time with William Bartram, observing birds in Bartram's garden and discussing natural history. Lawson, encouraged by Wilson's devotion to *American Ornithology* and convinced that his own reputation rested on the success of the books, finished the plates for volume 4, which was released on 12 September 1811.

The spring and summer of 1811 marked the beginning of Wilson's friendship with George Ord, the wealthy owner of a ship chandlery business and an avid student of natural history. Wilson first visited Cape May during the winter of 1810–1811 to study waterbirds, but he made six trips in all and was accompanied often by Ord, who became the editor of Wilson's posthumously published ninth volume, author of *Sketch of the Life of Alexander Wilson* (1828), and a jealous guardian of Wilson's reputation. Unfortunately and somewhat ironically, Ord's undiplomatic defense of Wilson alienated Audubon and the ensuing acrimonious debate persisted for many years, much to the detriment of Wilson's reputation.

In the autumn of 1811, Wilson devoted himself to volume 5, which was published on 12 February 1812 after unexpected troubles:

> To William Bartram — Favoured by Major Carr
>
> Philadelphia, 12 February 1812
>
> It has not been from neglect or forgetfulness I assure you that I have not been to see you since we parted in October. Since Mr. William Duncan left me as colourist I have been obliged to sit closely at that employment myself and what with Drawing, Writing my 5th volume

> (which is now nearly printed) and correcting the proofs I have had few moments to spare.[83]

Since 12 February 1811, Wilson had completed volumes 3, 4, and 5, a remarkable feat when you consider that he not only wrote and illustrated the text, but also oversaw all aspects of production and, as indicated in the letter just cited, colored the plates. Despite these demands on his time, Wilson's work on volume 6 was far advanced by spring 1812.

Wilson spent May of that year at Cape May, New Jersey, collecting material on water birds for volumes 7, 8, and 9, but the threat of financial ruin preoccupied him during the spring and summer of 1812. Subscribers paid for their books at the end of the year in which they received them, and because volume 5 had just been published in February, volume 6 was at the printer's, and volume 7 was in preparation, the expenses incurred by Bradford, Wilson's publisher, had exceeded the cash received from subscribers. By August the situation had become critical and Wilson was forced to undertake another business trip.

> To Sarah Miller　　　　Philadelphia, 7 August 1812
>
> . . . Mr. B. and I having all our accounts to settle, . . . I must now either run the risk of losing all or make one last and very long and expensive journey to collect what is due, and see how accounts are with the agents. There is no other choice left between this and absolute ruin . . . Mr. B. has positively refused to advance anything until he receives it and I have as positively told him that I will proceed no further with the work until I am paid for what I have done.[84]

The "journey" referred to was to New York and New England. Wilson left sometime in September, traveling up the Hudson River to Albany, then overland to Lake Champlain and Burlington, Vermont, across the Green Mountains to the Connecticut River and south to Haverhill. There he was arrested as a Canadian spy. One can only imagine the flashback to Paisley and his long days in the Tollbooth prison. But he was freed after convincing the judge of his natural history and business pursuits. He then visited Portsmouth and Portland before writing to Ord of the physical effects of months of hard work and long hours:

To George Ord Boston, 13 October 1812

I hope, my dear sir, that you have been well since I left you, I have myself been several times afflicted with a violent palpitation of the heart, and want to try whether a short voyage by sea will not be beneficial.[85]

Wilson was able to travel by sea from Boston to New York, but soon after his return to Philadelphia his colorists quit, leaving him with 4,050 plates to hand color for volume 7. Thus during the winter of 1812–1813 Wilson not only completed the text for volume 7; he also colored the plates. Ord notes in his biography of Wilson that he was responsible for all costs associated with the preparation of the text and drawings, which would have included travel to obtain information and specimens. These costs had exhausted Wilson's limited savings, and he had undertaken the job of colorist as his sole means of support. Bradford paid the colorists twenty-five cents a plate.

Despite these setbacks, Wilson signed the preface to volume 7 on 1 March 1813 and a month later wrote:

To William Bartram Philadelphia, 21 April 1813

Next week I shall publish my seventh volume; and shall send you your copy with the earliest opportunity. I am now engaged with the ducks, all of which, that I am acquainted with, will be comprehended in the eighth volume.

. . .

I long most ardently to breathe once more the fresh air of the country, and gaze on the lovely face of Nature. Will it be convenient for the family to accommodate me (as I shall be alone) this summer? Please to let me know.

In this, his last letter to William Bartram, Alexander Wilson turns again to his friend. To Wilson, William Bartram was far more than an important scientist, teacher, and colleague; he was the calm strength to which Wilson returned in spirit, if not in person, when he needed renewal for the trials ahead. Throughout their friendship, Alexander Wilson, with Bartram's support, faced the seemingly insurmountable difficulties of researching, writing, illustrating, editing, and publishing *American Ornithology* with remarkable determination and resiliency.

The letter continues with good news:

> The Philosophical Society of Philadelphia have done me the honor to elect me a member, for which I must certainly, in gratitude, make them a communication on some subject, this summer.[86]

Wilson's election on 16 April 1813 to the American Philosophical Society, the first scientific organization in the United States, was in recognition of the scientific importance of his *American Ornithology.* At the time Jefferson was president of the society, which must have greatly added to Wilson's pleasure at his election.

In May, Wilson and Ord went to Cape May where they remained for four weeks collecting material for the eighth volume. Upon their return Wilson did not, as requested in his letter to Bartram, spend the summer with his old friend, but devoted himself to preparing the text and drawings for volume 8. Wilson's last surviving letter was written to an unnamed friend in Paisley:

> Philadelphia, 6 July 1813
>
> . . . I am, myself, far from being in good health. Intense application to the study has hurt me much. My 8th volume is now in the press and will be published in November. One volume more will complete the whole which I hope to be able to finish in April next.[87]

On the morning of August 13, Wilson listed the species still to be drawn:[88]

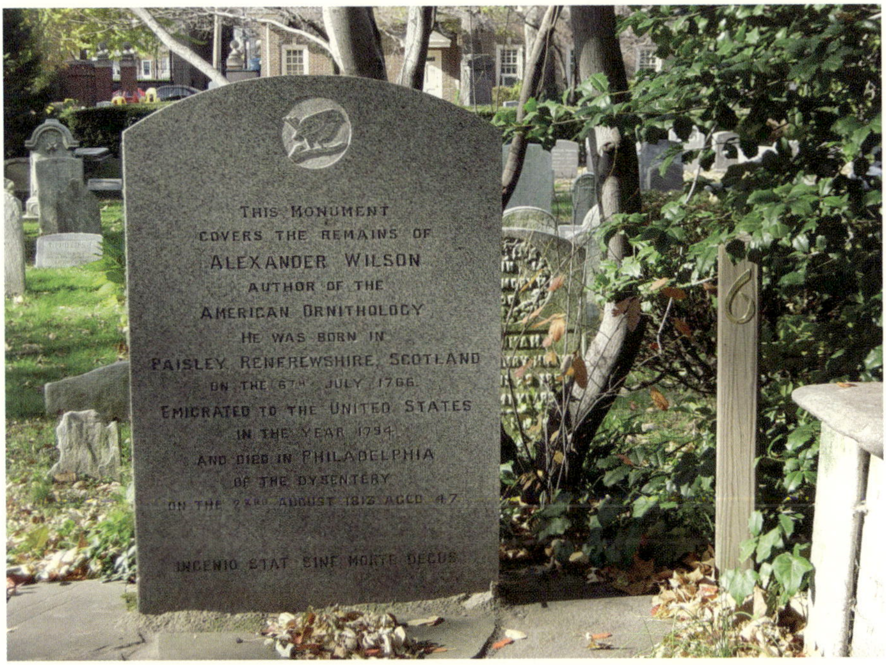

Figure 2.12. Alexander Wilson's grave in the Jones family plot at Old Swedes Church, Philadelphia. (Photograph by Edward H. Burtt, Jr.)

Gannet
Young
Frigate Pel.
Cormorant
Brown Booby
Great White P.
Swan
Albatross

Turkey
Blue Hawk
Razor Bill
Crested Grebe
Little G.
Speckled Diver
Shearwater Petrel

Tropic Bird
Puffin
Black-backed Gull
Skua G.
Kittiwake G.
Herring G.
Common G.

Hours later he was taken ill with what appears to have been dysentery. His exhausted condition contributed to the rapid progress of the disease and Wilson, recognizing the seriousness of his situation, dictated and signed his will on 16 August 1813. The following announcement appeared in the *United States Gazette of Philadelphia* on the evening of 23 August 1813:

> Died this morning, Mr. Alexander Wilson, author of the American Ornithology. His friends and acquaintances are invited to attend his funeral from Mr. William Jones', 233 Spruce St., Phila., at 9 o'clock tomorrow morning.

At the request of William Jones, Wilson's landlord of many years, the funeral was held on 24 August 1813 in the Swedish Lutheran Church, known in the neighborhood then and still as Old Swedes Church. The Reverend Mr. Clay officiated and Alexander Wilson's body was laid in the Jones family plot in the churchyard. A simple stone marks the grave (Figure 2.12)—the final resting place of Alexander Wilson: poet, author of *American Ornithology,* and Father of American Ornithology.

CHAPTER THREE

Illustrating *American Ornithology*

Toward the end of 1805, maybe as late as 1806, Alexander Wilson took his first tentative steps toward publication of his *American Ornithology* when he concluded that he had mastered the skills needed to complete his portrayal of America's birds. He was a superb marksman, skilled in wilderness survival, able to endure hardship, familiar with the natural history literature, and a skilled observer who could describe in words and pictures what he saw. His drawing had progressed to the point where his birds were not only accurate in detail, but also had an intensity, a liveliness that had been lacking in his earlier sketches. Reproducing his drawings on a copper plate for printing was the only aspect of publication that was still outside his skill set and, like Catesby and Edwards before him, he meant to master the etching process so as to keep the description and interpretation of birds entirely under his control.[1]

What follows are the pencil sketches, pen and ink drawings, and draft paintings known to us, for the plates in *American Ornithology*. The drawings and paintings demonstrate the skill and insight with which Wilson observed birds. They also show that as his experience increased, so did the sensitivity of his portraits. Some of this transformation can be seen in the illustrations for his books, but other aspects are evident only in the work of the moment, the sketch made as he watched the bird in its native habitat. To fully appreciate the difference between the sketches and the printed illustrations, it is helpful to understand how these drawings and paintings were translated into the copper plate engravings that were then printed and hand-colored.

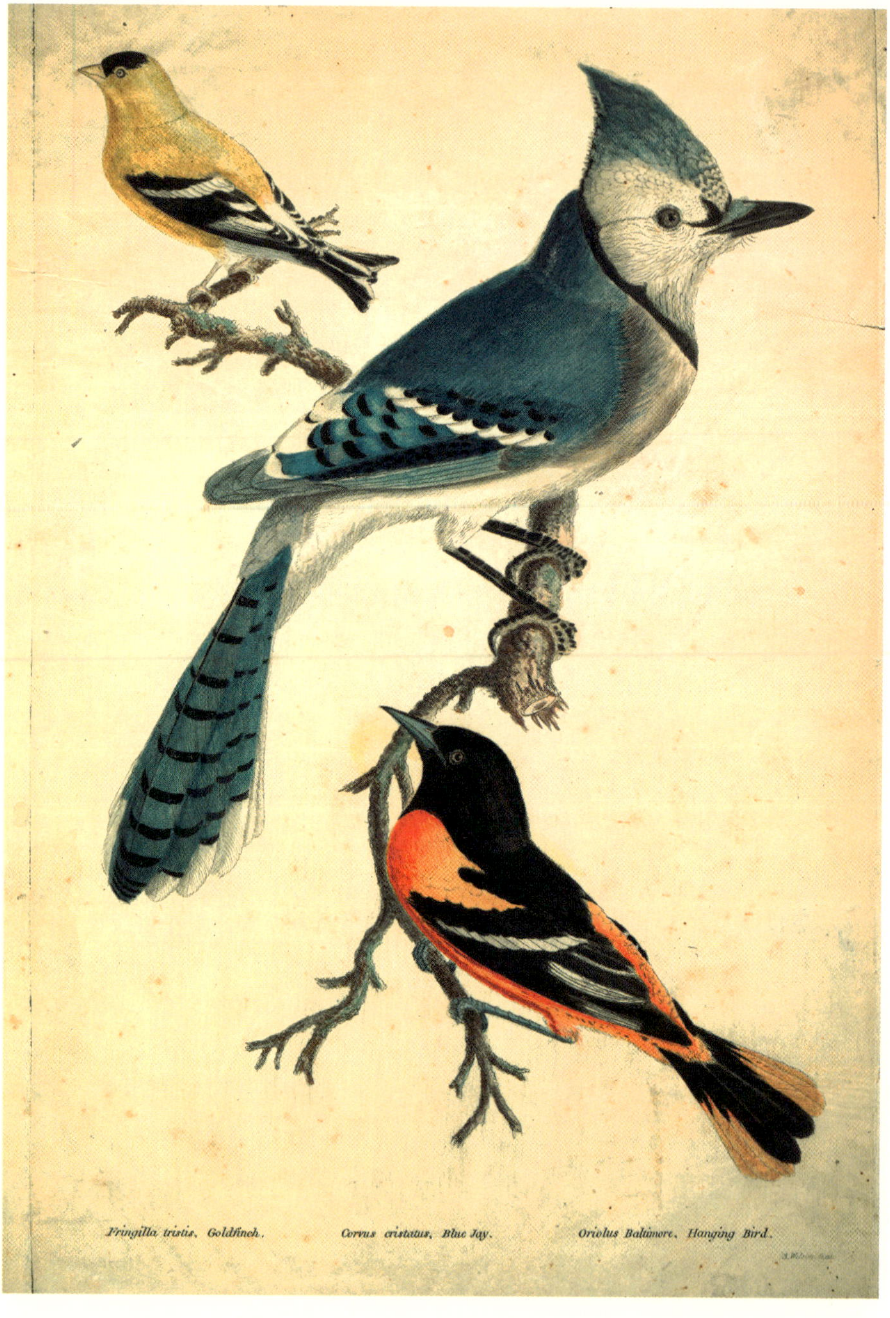

Figure 3.1. Wilson engraved this plate after instruction from Alexander Lawson and sent the print to William Bartram for his comments. (From the Academy of Natural Sciences of Philadelphia Library and Archives Collection, ANSP 79).

Preparing and Printing the Illustrations

Alexander Lawson, who had encouraged Wilson to draw birds, had a flourishing business engraving medical and chemical illustrations, maps, exploded views of engineering projects, and architectural views and plans. Quite naturally, Wilson turned to Lawson when he wished to learn to engrave his drawings and print his own illustrations. Lawson loaned him the tools and may have instructed him. Wilson engraved two plates, then printed and colored the first—an American Goldfinch, Blue Jay, and Baltimore Oriole (Figure 3.1)—and sent it to Bartram for comment.[2]

Whether due to his own evaluation or Bartram's gentle criticism, Wilson decided that Alexander Lawson's technique was superior, that Lawson was the right person to engrave the illustrations for *American Ornithology* (Figure 3.2). In the introduction to volume 1, Wilson offers two reasons for his decision: "Many years of application are necessary to enable a person, whatever may be his talents or diligence, to handle the graver with the facility and effect of the pencil, while the time thus consumed, might be more advantageously employed in finishing drawings, and collecting facts for the descriptive part, which is the proper province of the Ornithologist."[3] Wilson realized that his goal, to describe all the birds of the United States, required that he devote his time to the discovery and description of birds. Facilitating his decision to leave the engraving to Lawson was the realization that Lawson's skill, artistic ability, and knowledge of natural history would enable him to improve rather than simply copy Wilson's drawings.[4] As Wilson explained, "Every person who is acquainted with the extreme accuracy of eminent engravers, must likewise be sensible to the advantage of having the imperfections of the pencil corrected by the excellence of the graver. Every improvement of this kind the author has studiously availed himself of; and has frequently furnished the artist with the living or newly-killed subject itself to assist his ideas."[5]

For Wilson, the authenticity of his illustrations was central to his role as a pioneer ornithologist, and it was his close collaboration with his engraver that made him confident that his plates were faithful representations of nature. As a naturalist, Wilson was responsible for accurately describing the bird in

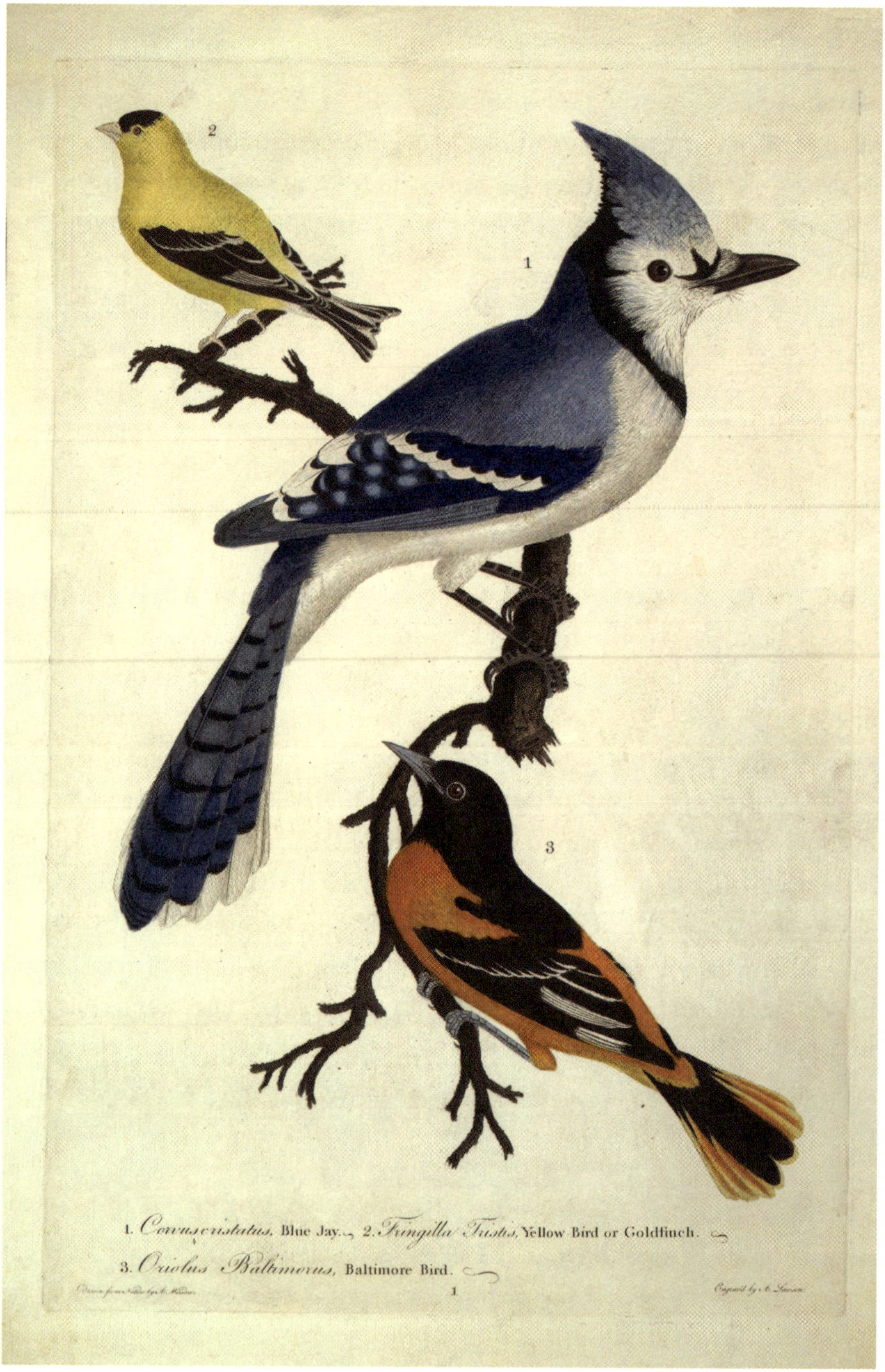
2
1
3
1. Corvus cristatus, Blue Jay. 2. Fringilla Tristis, Yellow Bird or Goldfinch.
3. Oriolus Baltimorus, Baltimore Bird.
1

word and illustration. As an engraver, Lawson was responsible for translating Wilson's drawings and paintings into the printed illustration. Their collaboration and respective roles were emphasized at the bottom of each plate where "Drawn from Nature by A. Wilson" appeared in the lower left corner and "Engraved by A. Lawson" appeared in the lower right corner.[6]

The process of creating and printing illustrations was a very different process at the beginning of the nineteenth century than it is today. Illustrations were composed in several ways. Wilson could prepare the illustration exactly as it would be transferred to the copper plate for engraving. An example of this approach is the drawing of the Orchard Oriole (Figure 3.3).[7] Alternatively, Wilson could draw a bird, for example the pencil sketch of the Canvasback (Figure 3.4). Later he would work it into a draft plate (Figure 3.5). Another method was to draw a bird, then cut the drawing from the larger sheet and glue it onto the larger drawing. The glue marks just above and behind the Canvasback in Figure 3.5 are evidence of this technique.

The missing bird in Figure 3.5 is a preening Long-tailed Duck as shown in Figure 3.6, which is a copy of the final plate as it appears in *American Ornithology,* volume 8. That Wilson often cut and pasted drawings of single birds into larger compositions is suggested by the irregular edges of many drawings such as that of the Mottled Owl (Eastern Screech-Owl; Figure 3.7). Still a fourth method he may have used was to cut the drawings from the larger sheet, but compose the illustration by placing several cutouts on the copper plate simultaneously or by adding and transferring them one at a time. How closely Wilson and Lawson collaborated on composition of the plates is unknown, but once the layout was finalized, transfer to the copper plate began. The back of each sketch was dusted with iron oxide (Figure 3.8). Daubs of glue were applied to the back of the drawings and the drawings positioned on the copper plate. (The small, brown spots visible in the

---

Figure 3.2. Lawson's engraving of the same painting that Wilson engraved previously (compare Figure 3.1) shows stronger lines with less wavering. Lawson also uses lines of varying depth to achieve lines that are darker or lighter, thicker or more delicate. (From the Academy of Natural Sciences of Philadelphia Library and Archives Collection, ANSP 79 [Drexel University].)

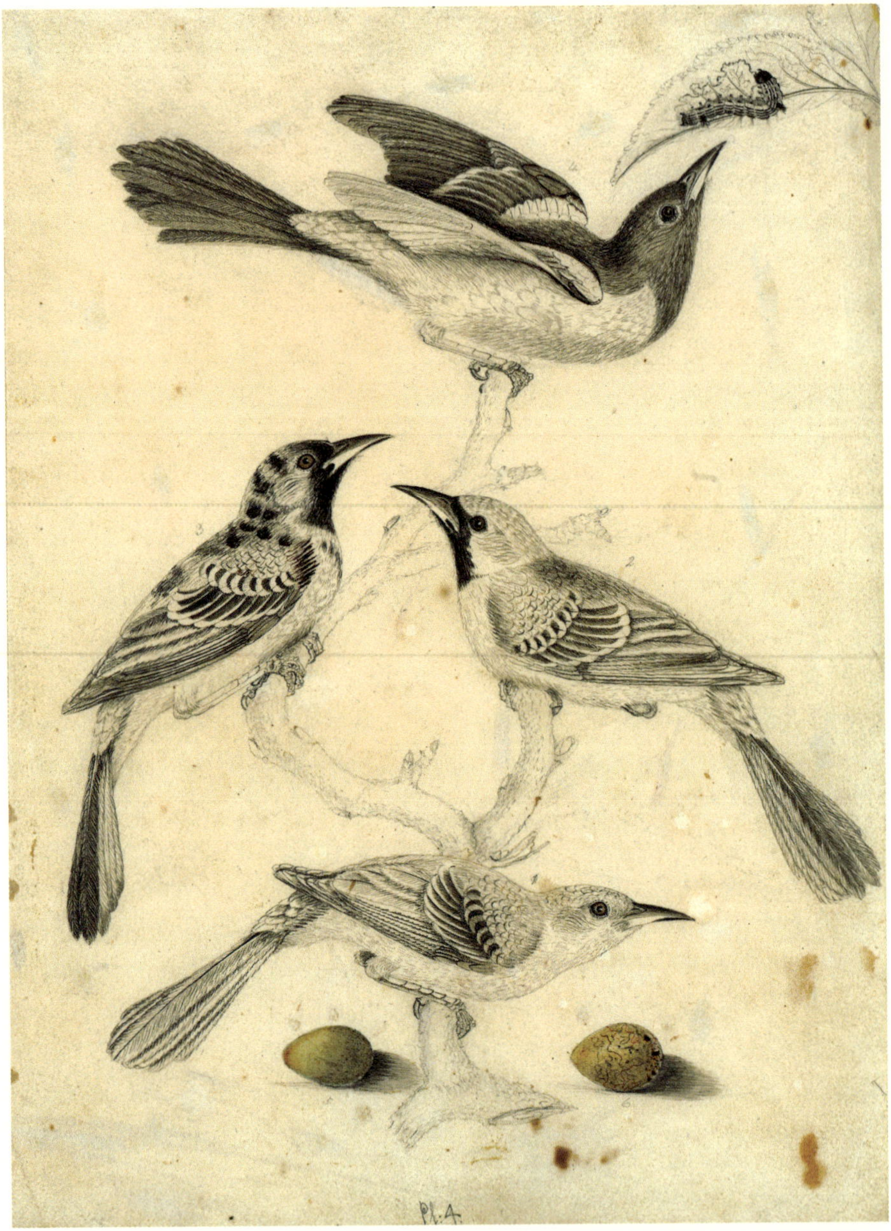

Figure 3.3. This illustration of an Orchard Oriole in pen and ink over pencil, with fully painted eggs, is an example of a complete drawing on a single sheet of paper. At the bottom Wilson has indicated that it is to be plate 4 (in volume 1 of *American Ornithology*), and indeed this drawing matches exactly the plate that appears in volume 1. (From the archives of the Ernst Mayr Library of the Museum of Comparative Zoology, Harvard University.)

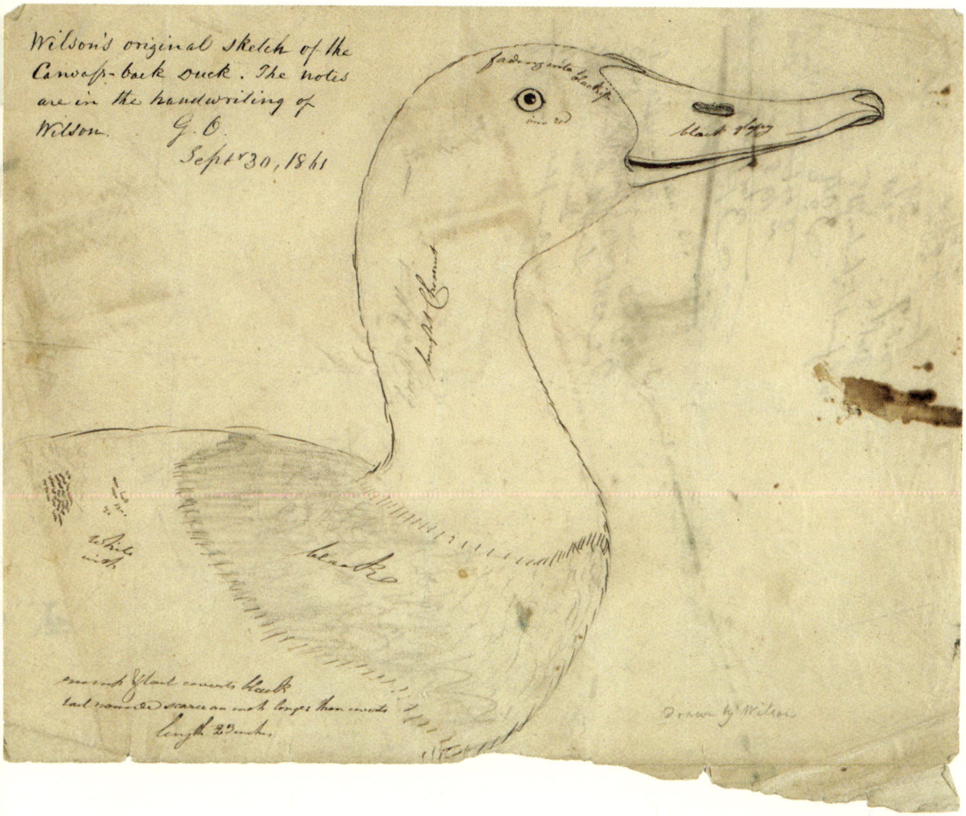

Figure 3.4. A pencil sketch of a Canvasback with Wilson's notes regarding the color of the feathers and of the bare tissues (including the iris). The latter are especially likely to fade after death. In the upper left is a note by George Ord ("G.O."). (From the archives of the Ernst Mayr Library of the Museum of Comparative Zoology, Harvard University.)

sky and across the bottom edge of Figure 3.5 show where the glue has soaked through the paper as the glue and paper aged.) The lines of the drawing were redrawn, thereby transferring the orange iron oxide powder from the back of the drawing to the surface of the copper plate. Finally the drawing was removed and Lawson used his tools to cut into the surface of the copper plate along the orange lines transferred from the back of the drawing. This step required both strength and exceptional control of the width and depth of the

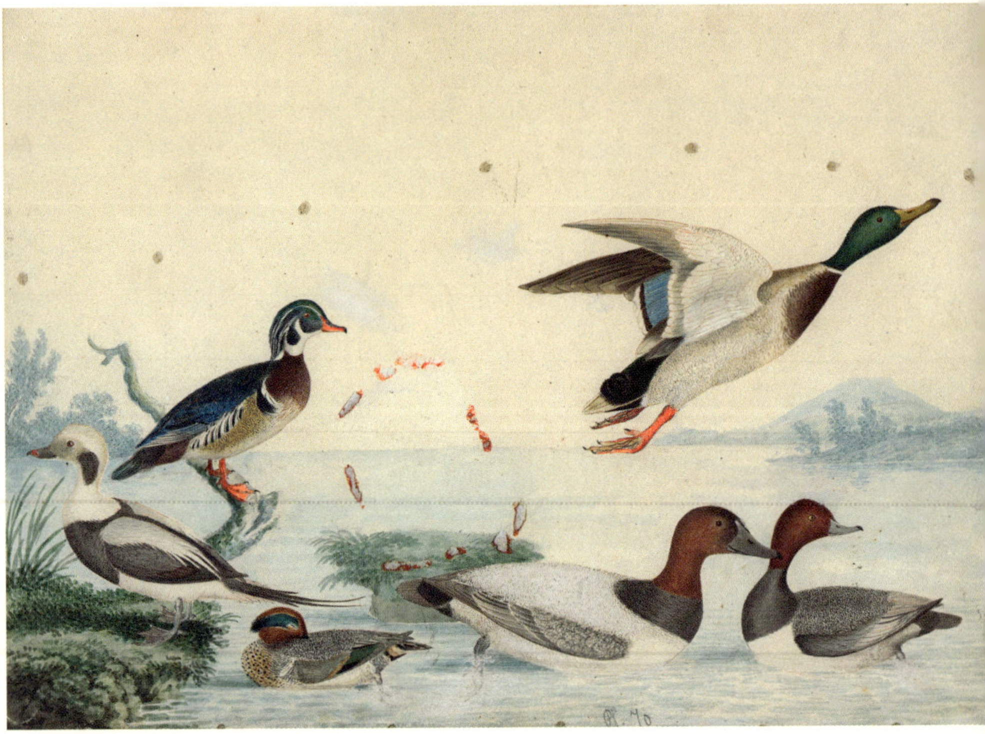

Figure 3.5. The painting for plate 70 of *American Ornithology* with a Canvasback on the right, next to a Redhead Duck. Notice how much better the proportions are in the preceding sketch (Figure 3.4) than in the painting. (From the archives of the Ernst Mayr Library of the Museum of Comparative Zoology, Harvard University.)

cut. A deeper cut held more ink, leading to a darker line. Shadows and three-dimensional effects were created by the darkness of the lines, the proximity of lines, and the loose or tight cross-hatching of lines.

Once the cutting was complete, the engraver would clean the plate to remove any copper shavings, dust, or detritus that had adhered to the surface or lingered in the cuts (Figure 3.9). A thick oil-based ink was applied, and a rounded tool was used to press the ink into the lines. The printer might also heat the plate to make the ink more fluid and help it settle into the cut lines more readily and evenly. Once the lines were filled, the excess ink was

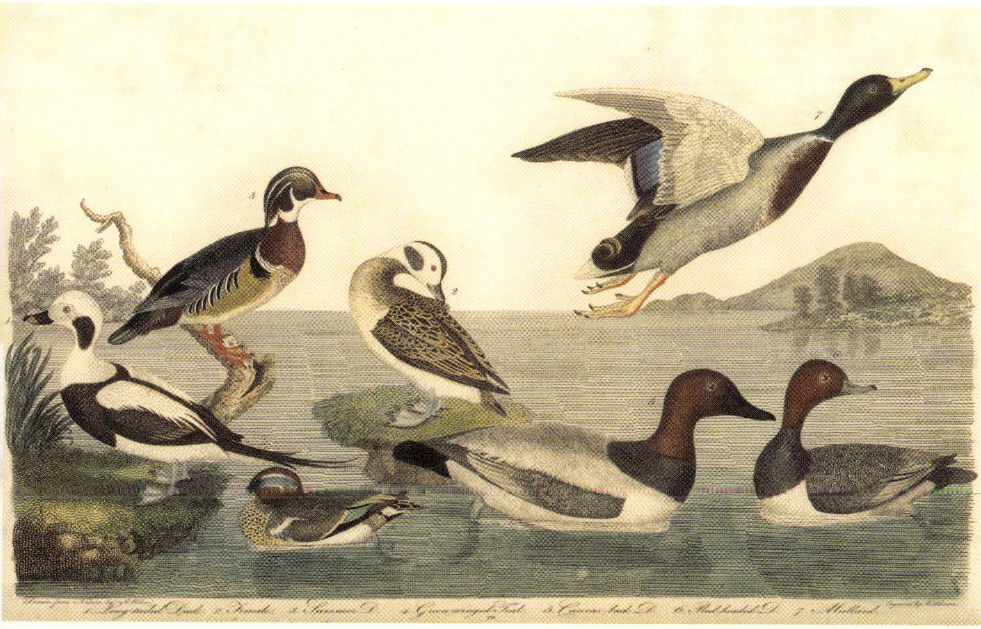

Figure 3.6. The final printed plate showing a Long-tailed Duck resting in the position indicated by the glue spots in Figure 3.5. (From the archives of the Ernst Mayr Library of the Museum of Comparative Zoology, Harvard University.)

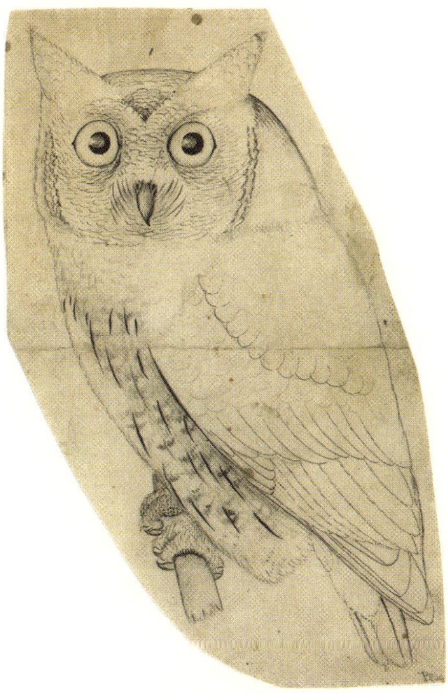

Figure 3.7. This Eastern Screech-Owl (Wilson's Mottled Owl) was drawn on a larger sheet of paper and then cut out so that it could be used in the composition of a plate featuring several birds. The same "cut and paste" process probably accounts for the presence of the Long-tailed Duck in Figure 3.6. (From the archives of the Ernst Mayr Library of the Museum of Comparative Zoology, Harvard University.)

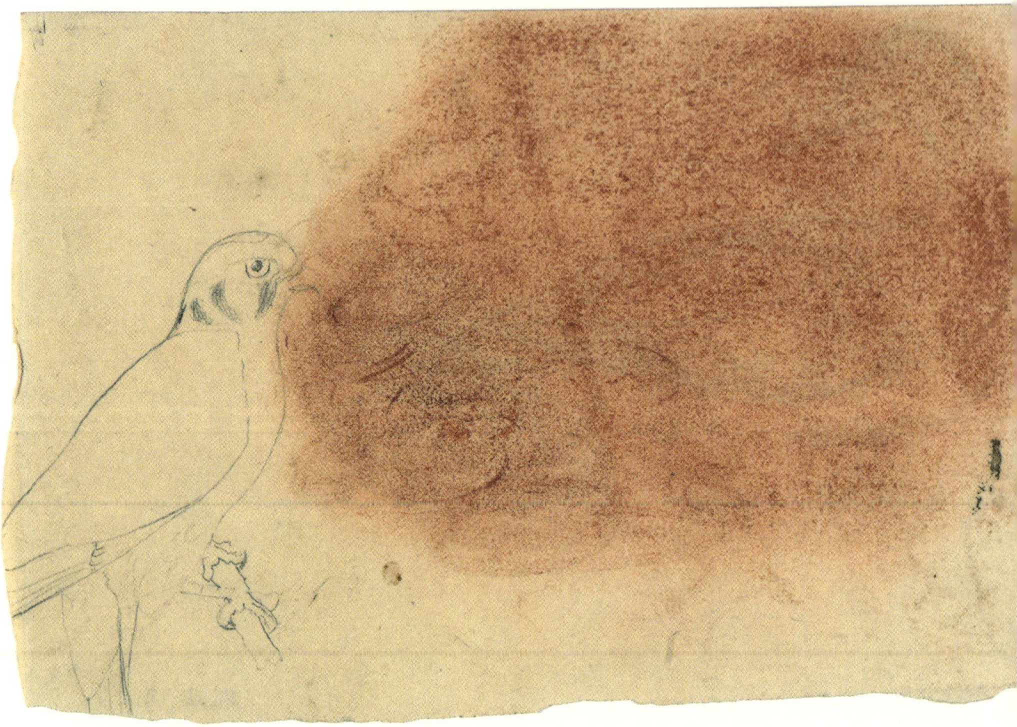

Figure 3.8. The pencil drawing of an American Black Duck (Figure 3.87) was done on the back side of what appears to be a preliminary drawing of the American Kestrel in Figure 3.22 (*a*). The iron oxide sprinkled on the kestrel side of the scrap was used to transfer the drawing of the duck to the final painting (Figure 3.84) from which the copper plate was engraved. (From the archives of the Ernst Mayr Library of the Museum of Comparative Zoology, Harvard University.)

Figure 3.9. One of the copper plates engraved by Alexander Lawson and used to print the plates of *American Ornithology*. Note that the Northern Flicker (*upper right*), Dickcissel (*Spiza americana, middle left*), and Eastern Bluebird (*Sialia sialis, bottom*) are line drawings with no stippling or other detail that could give them form. Stippling became possible about twenty years later when lithography was developed. Wilson could have used cross-hatching, but his emphasis was on shape and color as a means of identification: his goal was to help readers identify birds. (From the Academy of Natural Sciences of Philadelphia Library and Archives Collection, ANSP 79 [Drexel University].)

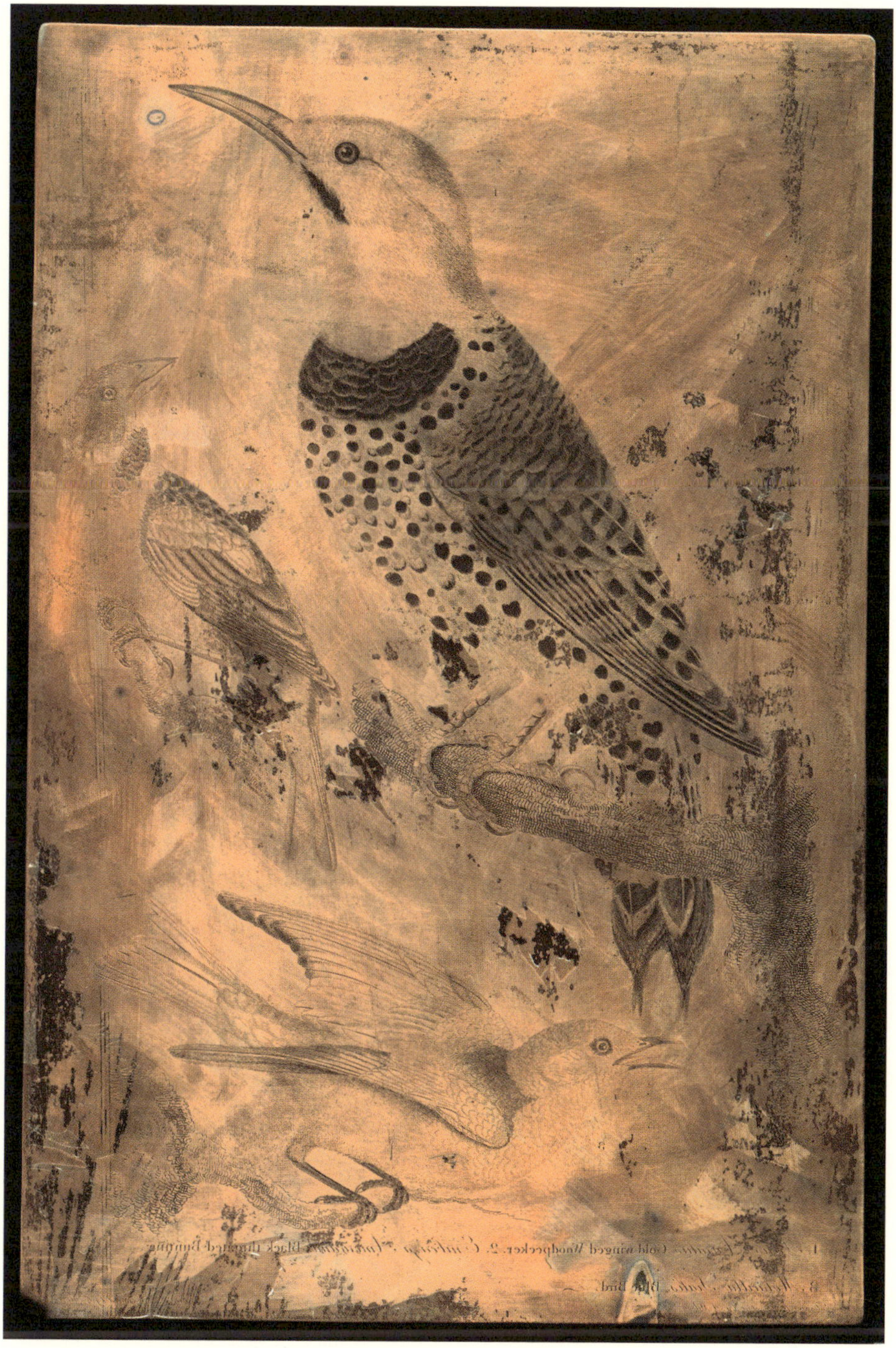

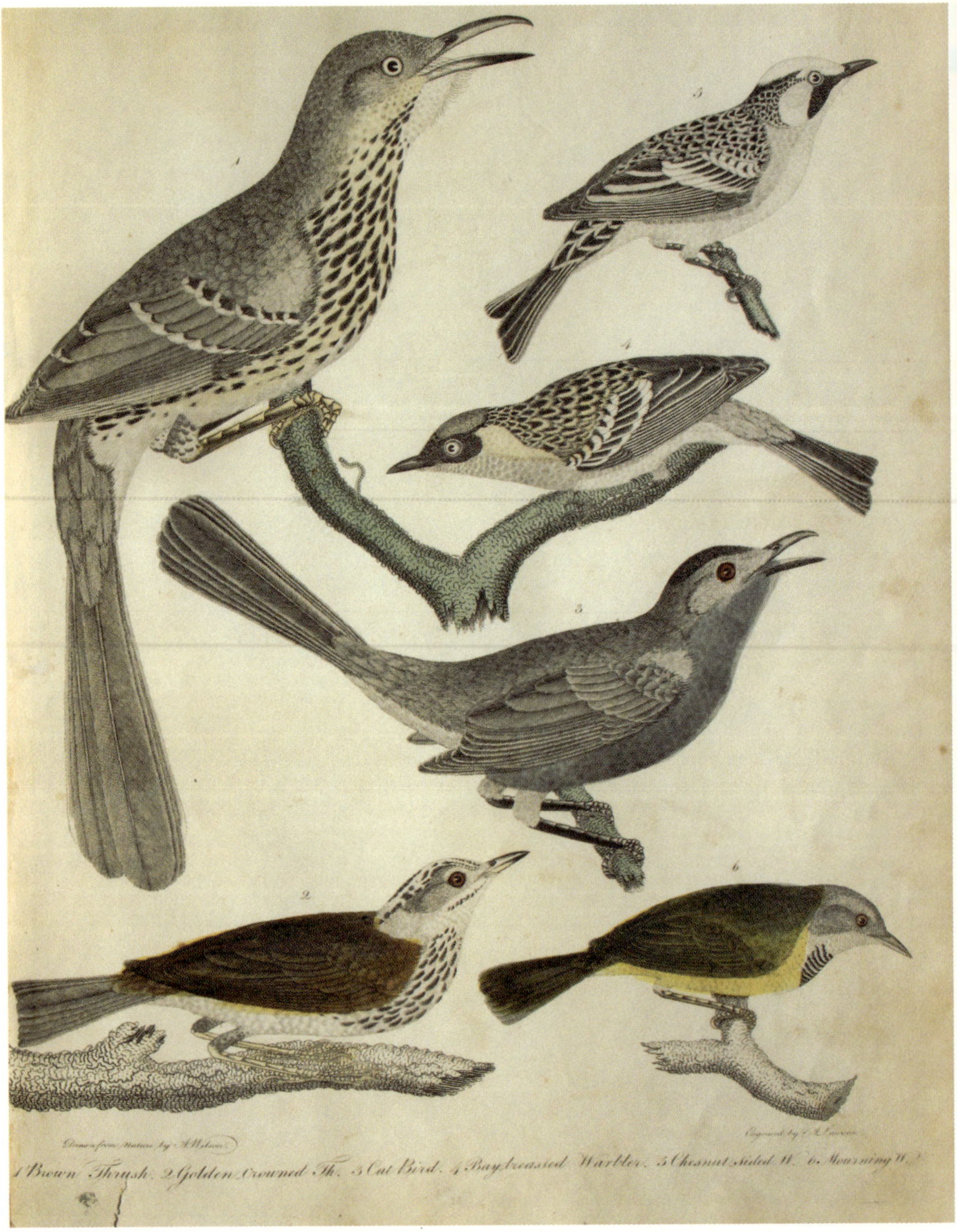

Figure 3.10. A printed, partially colored proof plate for *American Ornithology*. The print is detailed and the color is a semi-transparent wash over the image. The color could be intensified by repeated washes, which added a velvety texture to the plumage without obscuring the details. Charles Robert Leslie, one of the colorists, was a master of this technique. (From the Blanton Family Collection)

scraped off with a soft-edged implement, such as a piece of cardboard, that would not scratch the surface. The plate was then wiped with a soft cloth to be sure that the surface was clean, so that the spaces between the lines would not have spots or streaks of ink. Next the paper was dampened to improve the transfer and penetration of the ink, and the damp paper was laid on the plate and a thick pad and a thin pad were laid on top of the paper. The copper plate, paper, and pad sandwich was then placed in the press with the copper plate on the bottom, inked side up, and pressure was applied and then released. Afterward, the pad was removed from above the print and the print lifted off and set aside to dry. The printer then checked the print and the plate for stray ink marks. If any were found, the surface was wiped clean before a fresh sheet of paper was placed on the plate, the pads were repositioned, the sandwich was placed in the press, and another print was made.

Printing the illustrations required great skill and a lot of hard, physical work. Volume 1 had ten full-page illustrations, the remaining volumes had nine per volume, and there were 450 subscriptions, which meant that 4,500 or 4,050 illustrations had to be printed for each volume. The copper plate had to be cleaned every few prints and new ink applied. Imperfections caused by the pressure of the press had to be corrected. Depending on the damage from the repeated pressure, a new plate might have to be prepared.

Once printed, each illustration had to be hand-colored. Figure 3.10 shows a print that has been partially colored. The eyes of the Gray Catbird (*Dumetella carolinensis*), Ovenbird (*Seiurus aurocapillus*), and Mourning Warbler (*Geothlypis philadelphia*) have been painted and a wash has been applied to parts of each species.[8] Much has yet to be done. Areas that have not been colored will be, but just as importantly, some of the areas that have received a wash will receive additional layers of semi-transparent paint to give them a velvety, feathery texture without obscuring the underlying detail. Only when fully colored could the print be inserted into the text and the finished volume bound.

The written comments above the Horned Lark in Figure 3.11 indicate that Wilson himself checked the colorists' work. The handwriting of this comment appears to be that of Anne (Nancy) Bartram, who may have been

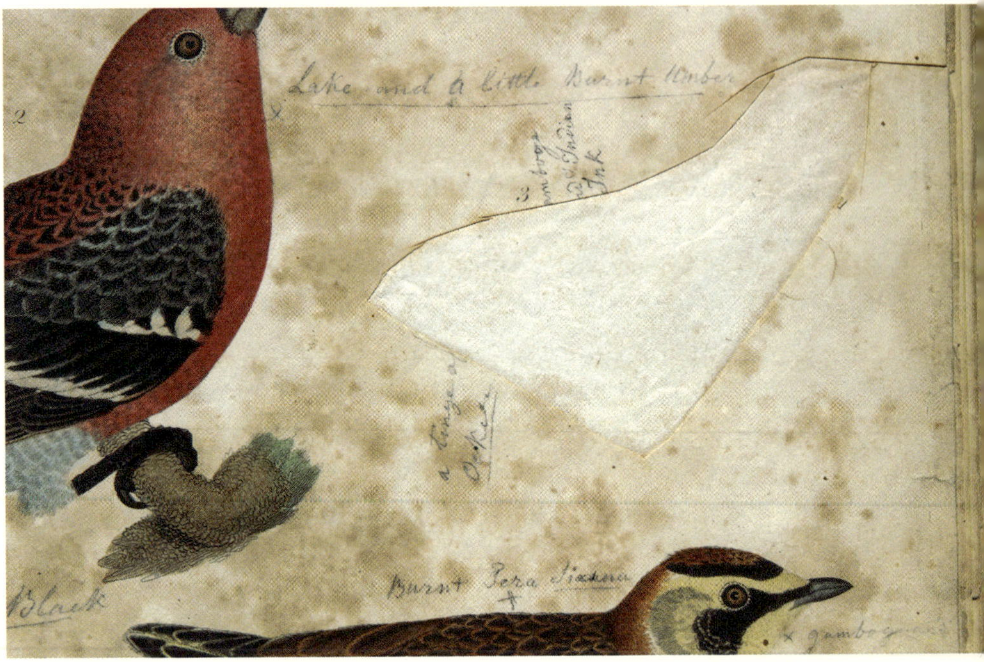

Figure 3.11. In this closeup from a page proof, handwritten directions above the back of a Shore Lark (Horned Lark,) indicate "Burnt Tera Sienna" and below the Pine Grosbeak, "umber over Black." The handwriting is probably that of Anne Bartram and may have been notes she made during a conversation with Alexander Wilson about the colors. (From the Blanton Family Collection)

writing down Wilson's suggestions for her own use or so that she could instruct the other colorists. Whenever possible, Wilson supplied specimens to the colorists so that they could compare the colors they were using with those of the real bird. This may explain the Xs through the Black-capped Chickadee (*Poecile atricapillus*) and Tufted Titmouse (*Baeolophus bicolor*) in Figure 3.12. The gray of each species is too blue. This is most noticeable in the titmouse. Clearly, Wilson and the colorists were concerned about the accuracy of the coloration; in the margin of the plate below and to the right of the titmouse are three brush strokes, each a slightly different gray (Figure 3.13). The topmost one seems the closest to the correct color and the closest

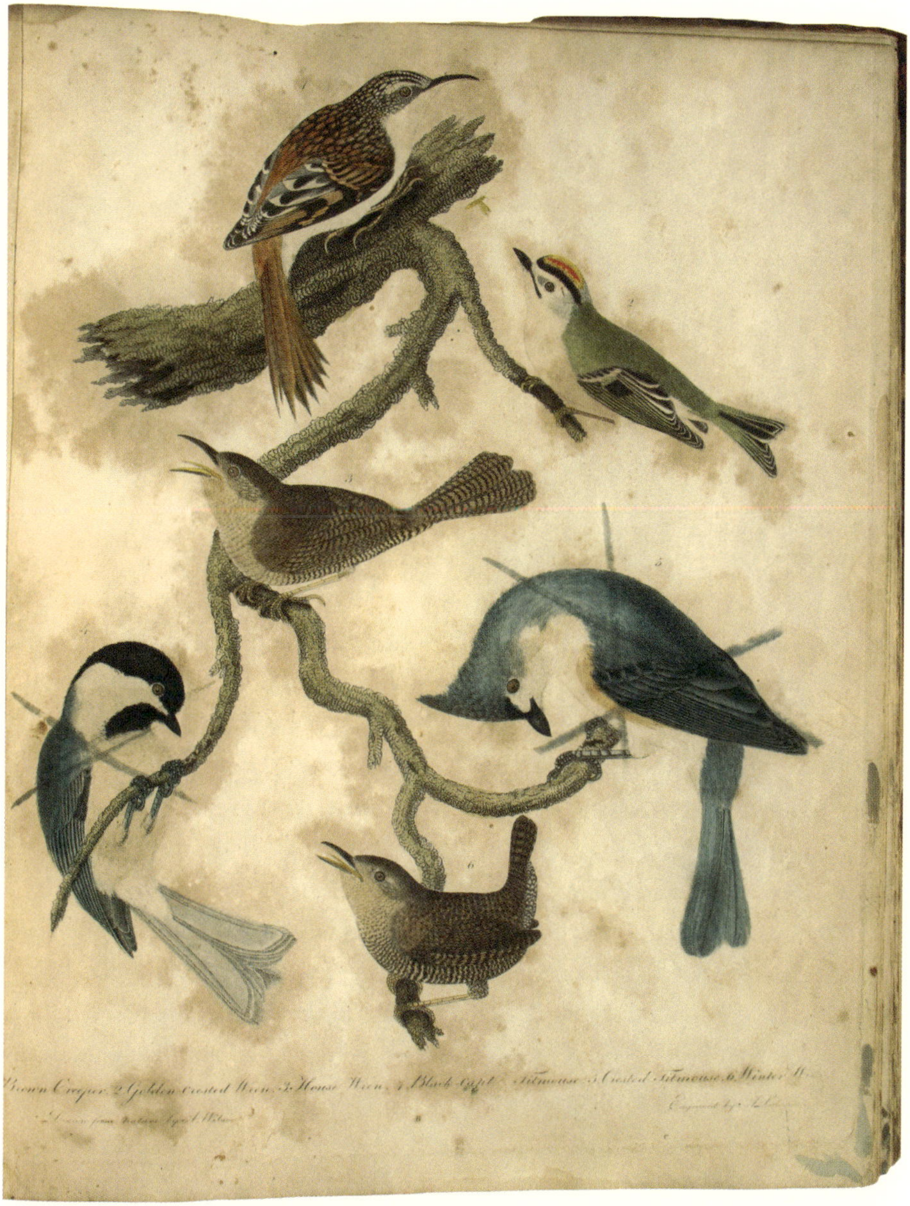

Figure 3.12. The Tufted Titmouse and Black-capped Chickadee in this illustration appear too blue and have been crossed out. Clearly Wilson took great care to ensure that the coloring appeared true to life, as promised in his introduction. (From the Blanton Family Collection)

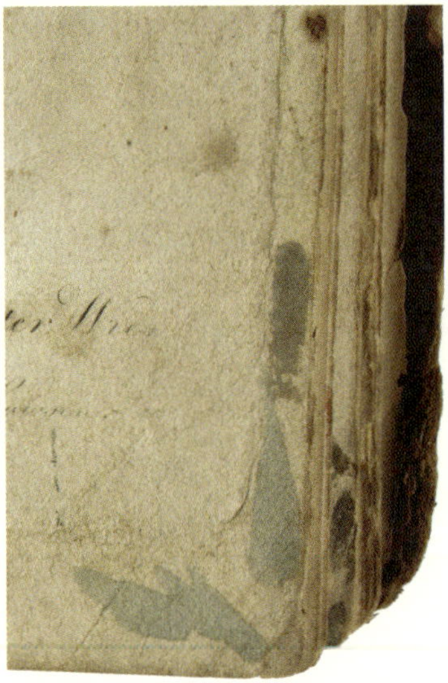

Figure 3.13. In this closeup of the corner of the proof page shown in Figure 3.12, there are brush strokes in different shades of gray, showing where the colorist attempted to match the colors of a fresh specimen. (From the Blanton Family Collection)

to the color we have seen in museum specimens and modern field guides. If you look closely at Figure 3.13 you will notice that the brush strokes of these test colors are between the edge of the paper and a faint line that resulted from the pressure of the press against the paper along the edge of the plate. This meant that the test colors and brush strokes would be cut away when the page was trimmed to size.

Wilson's attention to detail is evident in his use of colored inks. Had the plates been inked and printed with the black ink used for the printed page, the birds would have appeared as black outlines filled with color, much like a child's coloring book. Wilson avoided this effect by the occasional use of colored inks that matched the color of the bird, as in the crest of the Ivory-billed Woodpecker (Figure 3.14). Often dark lines were needed and black ink was indeed used, but the occasional use of colored inks created a picture that had

---

Figure 3.14. This finished print of an Ivory-billed Woodpecker shows the colored lines created by an application of red ink to the plate in the area of the crest. This technique enabled Wilson to indicate the presence of feathers without disrupting the color with lines of black ink. (From the Academy of Natural Sciences of Philadelphia Library and Archives Collection, ANSP 79 [Drexel University].)

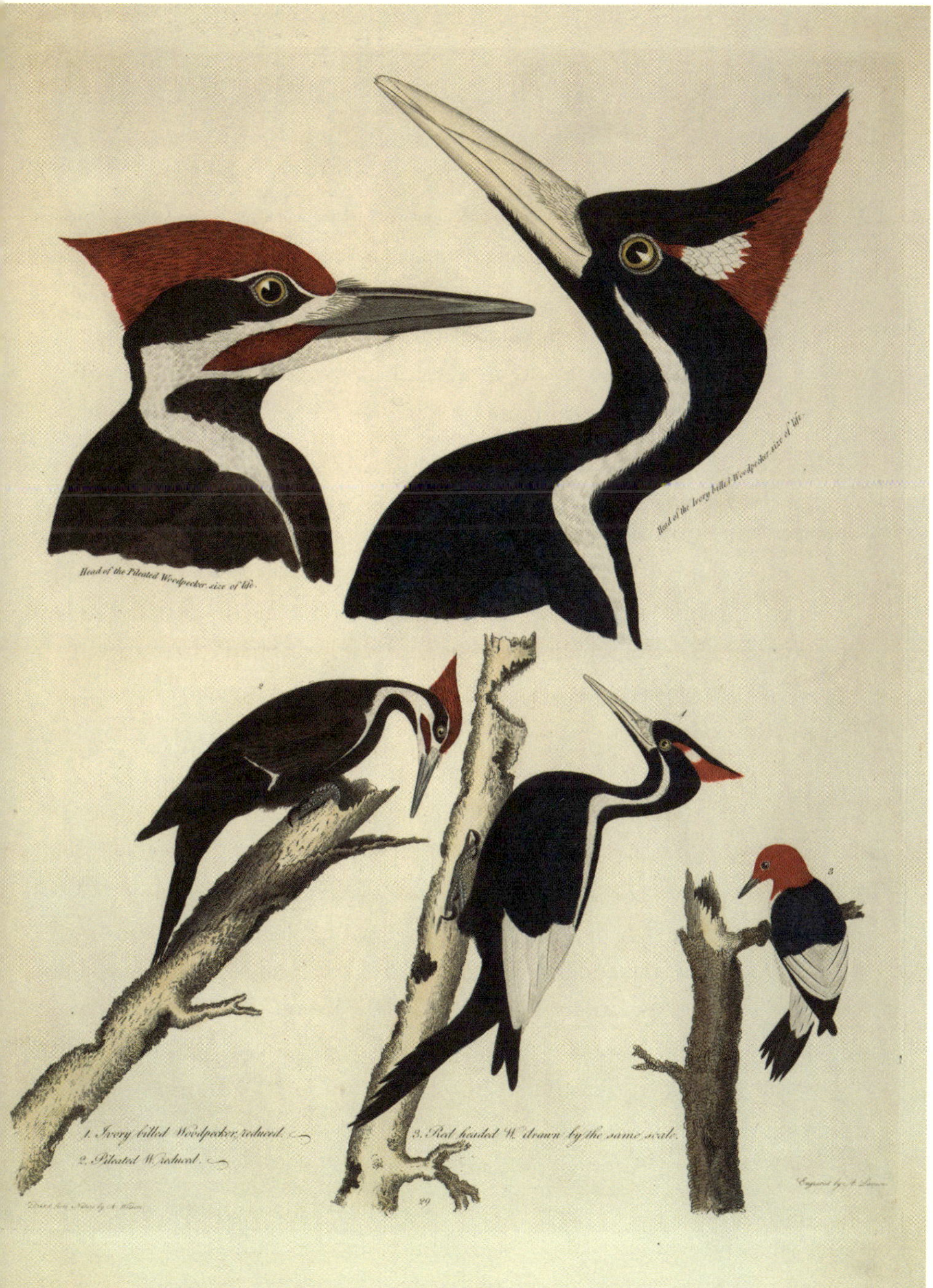
Head of the Pileated Woodpecker size of life.
Head of the Ivory billed Woodpecker size of life.
1. Ivory billed Woodpecker, reduced.
2. Pileated W. reduced.
3. Red headed W. drawn by the same scale.

the sharp edges of a printed line, yet edges that matched the interior colors applied by the colorists.

Initially Wilson turned to William Bartram and his niece, Anne (Nancy) Bartram—from whom he had learned to draw—for help with the coloring:

> By the impressions of my two first plates that accompany this you will see that I have a request to make to Miss Bartram if the state of her health will permit. We want well-colored Specimens of the plates to be sent to Boston Charleston New York etc. and as my time will not permit me to do them myself I have presumed to apply to her to colour the . . . impressions that are sent with this according to the Specimens that accompany them, for which I shall make any return. Perhaps Mary Leech might be set to some parts of them with safety which would lessen the drudgery . . . In washing the Blue Jay the most difficult part of the process is to lay on the colour without being streaked (which you will see I have not succeeded in) and in giving the true tint which I think is nearly approached in the specimen. Nothing but a wash is necessary as the engraving must be seen thro' the colour.[9]

As the first volume approached completion, Wilson began to hire colorists. Among the first was Mary Leech, an artistically inclined teenage student in his school (Figure 3.15).[10] She was the younger sister of Isaac Leech, who had accompanied Wilson on his trek to Niagara Falls, and the older sister of Wilson Leech, Alexander Wilson's namesake. Trained artists were also available to color illustrations, and even though Wilson's publisher was offering only twenty-five cents a page, Wilson was able to hire Alexander Rider, a Swiss-born artist. Rider applied the paints too heavily, however, thereby obscuring details of the engravings, and his use of color was neither accurate nor consistent. Although Wilson retained Rider, who later became an important zoological illustrator, he decided that his best option was to hire young men and women with artistic talent whom he could supervise closely to ensure uniform quality of the plates. For this reason, much of the work

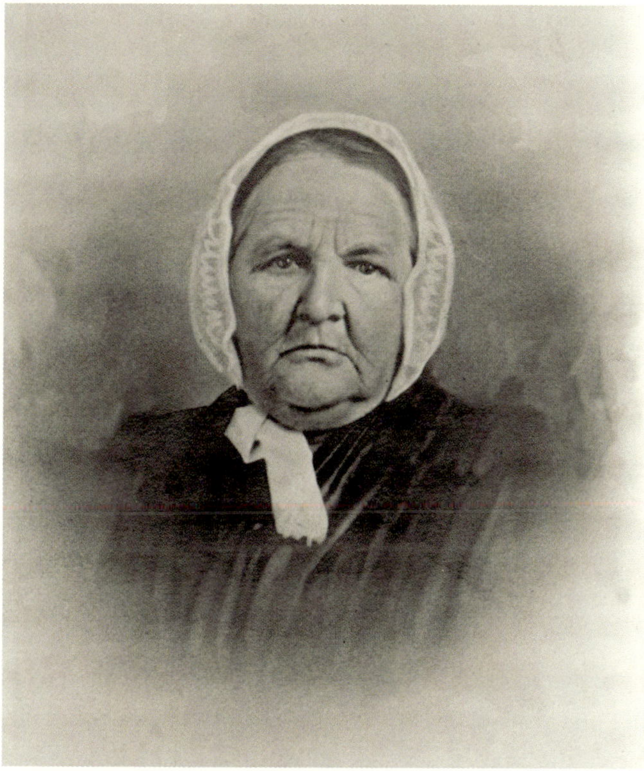

Figure 3.15. A daguerreotype of Mary B. Leech, known to her family as Molly, taken around 1850, shortly before she moved to Ohio where she died in 1870 at the age of seventy-eight. A student in Wilson's school at Gray's Ferry, she became one of his colorists in 1807 at the age of fifteen. (Courtesy of the Leech-Fetters-Webb family.)

went to Ann Bartram and Mary Leech, though Henry Hopkins, a young clerk of Wilson's acquaintance, also colored some illustrations.

There were other colorists. By a stroke of great good luck, Charles Robert Leslie, an exceptionally artistic teenager, was an apprentice in Bradford's bookstore until Bradford, Wilson's publisher, reassigned him to coloring plates. Wilson and a number of Philadelphia businessmen bought up the remaining years of Leslie's apprenticeship after Leslie had worked three years as a colorist. They arranged for him to go to England to study under Benjamin West. Leslie went on to become one of the most famous of Victorian artists and court painter to Queen Victoria herself. Leslie's younger sister, Eliza, also became a colorist, along with Anna Peale, daughter of Charles Wilson Peale, who compounded the paints used by the colorists. For a short time, William Duncan, Wilson's nephew, became a colorist as well. After the War of 1812 was declared, however, the U.S. economy became difficult and Wilson's cash flow from subscribers stopped. In the winter of 1812–1813 he had to release all his colorists and take over the coloring himself. During that winter he colored the plates for 450 copies of each of two volumes, which featured nine plates per volume and an average of four birds per plate. That is, he hand-colored approximately 32,400 bird illustrations.

The Artwork: Contemporary and Historical Insights

We now present the artwork known to us on which Alexander Wilson based the illustrations for his *American Ornithology.* They vary from pencil, to pen and ink, to mixed media. A few contain penciled notes on colors—particularly the colors of the birds' soft parts, which change quickly after death—and many are partially or completely colored. Some of Wilson's sketches are very similar to the plate for which they were the template. Others offer just a few lines to indicate the general features of the species. (Such sketches may have been done in the field, as Wilson himself suggests in his introduction.) And still other drawings bear no resemblance to any of Wilson's published illustrations. This raises a point worth emphasizing. Wilson drew a lot of birds, not just those that appear in the plates of *American Ornithology.* He practiced. He worked hard to be true to nature. Some of the drawings presented here we have had to piece together from scraps left by Wilson when he cut a larger drawing in order to use the reverse side to draw a smaller species. Wilson was working on a very tight budget and paper was expensive, so he used it very carefully. Moreover, his goal was not to make a fine drawing or painting, but rather to create an illustration that would give life to his description of the species. He made little effort to keep the original artwork that was used to create the engravings, so there are relatively few extant drawings from his early volumes. The surviving drawings thus feature few songbirds and more waterfowl.

Here we provide the current common and scientific names for each species pictured, the common and scientific names as given in *American Ornithology* and the original name given to the species when it was first described. When the name of the person who first described the species is enclosed in parentheses in the current listing, that person's species name is still used, but the original genus name has been changed (as you will see if you compare the current and original scientific names). Note that most of the names derive from the tenth edition of Carl Linnaeus's *Systema Naturae,* published in 1758. This edition of Linnaeus's system of names—whereby every known species received a genus name, which began with an uppercase letter, and a species name, which usually, but not always, began with a lowercase letter—is considered the worldwide starting point for scientific names of species. What name a species receives and who is credited with discovering and naming the species is determined by who first published both a name and a complete description of the newly discovered organism after the tenth edition of the *Systema Naturae* was published. That means that all species named and described in the tenth edition of Linnaeus's *Systema Naturae* are attributed to him. Why should Linnaeus receive the credit when, as was often the case, the species was seen and described by an earlier naturalist, Mark Catesby for example, who risked life and limb to venture into the unknown to collect and describe plants and animals? Simply put, Linnaeus presented a logical system for generating names that allowed species to be easily cataloged. Furthermore, he succeeded in compiling most of the information then available on species throughout the world. Thus his *Systema Naturae* provided both an organizational framework of existing names and a logical system for adding names as new species were discovered and described. No other naturalist or scholar had accomplished either task.

Although many biologists of the late eighteenth century adopted Linnaeus's system, some resisted. Those like Johann Freidrich Gmelin and Pieter Boddaert, who did use the Linnaean system, are credited with naming new species. Those like William Bartram, however, who did not follow Linnaeus's system, are not credited with naming new species even though Wilson often referred to names given by Bartram, his mentor, when listing the previ-

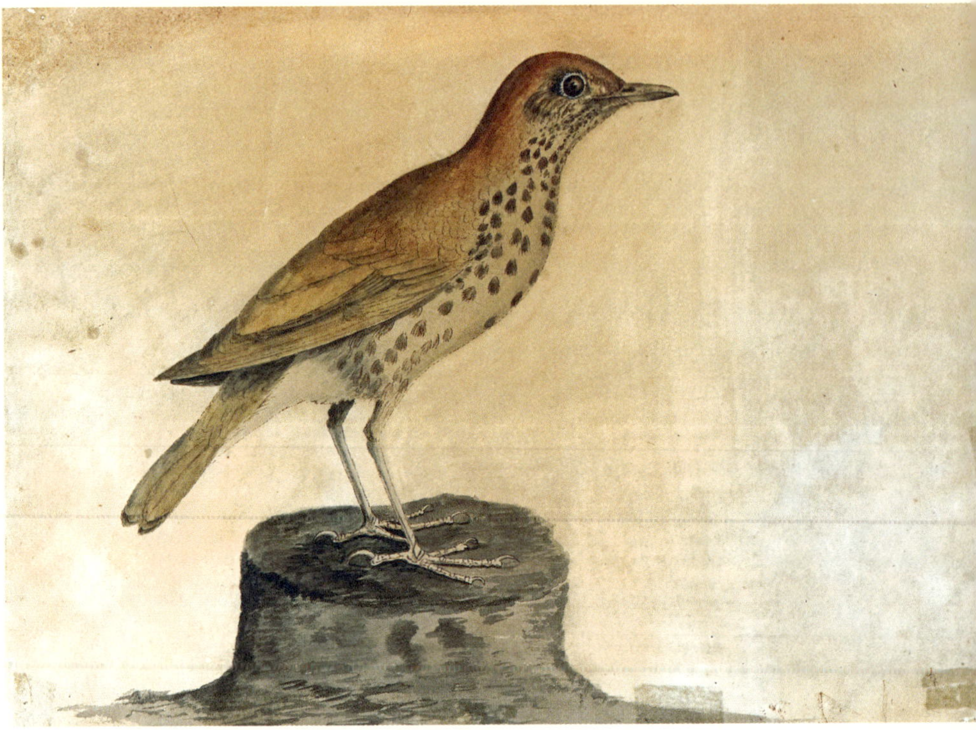

Figure 3.16. Wood Thrush by Wilson. (From the archives of the Paisley Museum, Renfrewshire Council, Scotland.)

ous names of species. Alexander Wilson was the first American naturalist to apply the Linnaean system to the naming of American birds.

When possible we include information on when and where Wilson drew the species, when it was published, and who engraved the plate based on the artwork. We include anything that is known about the history of the drawing. We have used our background as scientists and Davis's background as a scientific illustrator to comment on outstanding or unusual artistic and biological aspects of the portrait itself. Where space allows, we have also included relevant excerpts from Wilson's written account of the species.

Current name: Wood Thrush, *Hylocichla mustelina* (Gmelin)[11]
Wilson's name: Wood Thrush *Turdus melodus*
Original name: *Turdus mustelina* Gmelin

Commentary

The Wood Thrush in the illustration is not a draft of the Wood Thrush in *American Ornithology,* which is perched almost horizontally on a small branch. The correct overlapping of the wing feathers and the booted pattern of scales on the legs and feet are detailed and correct, a characteristic of Wilson's drawings that is not true of earlier illustrators. The feet are too far back and the tail seems short, giving the bird a slightly off-balance look. Stumps and sticks as perches are typical of Wilson's early drawings, as they are of drawings by Bartram and most eighteenth-century illustrators, but Wilson's artwork becomes more ecologically oriented with time. This drawing is from the Paisley Museum Library archives. Wilson must have sent the drawing to his parents.

Wood Thrush

At whatever time the Wood Thrush may arrive, he soon announces his presence in the woods. With the dawn of the succeeding morning, mounting to the top of some tall tree that rises from a low thick-shaded part of the woods, he pipes his few but clear and musical notes in a kind of ecstasy; the prelude, or symphony to which, strongly resembles the double-tongueing of a German flute, and sometimes the tinkling of a small bell; the whole song consists of five or six parts, the last note of each of which is in such a tone as to leave the conclusion evidently suspended; the finale is finely managed, and with such charming effect as to soothe and tranquillize the mind, and seem sweeter and mellower at each successive repetition. Rival songsters, of the same species, challenge each other from different parts of the the wood, seeming to vie for softer tones and more exquisite responses. During the burning heat of the day, they are comparatively mute; but in the evening the same melody is renewed, and continued long after sunset.[12]

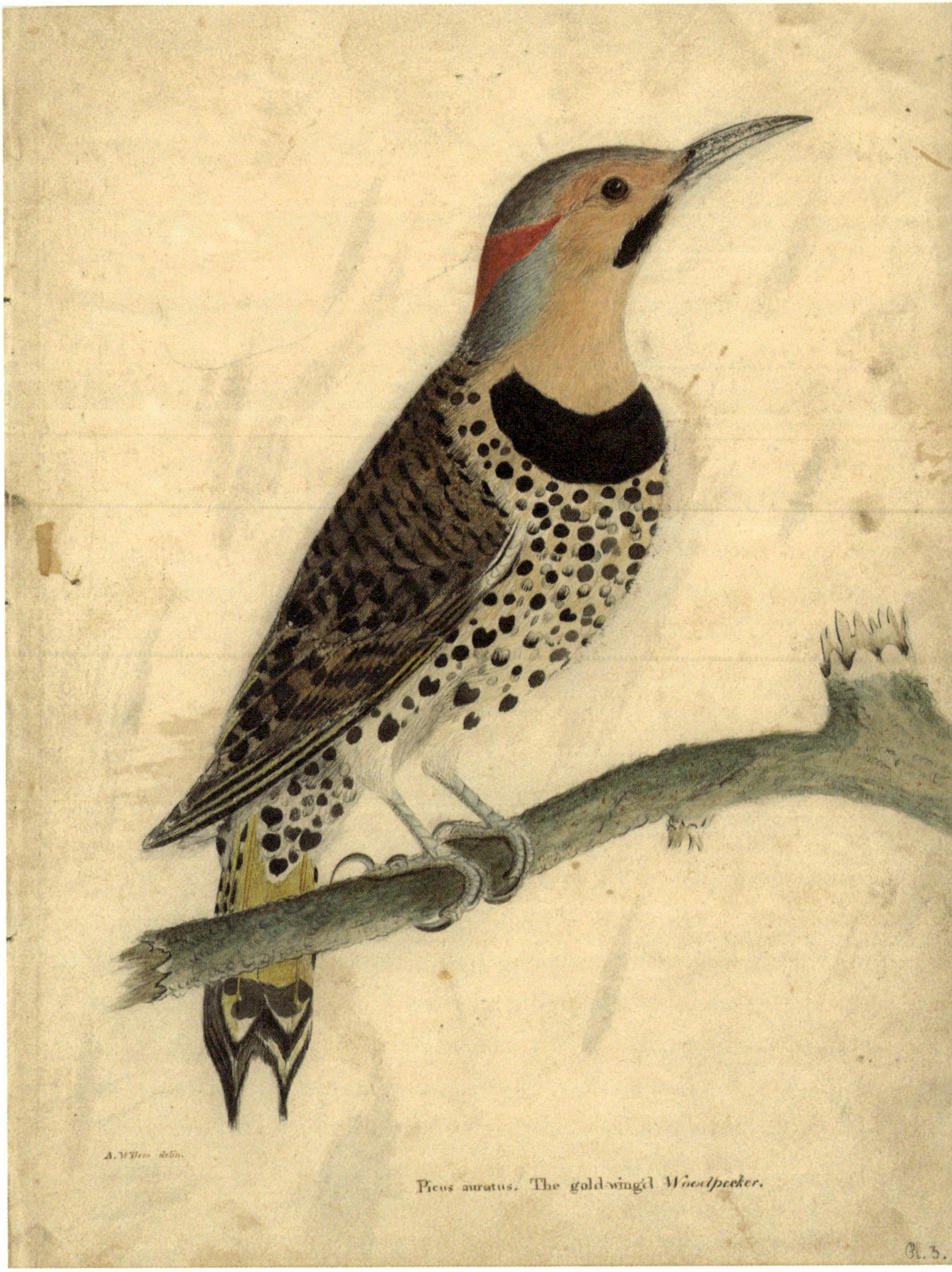

Figure 3.17. Northern Flicker. Note the "A. Wilson" in the lower left corner. (From the archives of the Ernst Mayr Library, Museum of Comparative Zoology, Harvard University.)

Current name: Northern Flicker, *Colaptes auratus* (Linnaeus)
Wilson's name: Golden-winged Woodpecker, *Picus auratus*
Original name: *Cuculus auratus* Linnaeus 1758

Commentary

The flicker is perched on a rudimentary branch in a stylistic pose such as was commonly used by previous illustrators—for example, Mark Catesby or William Bartram—and it is a portrait of a single bird—again characteristic of previous illustrators. (In later plates, Wilson turned to illustrating both sexes when the genders differed in color pattern, and painted the birds within their habitat.) Again Wilson's rendering of the legs and feet is superior to that of previous illustrators. The tail is turned to illustrate the feather pattern and the head is tilted to full profile. The figure of the bird has been outlined in pencil and then filled in with watercolor. Note the hand-printed designation "gold wing'd *Woodpecker.*"

Golden-winged Woodpecker (Northern Flicker)

> The food of this bird varies with the season. As the common cherries, bird cherries, and berries of the sour gum successively ripen he regales plentifully on them, particularly on the latter; but the chief food of this species, or that which is most usually found in the stomach, is wood-lice, and the young and larvae of ants, of which he is so immoderately fond, that I have frequently found his stomach distended with a mass of these, and these only, as large nearly as a plumb. For the procuring of these insects nature has remarkably fitted him. The bills of Woodpeckers, in general, are straight, grooved or channelled, wedge-shaped, and compressed to a thin edge at the end, that they may easier penetrate the hardest wood; that of the Golden-winged Woodpecker is long, slightly bent, ridged only on the top, and tapering almost to a point, yet still retaining a little of the wedge form there. Both, however, are admirably adapted for the peculiar manner each has of procuring its food.[13]

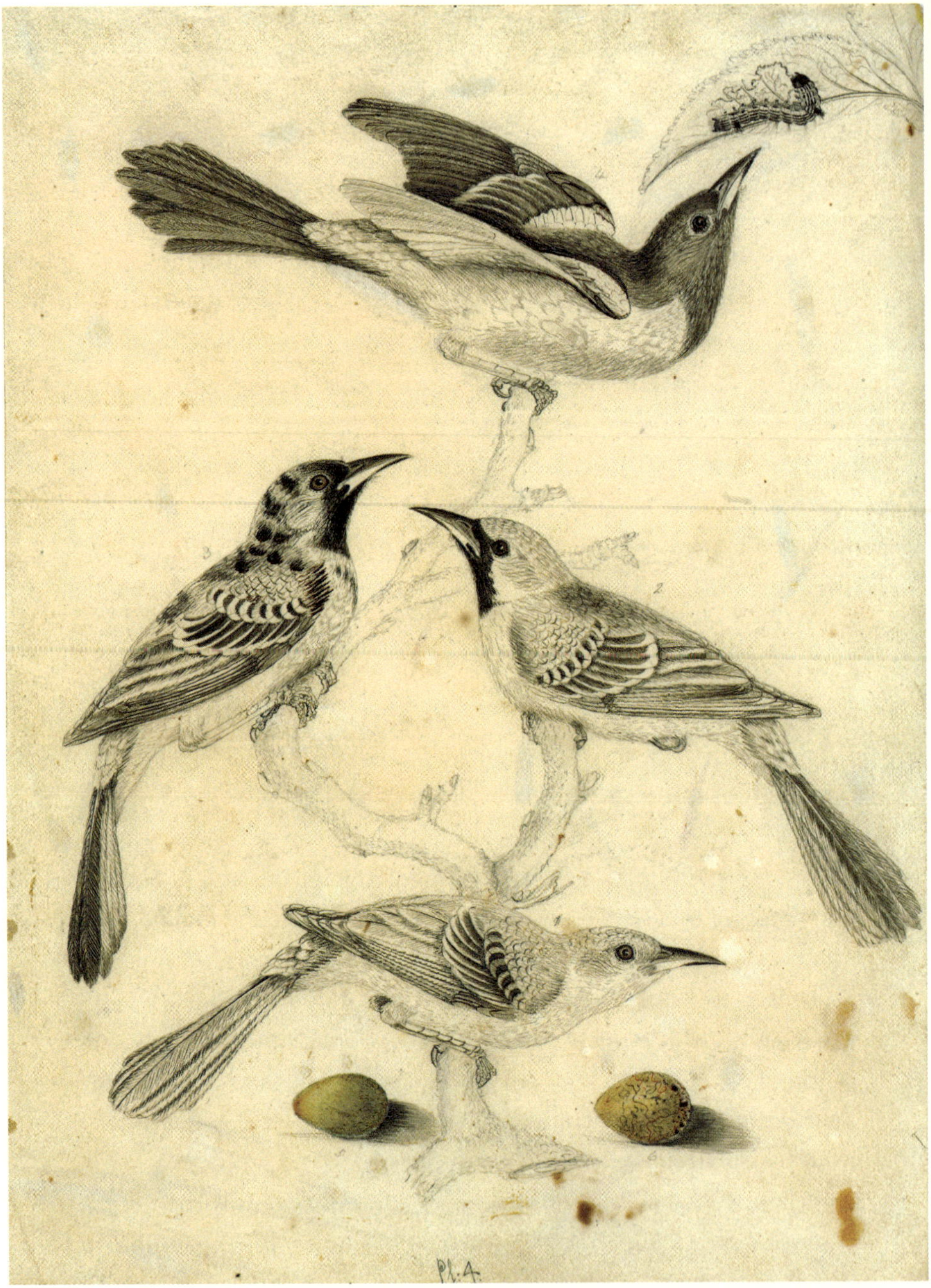

Figure 3.18. Orchard Oriole. (From the archives of the Ernst Mayr Library, Museum of Comparative Zoology, Harvard University.)

Current name: Orchard Oriole, *Icterus spurius* (Linnaeus)
Wilson's name: Orchard Oriole, *Oriolus mutatus*
Original name: *Oriolus spurious* Linnaeus 1766

Commentary

This is an extremely detailed drawing in pencil and ink (done with a quill pen). The feather patterns, especially on the top bird, are individually drawn. The individual scutes of the legs, feet, and toes are individually rendered, a characteristic of Wilson's work and a great improvement over that of most previous illustrators. The heads are all in full profile and the tails turned to show the feather pattern, although this posture gives the birds an awkward and twisted appearance. Wilson faithfully illustrates the development of adult male plumage with three birds, and a fourth, the bird closest to the eggs, illustrates the female plumage. This is an important innovation because often different plumaged birds of the same species were incorrectly identified or described as different species. Another innovation is the caterpillar on the leaf above the male who is looking at it. Orchard Orioles feed on such caterpillars, making this the first bird portrait to include ecological information.

Orchard Oriole

> There are no circumstances relating to birds, which tend so much to render their history obscure and perplexing, as the various changes of color which many of them undergo. These changes are in some cases periodical; in others progressive; and are frequently so extraordinary, that, unless the naturalist has resided for years in the country which the birds inhabit, and has examined them at almost every season, he is extremely liable to be mistaken and imposed on by their novel appearance . . . The species we are now about to examine is a remarkable example of this; and as it has never, to my knowledge, been either accurately figured or described, I have devoted one plate to the elucidation of its history.[14]

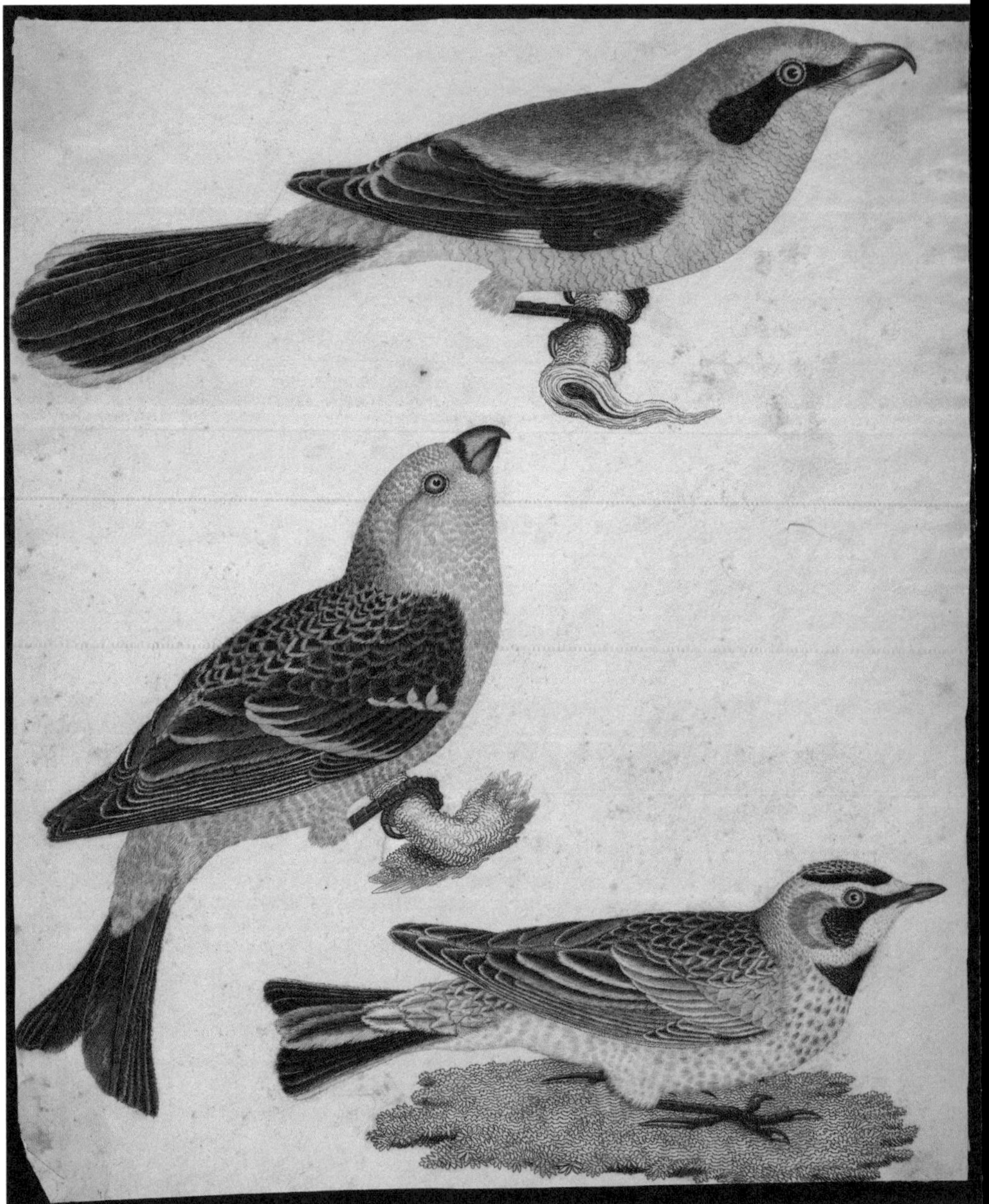

Figure 3.19. Northern Shrike (*top*); Pine Grosbeak (*middle*); Horned Lark (*bottom*). (From the Academy of Natural Sciences of Philadelphia Library and Archives Collection, ANSP 79 [Drexel University].)

Current name: Northern Shrike, *Lanius excubitor* Linnaeus
Wilson's name: Great American Shrike, or Butcher Bird, *Lanius excubitor*
Original name: *Lanius excubitor* Linnaeus 1758

Current name: Pine Grosbeak, *Pinicola enucleator* (Linnaeus)
Wilson's name: Pine Grosbeak, *Loxia enucleator*
Original name: *Loxia enucleator* Linnaeus 1758

Current name: Horned Lark, *Eremophila alpestris* (Linnaeus)
Wilson's name: Shore-lark, *Alauda alpestris*
Original name: *Alauda alpestris* Linnaeus 1758

Commentary

This fully rendered drawing is very like the final plate 5 in the *Ornithology* but it lacks the Ruby-crowned Kinglet that appears there. Wilson would often add a bird to a plate before it went to the engraver. Wilson kept a Pine Grosbeak in his room for six months to study its behavior and molt cycle. Indeed, this is one of many songbirds, hawks, and owls, even a hummingbird, that he kept for close study of the living bird. This approach was radically different from that of European ornithologists, who studied specimens and had little knowledge of the living bird.

Great American Shrike or Butcher Bird (Northern Shrike)

When we compare the beak of this species with his legs and claws, they appear to belong to two very different orders of birds; the former approaching, in its conformation, to that of the Accipitrine; the latter to those of the Pies; and, indeed, in his food and manners he is assimilated to both . . . In his manners he has more resemblance to the Pies than to birds of prey, particularly in the habit of carrying off his surplus food, as if to hoard it for future exigencies; with this difference, that the Crows, Jays, Magpies, &c., conceal theirs at random, in holes and crevices, where, perhaps, it is forgotten, or never again found; while the Butcher Bird sticks his on thorns and bushes, where it shrivels in the sun, and soon becomes equally useless to the hoarder. Both retain the same habits in a state of confinement, whatever the food may be that is presented to them.[15]

Pine Grosbeak

I have kept one of these Pine Grosbeaks, a male, for more than half a year. In the month of August those parts of the plumage which were red became of a greenish yellow, and continue so still. In May and June its song, though not so loud as some birds of its size, was extremely clear, mellow and sweet. It would warble out this for a whole morning together, and acquired several of the notes of the Red-bird *(L. cardinalis),* that hung near it. It is exceedingly tame and familiar, and when it wants food or water utters a continual melancholy and anxious note.[16]

Shore-Lark (Horned Lark)

There is a singular appearance in this bird, which I have never seen taken notice of by former writers, viz., certain long, black feathers, which extend, by equal distances beyond each other, above the eyebrow; these are longer, more pointed, and of a different texture from the rest around them; and the bird possesses the power of erecting them, so as to appear as if horned, like some of the Owl tribe. Having kept one of these birds alive for some time, I was much amused at this odd appearance, and think it might furnish a very suitable specific appellation, viz. *Alauda cornuta,* or, Horned Lark. These horns become scarcely perceivable after the bird is dead. The head is slightly crested.[17]

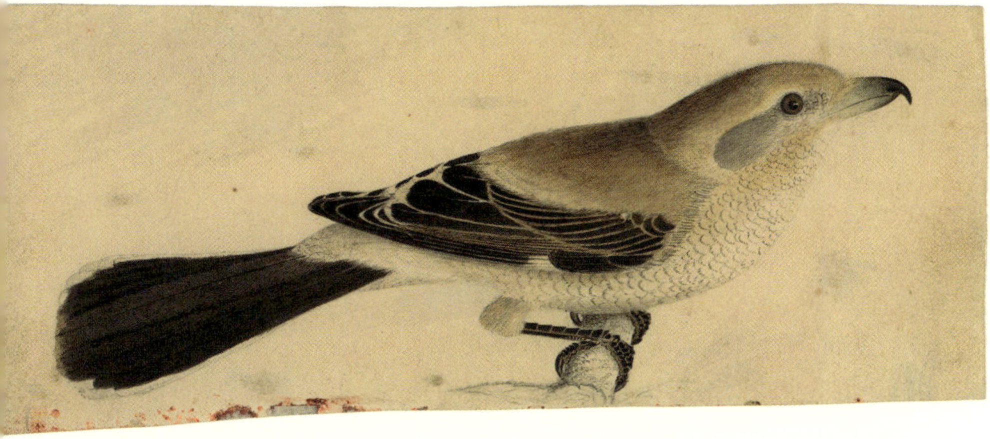

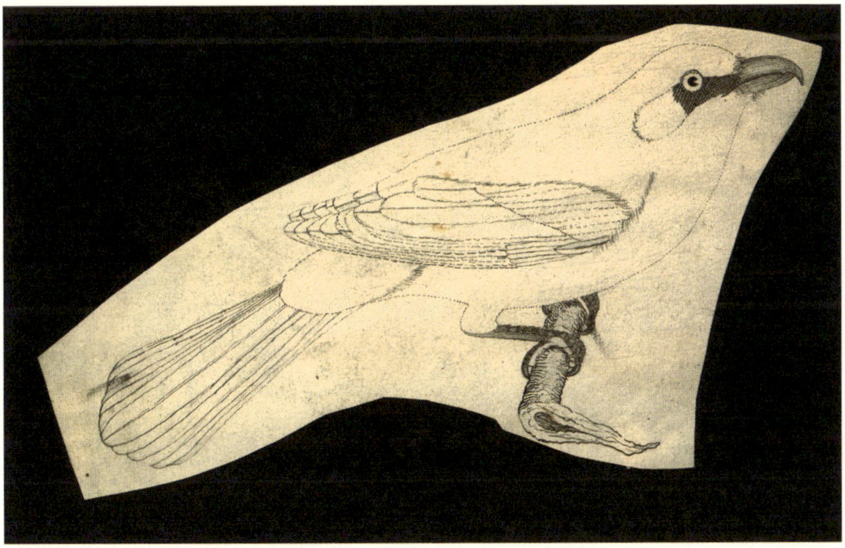

Figure 3.20. The uppermost Northern Shrike has been washed in gray-brown and sits closer to the perch and more horizontally than the Northern Shrike below, which is a pencil drawing and essentially the same as the one in the final painting (Figure 3.19), even to the shape and orientation of the perch. The line drawings of the House Wren (see next page) and shrike emphasize the care with which Wilson observed and rendered the anatomical details. (From the Academy of Natural Sciences of Philadelphia Library and Archives Collection, ANSP 79 [Drexel University].)

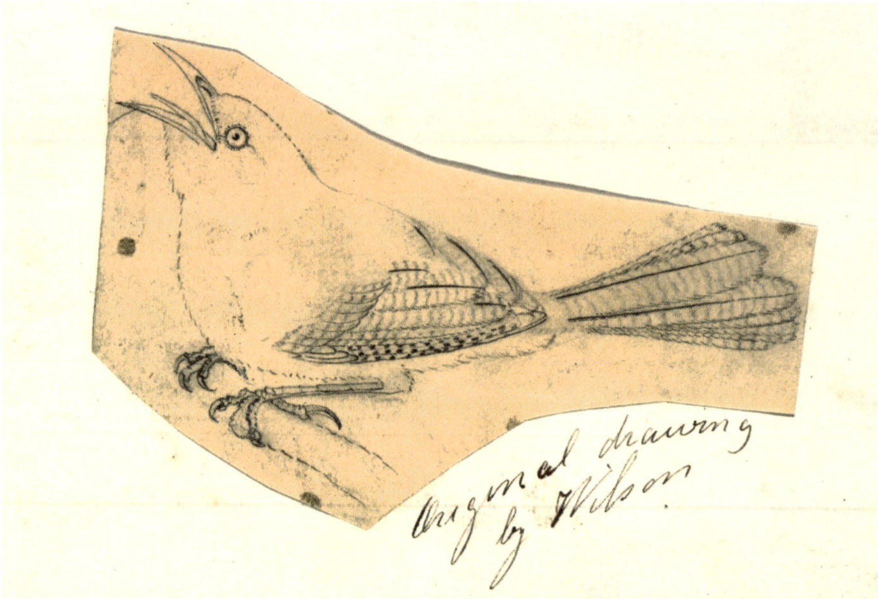

Current name: House Wren, *Troglodytes aëdon* Vieillot
Wilson's name: House Wren, *Sylvia domestica*
Original name: House Wren, *Troglodytes aëdon* Vieillot 1809

Commentary

Often Wilson's drawings were superior to the final engravings for the printed plates in *American Ornithology*, as is the case here with the detail of the eye and its surroundings in the colored version of the Northern Shrike (*top*). In both shrikes and the wren the tail has been twisted to show the feather pattern, a technique that detracts from the realism of the drawings, but shows the markings of the tail. Wilson often did several sketches of the same species, as exemplified by the three drawings of the Northern Shrike, two here and the one in Figure 3.19. The House Wren (*bottom*) is pictured singing, the first bird portrait to emphasize behavior.

Although Vieillot is credited with naming the House Wren, he did so in an advertisement for the publication of his book, which was not published until 1809, a year after Wilson had published his Latin name for the wren, along with a written description and an illustration based on this sketch. Curiously, both men chose the same common name. Because of the British blockade of Napoleonic France, Wilson and Vieillot knew little or nothing of each other's work.

House Wren

> Having seen no accurate description of this bird in any European publication, I have confined my references to Mr. Bartram and Mr. Peale; but though Europeans are not ignorant of the existence of this bird, they have considered it, as usual, merely as a slight variation from the original stock, (*M. troglodytes*) their own Wren; in which they are, as usual, mistaken; the length and bent form of the bill, its notes, migratory habits, long tail, and red eggs, are sufficient specific differences.[18]

Current name: Summer Tanager, *Tangara æstiva*
Wilson's name: Summer Red-bird, *Tanagra æstiva*
Original name: *Fringilla rubra* Linnaeus 1758

Commentary

This is a true pencil sketch. And it may be Wilson's first attempt at drawing a Summer Red-bird, although the sketch is so preliminary that identification is problematic. The wing feathers are detailed and the coverts—feathers lying over the bases of the major flight feathers—are correctly angled relative to the alignment of the flight feathers. The feet are incomplete. The feathers of the eyebrow and face are sketched in, but no markings are shown. The bill is large, a characteristic of the Summer Tanager, and has the curved edge of a tanager's upper mandible, although the bill appears shorter and deeper than is depicted for the species in Wilson's *American Ornithology*.

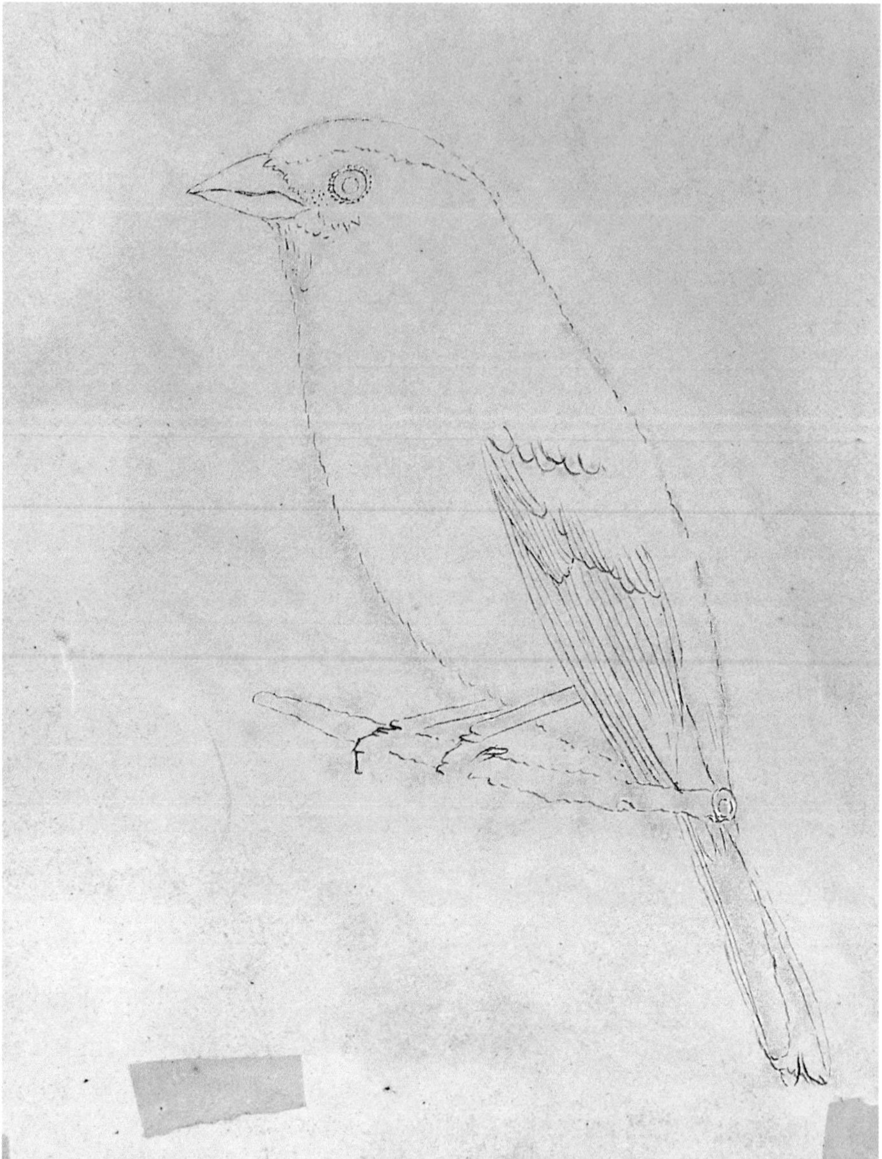

Figure 3.21. Summer Tanager (From the archives of the Paisley Museum, Renfrewshire Council, Scotland.)

Summer Red-bird (Summer Tanager)

> The male of the Summer Red-bird . . . is wholly of a rich vermilion color, most brilliant on the lower parts, except the inner vanes and tips of the wings, which are of a dusky brown; . . . In the month of August the young males are distinguished from the females by their motleyed garb; the yellow plumage below, as well as the olive green above, first becoming stained with spots of a buff color, which gradually brighten into red; these being irregularly scattered over the whole body, except the wings and tail, particularly the former, which I have often found to contain four or five green quills in the succeeding June.[19]

Current name: American Kestrel, *Falco sparverius* Linnaeus
Wilson's name: American Sparrow Hawk, *Falco sparverius*
Original name: *Falco sparverius* Linnaeus 1758

Current name: Acadian Flycatcher, *Empidonax virescens* (Vieillot)
Wilson's name: Small Green-crested Flycatcher, *Muscicapa querula*
Original name: *Platyrhynchos virescens,* Vieillot 1818

Commentary

In this drawing of the Kestrel, the tail is less rotated than in many of Wilson's early drawings so the pose is more realistic. Wilson often would only outline a branch and let the engraver fill in the details; nonetheless in this drawing the feet are gripping the branch. The eyeball appears to be the whole eyeball, and thus seems large relative to what can be seen in the intact bird. (Wilson dissected many birds and studied the details of their anatomy.) The Acadian Flycatcher is posed so that the width of the bill is evident and the detailed rendering of the head includes the rictal bristles. Wilson's intense interest in the natural history of the birds he studied and drew is evident here. The bill

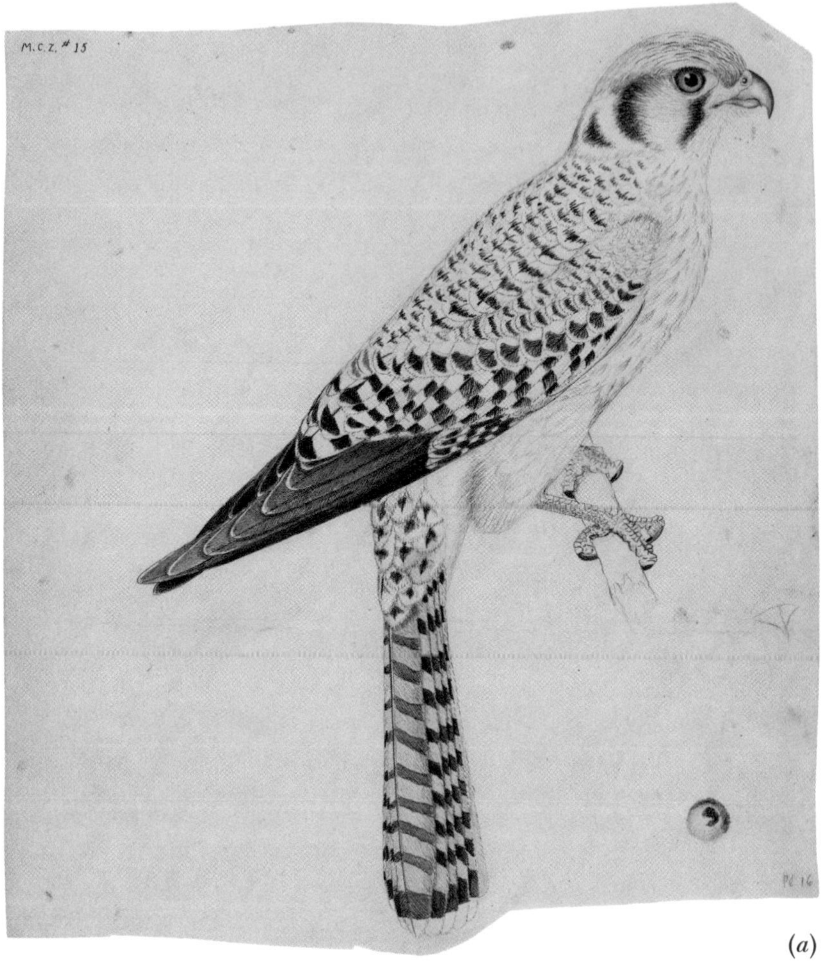

(a)

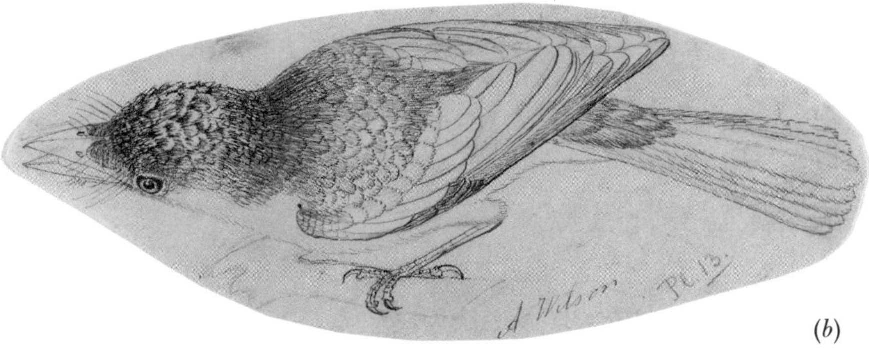

(b)

is the "business end" of a bird and a structure that reflects its mode of food procurement. Wilson's original name for the Acadian Flycatcher was already in use, so Vieillot's species name, *Platyrhynchos virescesns,* published in 1818, was employed instead.

American Sparrow Hawk (American Kestrel)

> In no department of ornithology has there been greater confusion, or more mistakes made, than among this class of birds of prey. The great difference of size between the male and female, the progressive variation of plumage to which, for several years, they are subject, and the difficulty of procuring a sufficient number of specimens for examination,—all these causes conspire to lead the naturalist into almost unavoidable mistakes. For these reasons, and in order, if possible, to ascertain each species of this genus distinctly, I have determined, where any doubt or ambiguity prevails, to represent both male and female, as fair and perfect specimens of each may come into my possession . . . The habits and manners of this bird are well known. It flies rather irregularly, occasionally suspending itself in the air, hovering over a particular spot for a minute or two, and then shooting off in another direction. It perches on the top of a dead tree or pole, in the middle of a field or meadow, and, as it alights, shuts its long wings so suddenly, that they seem instantly to disappear; it sits here in an almost perpendicular position, sometimes for an hour at a time, frequently jerking its tail, and reconnoitering the ground below, in every direction, for mice, lizards, &c.[20]

Figure 3.22. American Kestrel (*a*); Acadian Flycatcher (*b*). (From the archives of the Ernst Mayr Library, Museum of Comparative Zoology, Harvard University.)

Small, Green, Crested Flycatcher (Acadian Flycatcher)

This bird is but little known. It inhabits the deepest, thick-shaded, solitary parts of the woods[;] sits generally on the lower branches[;] utters, every half minute or so, a sudden, sharp squeak, which is heard a considerable way through the woods; and, as it flies from one tree to another, has a low, querulous note, something like the twitterings of Chickens nestling under the wings of the Hen. On alighting, this sound ceases, and it utters its note as before.[21]

Current name: Eastern Meadowlark, *Sturnella magna* (Linnaeus)
Wilson's name: Meadow Lark, *Alauda magna*
Original name: *Alauda magna* Linnaeus 1758

Current name: Eastern Screech-Owl, *Megascops asio* (Linnaeus)
Wilson's name: Mottled Owl, *Strix naevia*
Original name: *Strix Asio* Linnaeus 1758

Current name: Black-and-White Warbler, *Mniotilta varia* (Linnaeus)
Wilson's name: Black and White Creeper, *Certhia maculata*
Original name: *Motacilla varia* Linnaeus 1766

Commentary

The meadowlark is crouched as is the Black-and-white Warbler to reduce the space required for the portraits. The Meadowlark in a full erect stance would have taken up double the space. These unusual postures—which have been used to reduce the number of plates required—detract from the portrait's realism. The meadowlark appears to be surrounded by grass, which makes sense if it is crouched, but is also the first time Wilson indicates habitat in his

---

Figure 3.23. Eastern Meadowlark (*top*); Eastern Screech-Owl (*left*); Black-and-White Warbler (*right*). (From the archives of the Ernst Mayr Library, Museum of Comparative Zoology, Harvard University.)

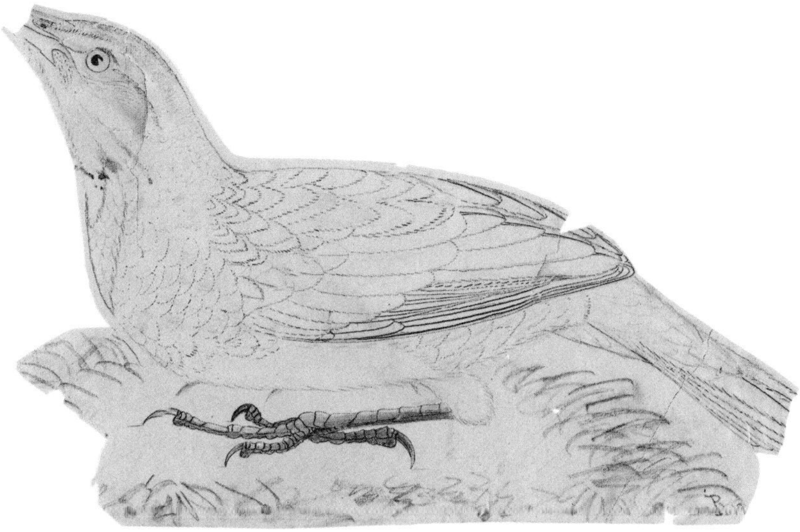

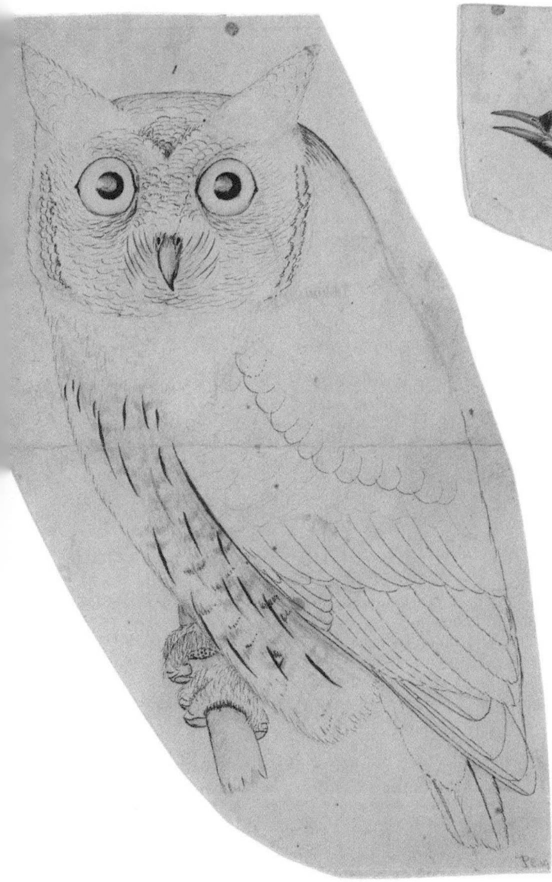

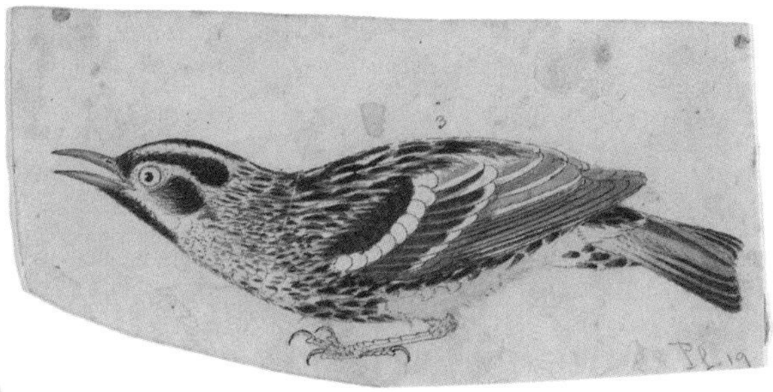

sketches. The Screech-Owl is not a finished drawing and may be a preliminary sketch that was ultimately replaced by a more complete rendering with colored wash, as was the case with the drawings of the Northern Shrikes.

Meadow Lark (Eastern Meadowlark)

Though this well-known species cannot boast of the powers of song which distinguish that "harbinger of day," the Sky Lark of Europe, yet in richness of plumage, as well as in sweetness of voice, (as far as his few notes extend,) he stands eminently its superior. He differs from the greater part of his tribe in wanting the long straight hind claw, which is probably the reason why he has been classed, by some late naturalists, with the Starlings. But in the particular form of his bill, in his manners, plumage, mode and place of building his nest, Nature has clearly pointed out his proper family.[22]

Mottled Owl (Eastern Screech-Owl)

Hollow trees, either in the woods or orchard, or close evergreens in retired situations, are the usual roosting-places of this and most of our other species [of owls]. These retreats, however, are frequently discovered by the Nuthatch, titmouse, or Blue Jay, who instantly raise the alarm; a promiscuous group of feathered neighbors soon collect round the spot, like crowds in the streets of a large city, when a thief or murderer is detected; and, by their insults and vociferation, oblige the recluse to seek for another lodging elsewhere. This may account for the circumstance of sometimes finding them abroad during the day, on fences and other exposed situations.[23]

Black and White Creeper (Black-and-White Warbler)

This nimble and expert little species seldom perches on the small twigs; but circumambulates the trunk and larger branches, in quest of ants and other insects, with admirable dexterity.[24]

Current name: Lewis's Woodpecker, *Melanerpes lewis* (Gray)
Wilson's name: Lewis's Woodpecker, *Picus torquatus*
Original name: *Picus Lewis* Gray 1849

Current name: Clark's Nutcracker, *Nucifraga columbiana* (Wilson)
Wilson's name: Clark's Crow, *Corvus columbianus*
Original name: Clark's Crow, *Corvus columbianus* Wilson 1811

Commentary

These drawings are of two species, which, along with the Western Tanager (*Piranga ludoviciana* [Wilson]), were collected by members of the Lewis and Clark expedition and given to Wilson by Meriwether Lewis upon his return to Philadelphia in 1807. Wilson provided the first compete descriptions, illustrations, and names for these species, which were the only specimens to survive the return journey of the expedition. The name *Picus torquatus,* which Wilson gave to Lewis's Woodpecker had been given in 1783 by Boddaert to the Ringed Woodpecker of the Amazon Basin and northeastern South America. Gray, writing in 1849, renamed the woodpecker, thereby eliminating the duplication of names and garnering credit for naming the species. He retained Wilson's original intent by using *lewis* as the species' scientific name.

In his drawings, Wilson frequently delineated short contour feathers with crescents of short broken lines, as in the head feathers of the nutcracker. He used longer lines for rendering the larger flight feathers, as in the primaries, secondaries, and coverts of both birds.

Lewis's Woodpecker

It was the request and particular wish of Captain Lewis, made to me in person, that I should make drawings of such of the feathered tribes as had been preserved, and were new. That brave soldier, that ami-

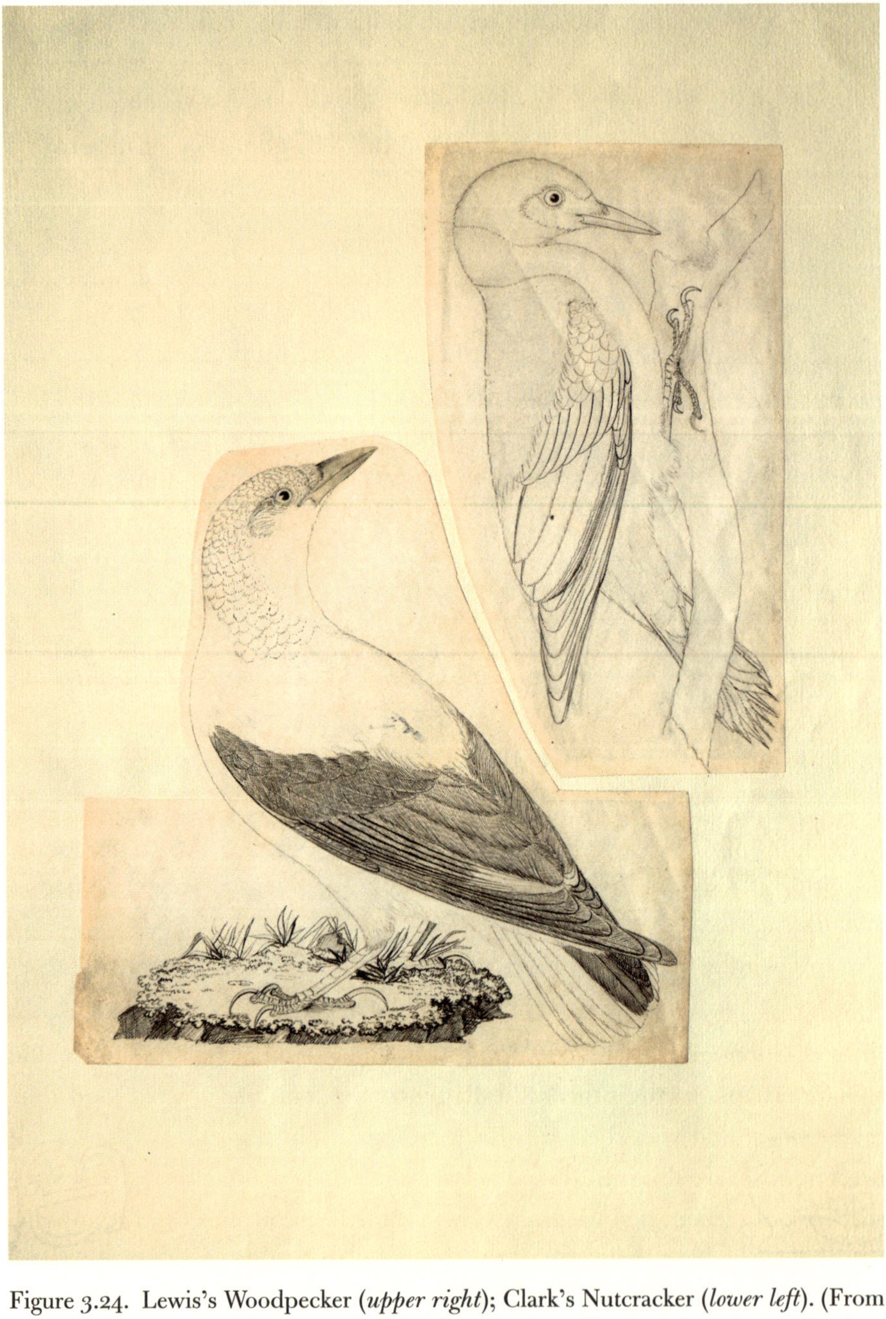

Figure 3.24. Lewis's Woodpecker (*upper right*); Clark's Nutcracker (*lower left*). (From the Academy of Natural Sciences of Philadelphia Library and Archives Collection, ANSP 79 [Drexel University].)

able and excellent man, over whose solitary grave in the wilderness I have since shed tears of affliction, having been cut off in the prime of his life, I hope I shall be pardoned for consecrating this humble note to his memory, until a more able pen shall do better justice to the subject.[25]

Clark's Crow (Clark's Nutcracker)

This species resembles, a little, the Jackdaw of Europe, *(Corvus monedula,)* but is remarkable for its formidable claws, which approach to those of the *Falco* genus, and would seem to intimate that its food consists of living animals, for whose destruction these weapons must be necessary.[26]

Current name: Common Grackle, *Quiscalus quiscula* Linnaeus
Wilson's name: Purple Grakle, *Gracula quiscala*
Original name: *Quiscalus quiscula* Linnaeus 1758

Current name: Rusty Blackbird, *Euphagus carolinus* Müller
Wilson's name: Rusty Grakle, *Gracula ferruginea*
Original name: *Euphagus carolinus* P. L. S. Müller 1776

Commentary

Wilson has the Purple Grakle doing something—pulling up a plant by the roots, perhaps. In the process he flattened the image somewhat, and thus saved space that would have been required to illustrate the bird in a full upright pose. He also introduced behavior into his portraits, an advance over the work of earlier artist/naturalists, whose birds sat on sticks and stumps.

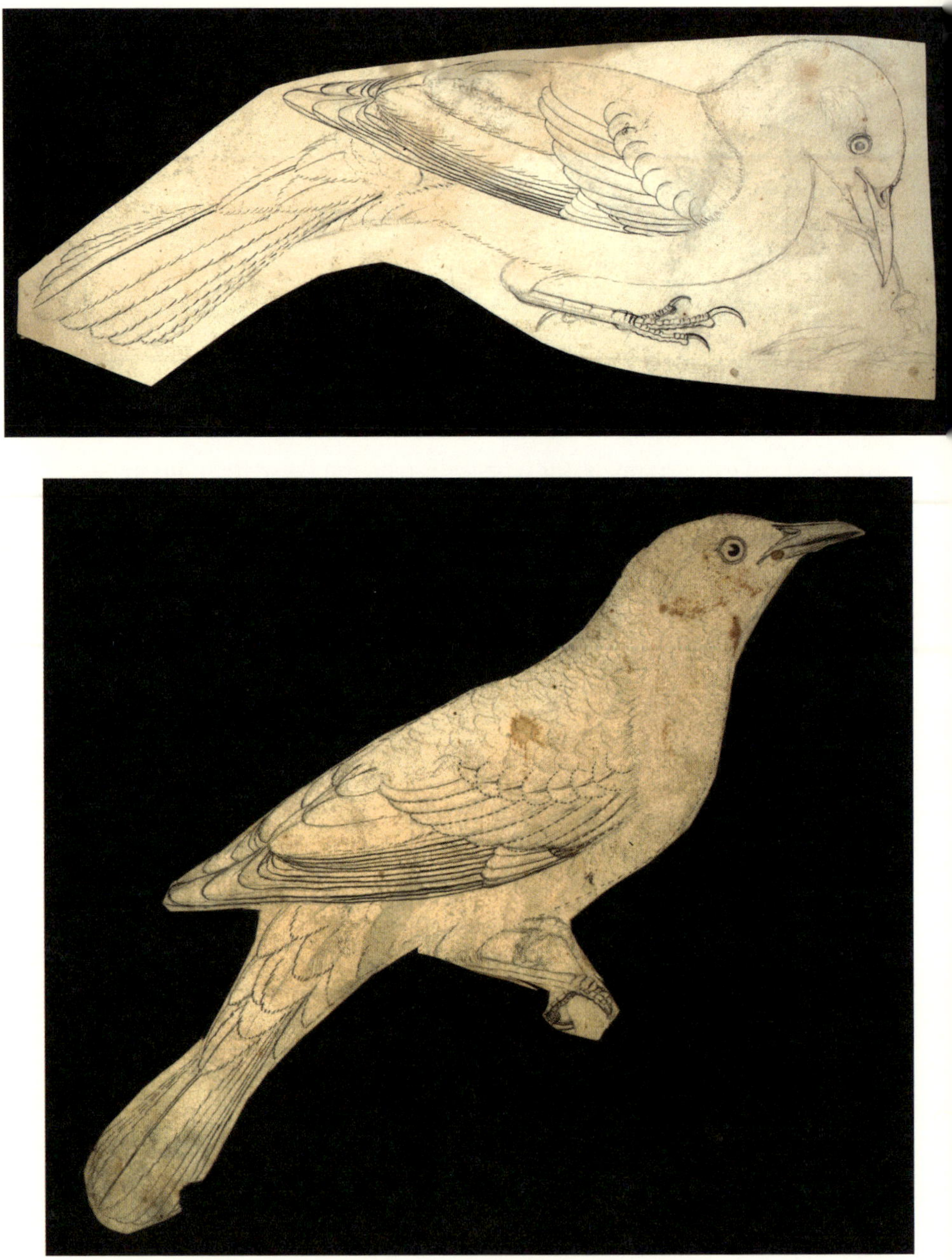

Figure 3.25. Common Grackle *(top)*; Rusty Blackbird *(bottom)*. (From the Academy of Natural Sciences of Philadelphia Library and Archives Collection, ANSP 79 [Drexel University].)

Purple Grakle (Common Grackle)

The lower parts of Virginia, North and South Carolina, and Georgia, are the winter residences of these flocks. Her numerous bodies, collecting together from all quarters of the interior and northern districts, and darkening the air with their numbers, sometimes form one congregated multitude of many hundred thousands. A few miles from the banks of the Roanoke, on the 20th of January, I met with one of those prodigious armies of Grakles. They rose from the surrounding fields with a noise like thunder, and, descending on the length of road before me, covered it and the fences completely with black; and when they again rose, and, after a few evolutions, descended on the skirts of the high-timbered woods, at that time destitute of leaves, they produced a most singular and striking effect; the whole trees for a considerable extent, from the top to the lowest branches, seemed as if hung in mourning; their notes and screaming the mean while resembling the distant sound of a great cataract, but in more musical cadence, swelling and dying away on the ear, according to the fluctuation of the breeze.[27]

Rusty Grakle (Rusty Blackbird)

Here is a single species described by one of the most judicious naturalists of Great Britain no less than five different times!—The greater part of these descriptions is copied by succeeding naturalists, whose synonyms it is unnecessary to repeat: so great is the uncertainty in judging, from a mere examination of their dried or stuffed skins, of the particular tribes of birds, many of which, for several years, are constantly varying in the colors of their plumage, and, at different seasons, or different ages, assuming new and very different appearances. Even the size is by no means a safe criterion, the difference in this respect between the male and female of the same species (as in the one now before us) being sometimes very considerable.[28]

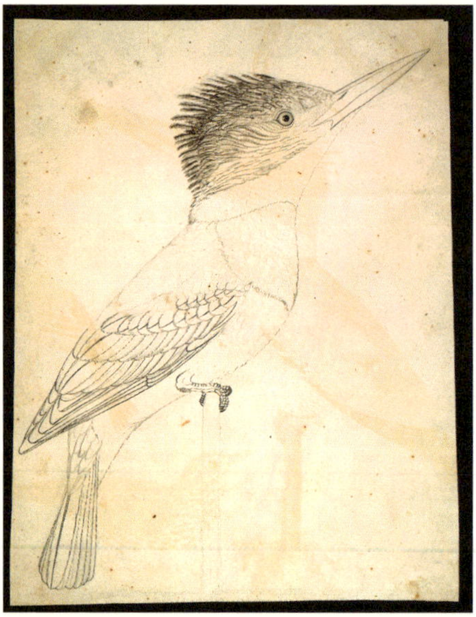

Figure 3.26. Belted Kingfisher (*top*); Fox Sparrow (*middle*); Savannah Sparrow (*bottom*). (From the Academy of Natural Sciences of Philadelphia Library and Archives Collection ANSP 79 [Drexel University].)

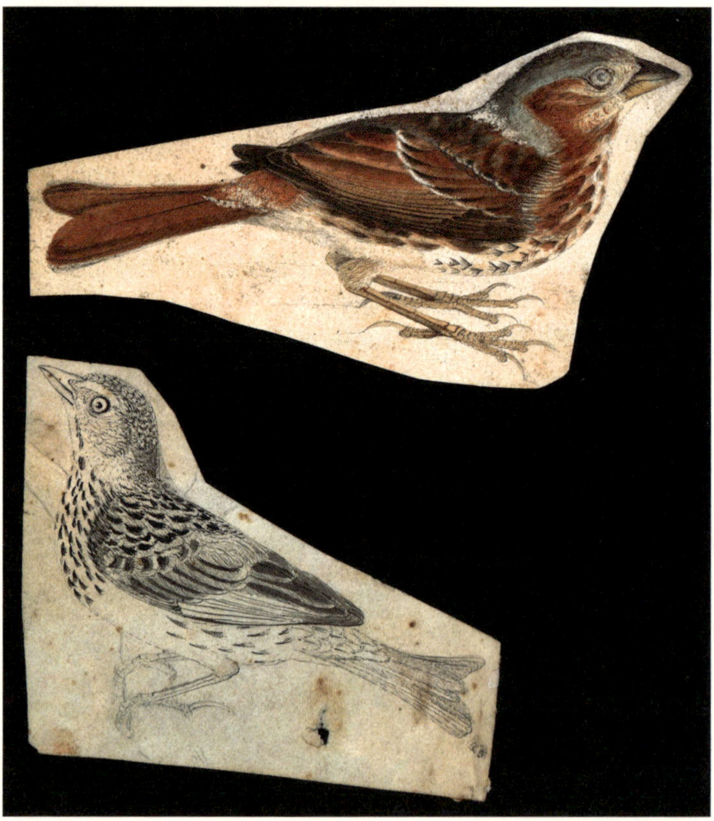

Current name: Belted Kingfisher, *Megaceryle alcyon* (Linnaeus)
Wilson's name: Belted Kingsfisher, *Alcedo alcyon*
Original name: *Alcedo alcyon* Linnaeus 1758

Current name: Fox Sparrow, *Passerella iliaca* (Merrem)
Wilson's name: Fox-colored Sparrow, *Fringilla rufa*
Original name: *Fringilla iliaca* Merrem 1786

Current name: Savannah Sparrow, *Passerculus sandwichensis* (Gmelin)
Wilson's name: Savannah Sparrow, *Fringilla savanna*
Original name: *Emberiza sandwichensis* Gmelin 1789

Commentary

These were preliminary drawings for birds figured in plates 22 and 23 of *American Ornithology*. Like he has done in so many other illustrations, Wilson has rotated the tails to show the forking or the feather pattern.

Belted Kingsfisher

This is a general inhabitant of the banks and shores of all our fresh water rivers, from Hudson's Bay to Mexico; and is the only species of its tribe found within the United States. This last circumstance, and its characteristic appearance, make it as universally known here as its elegant little brother, the Common Kingfisher of Europe, is in Britain. Like the lovelorn swains, of whom poets tell us, he delights in murmuring streams and falling waters; not, however, merely that they may soothe his ear, but for a gratification somewhat more substantial. Amidst the roar of the cataract, or over the foam of a torrent, he sits perched upon an overhanging bough, glancing his piercing eye in every direction below for his scaly prey, which, with a sudden, circular plunge, he sweeps from their native element, and swallows in an instant. His voice, which is not unlike the twirling of a watchman's rattle, is naturally loud, harsh, and sudden; but is softened by the sound of the brawling streams and cascades among which he generally rambles.[29]

Fox-colored Sparrow (Fox Sparrow)

This plump and pretty species arrives in Pennsylvania from the north about the 20th of October; frequents low, sheltered thickets; associates in little flocks of ten or twelve; and is almost continually scraping the ground, and rustling among the fallen leaves . . . They are rather of a solitary nature, seldom feeding in the open fields, but generally under thickets, or among tall, rank weeds on the edges of fields.[30]

Savannah Sparrow

This new species is an inhabitant of the low countries on the Atlantic coast, from Savannah, where I first discovered it, to the state of New York, and is generally resident in these places, though rarely found inland, or far from the sea-shore.[31]

Current name: Painted Bunting, *Passerina ciris* (Linnaeus)
Wilson's name: Painted Bunting, *Emberiza ciris*
Original name: *Emberiza Ciris* Linnaeus 1758

Current name: Prothonotary Warbler, *Protonotaria citria* (Boddaert)
Wilson's name: Prothonotary Warbler, *Sylvia protonotarius*
Original name: *Protonotaria citria* Boddaert 1783

Current name: Blue Grosbeak, *Passerina caerulea* (Linnaeus)
Wilson's name: Blue Grosbeak, *Loxia caerulea*
Original name: *Loxia caerulea* Linnaeus 1758

Current name: Worm-eating Warbler, *Helmitheros vermivorum* (Gmelin)
Wilson's name: Worm-eating Warbler, *Sylvia vermivora*
Original name: *Motacilla vermivora* Gmelin 1789

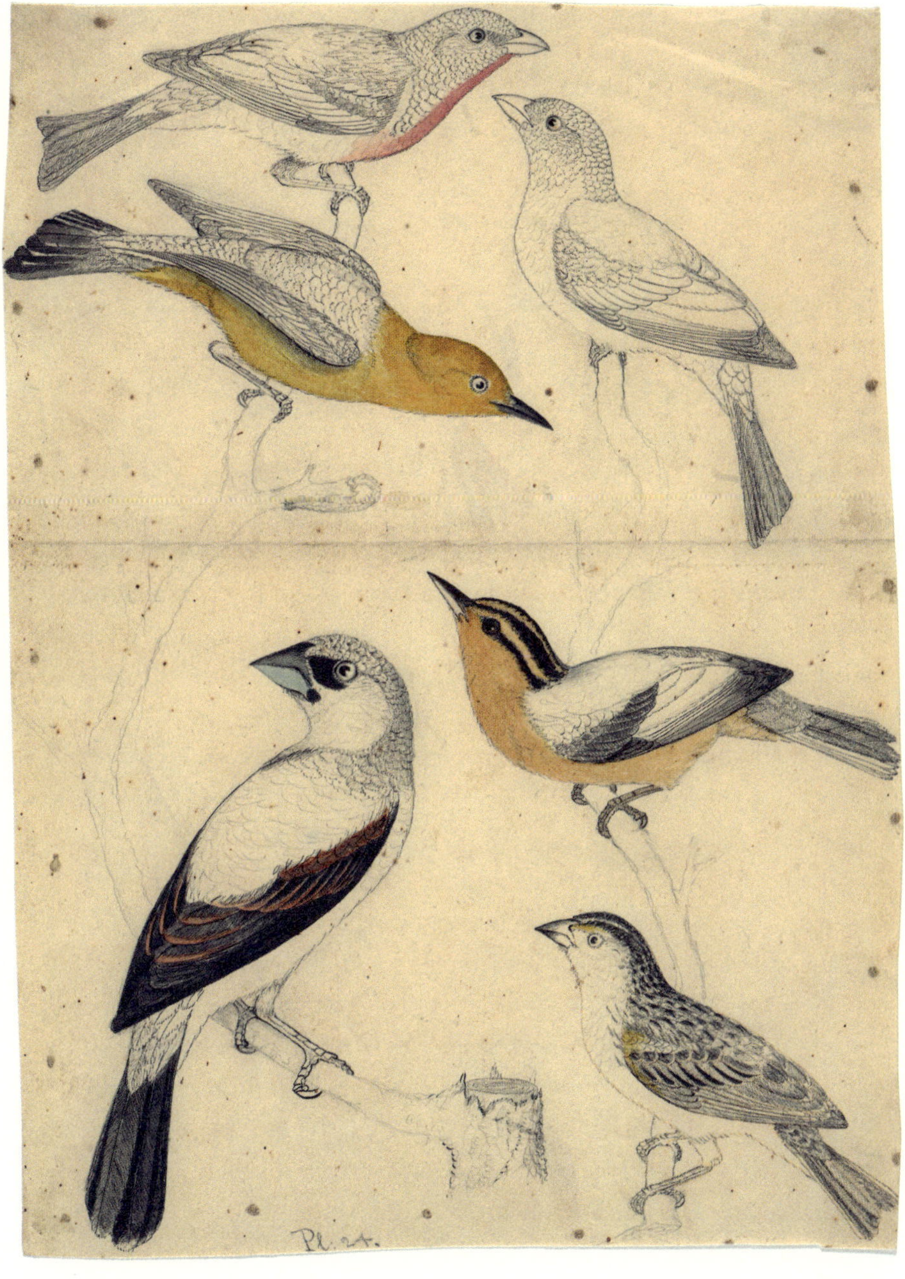

Figure 3.27. Painted Bunting (*top* and *upper right*); Prothonotary Warbler (*middle left*); Blue Grosbeak (*lower left*); Worm-eating Warbler (*middle right*); Grasshopper Sparrow (*bottom right*). (From the archives of the Ernst Mayr Library, Museum of Comparative Zoology, Harvard University.)

Current name: Grasshopper Sparrow, *Ammodramus savannarum* (Gmelin)
Wilson's name: Yellow-winged Sparrow, *Fringilla passerina*
Original name: *Fringilla Savannarum* Gmelin 1789

Commentary

The posture of the Blue Grosbeak, with its head turned sharply to the left, is reminiscent of the work of William Bartram and George Edwards. The postures of the warblers are unusual, but not unrealistic. The Prothonotary Warbler appears to be lunging forward toward an insect, and the Worm-eating Warbler may be studying the underside of a leaf for aphids. Whatever these birds are doing, the bills are excellently proportioned and reflect Wilson's understanding of the importance of this structure in the natural history of birds. The beak of a grosbeak, as the name implies, is large and well adapted to cracking large seeds. The bills of the warblers, in contrast, are thin and pointed, and make excellent tools for grasping the small insects that are the principal food of this group of birds.

Painted Bunting

The young birds of both sexes, during the first season, are of a fine green olive above, and dull yellow below. The females undergo little or no change, but that of becoming of a more brownish cast. The males, on the contrary, are long and slow in arriving at their full variety of colors. In the second season, the blue on the head begins to make its appearance, intermixed with the olive green; the next year, the yellow shows itself on the back and rump, and also the red, in detached spots, on the throat and lower parts. All these colors are completed in the fourth season, except, sometimes, that the green still continues on the tail. On the fourth and fifth season, the bird has attained his complete colors, and appears then as represented in the plate. No dependence, however, can be placed on the regularity of this change in birds confined in a cage, as the want of proper food, sunshine, and variety of climate, all conspire against the regular operations of nature.[32]

Prothonotary Warbler

This is an inhabitant of the same country as the preceding species, and also a passenger from the south, with this difference, that the bird now before us seldom approaches the house or garden, but keeps among the retired, deep, and dark, swampy woods, through which it flits nimbly in search of small caterpillars, uttering every now and then a few screaking notes, scarcely worthy of notice.[33]

Blue Grosbeak

During its stay with me, I fed it on Indian corn, which it seemed to prefer, easily breaking with its powerful bill the hardest grains. They also feed on hemp seed, millet and the kernels of several kinds of berries. They are timid birds, watchful, silent, and active, and generally neat in their plumage. Having never yet met with their nest, I am unable to describe it.[34]

Worm-eating Warbler

This bird is remarkably fond of spiders, darting about wherever there is a probability of finding these insects. If there be a branch broken, and the leaves withered, it shoots among them in preference to every other part of the tree, making a great rustling, in search of its prey. I have often watched its manoeuvres while thus engaged, and flying from tree to tree in search of such places.[35]

Yellow-winged Sparrow (Grasshopper Sparrow)

This small species is now for the first time introduced to the notice of the public. I can, however, say little towards illustrating its history, which like that of many individuals of the human race, would be but a dull detail of humble obscurity.[36]

Current name: Carolina Parakeet, *Conuropsis carolinensis* (Linnaeus)
Wilson's name: Carolina Parrot, *Psittacus carolinensis*
Original name: *Psittacus Carolinensis* (Linnaeus 1758)

Commentary

This fragment of the Carolina Parakeet drawing has been cut from the original and the obverse side used for another drawing. After the obverse drawing was made, this original side was covered with iron oxide for transfer of the drawing on the reverse side to the engraving surface. This Carolina Parakeet drawing fits perfectly over plate 26 in *American Ornithology;* see Figure 3.29.

Carolina Parrot (Carolina Parakeet)

Anxious to try the effects of education on one of the Carolina Parrots which I procured at Big Bone Lick[,] . . . I wrapped it closely in a silk handkerchief, tying it tightly around and carried it in my pocket. When I stopped for refreshment, I unbound my prisoner and gave it its allowance [of cockleburs], which it generally dispatched with great dexterity, unhusking the seeds from the bur in a twinkling; in doing which, it always employed the left foot to hold the bur, as did several others that I kept for some time. I began to think that this might be peculiar to the whole tribe, . . . but by shooting a number afterwards while engaged in eating mulberries, I found sometimes the left, sometimes the right, foot stained with the fruit, the other

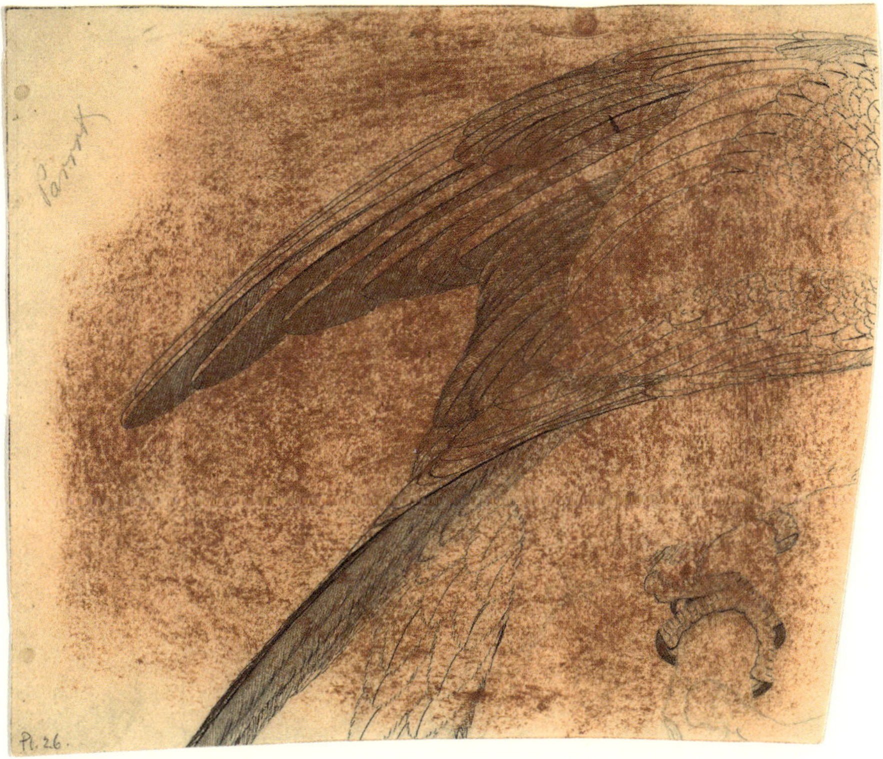

Figure 3.28. The wings, body, and feet cut from Wilson's drawing of a Carolina Parakeet. (From the archives of the Ernst Mayr Library, Museum of Comparative Zoology, Harvard University.)

always clean; from which and the constant practice of those I kept, it appears, that, like the human species in the use of their hands, they . . . are either left or right-footed . . . In passing through the Chickasaw and Chactaw nations, the Indians, wherever I stopped to feed, collected around me, men, women and children, laughing and seeming wonderfully amused with the novelty of my companion. The Chickasaws called it in their language *"Kelinky;"* but when they heard me call it Poll, they soon repeated the name; and, wherever I chanced to stop among these people, we soon became familiar with each other through the medium of Poll.[37]

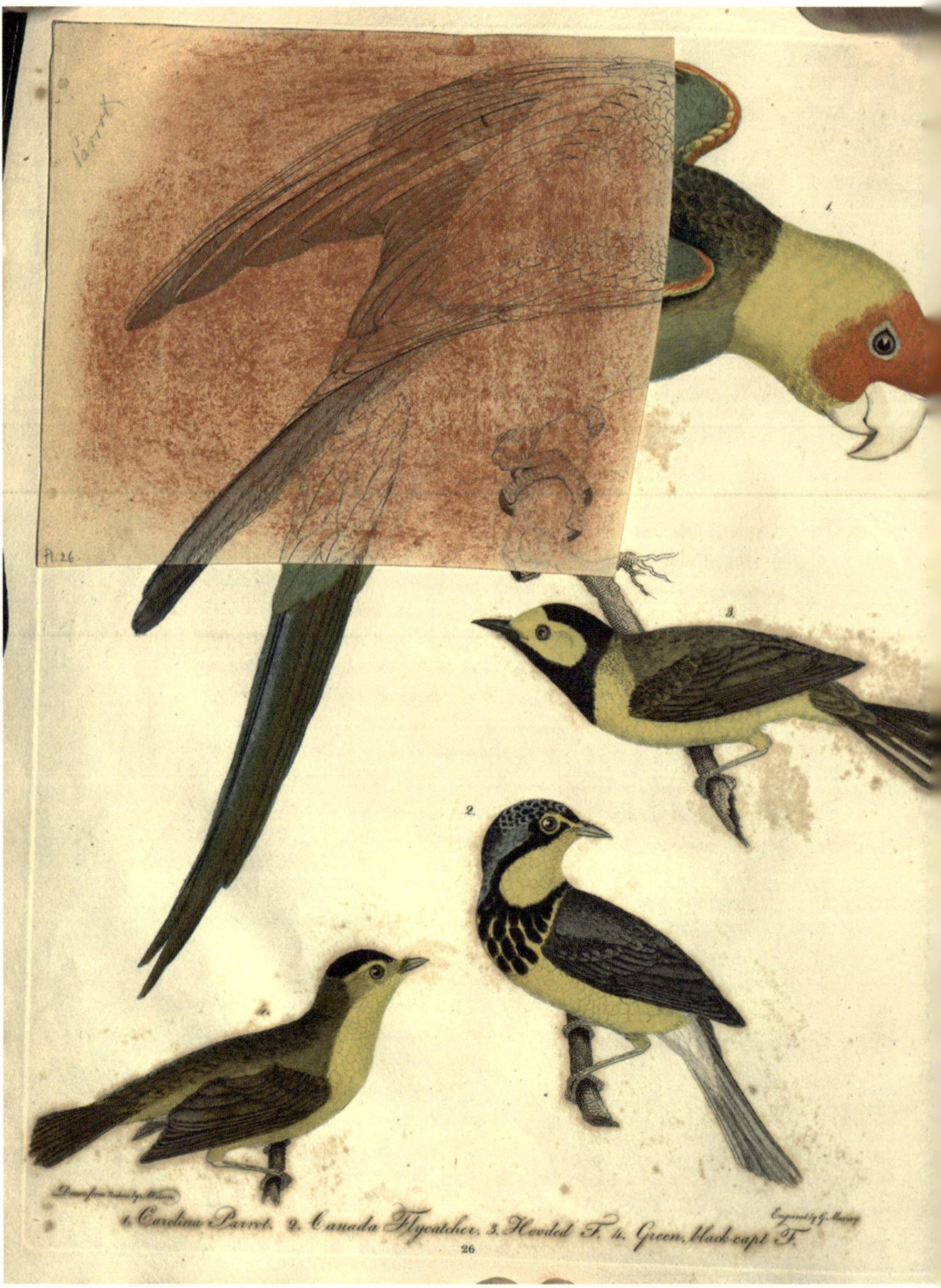
Pl. 26.
Drawn from Nature by A. Wilson
1. Carolina Parrot. 2. Canada Flycatcher. 3. Hooded F. 4. Green, black-capt F.
Engraved by G. Murray
26

Current name: Hooded Warbler, *Setophaga citrina* (Boddaert)
Wilson's name: Hooded Flycatcher, *Muscicapa cucullata*
Original name: *Muscicapa citrina* Boddaert 1783

Current name: Wilson's Warbler, *Cardellina pusilla* (Wilson)
Wilson's name: Green Black-capped Flycatcher, *Muscicapa pusilla*
Original name: *Muscicapa pusilla* Wilson 1811

Current name: Canada Warbler, *Cardellina canadensis* (Linnaeus)
Wilson's name: Canada Flycatcher, *Muscicapa canadensis*
Original name: *Muscicapa Canadensis* Linnaeus 1766

Commentary

This figure shows how perfectly the fragment of the Wilson drawing of a Carolina Parakeet (Figure 3.28) superimposes on the parakeet in plate 26 of *American Ornithology*. Wilson's arrangement of the four birds with the life-sized Carolina Parakeet framing the three smaller warblers speaks to his exquisite artistic sense. For many years the warblers were all in the genus *Wilsonia.*

Hooded Warbler

> This species is seldom seen in Pennsylvania and the northern states; but through the whole extent of country south of Maryland, from the Atlantic to the Mississippi, is very abundant. It is, however, most partial to low situations, where there is plenty of thick underwood; abounds among the canes in the state of Tenessee [*sic*], and in the Mississippi territory; and seems perpetually in pursuit of winged insects; now and then uttering three loud not unmusical and very

Figure 3.29. The plate from *American Ornithology* showing the Carolina Parakeet (*top*) with the scrap placed in its original position in the drawing (Figure 3.28). Also shown are the Hooded Warbler (*middle right*), Wilson's Warbler (*bottom left*) and Canada Warbler (*bottom right*). (From the archives of the Ernst Mayr Library, Museum of Comparative Zoology, Harvard University.)

lively notes, resembling *twee, twee, twitchie,* while engaged in the chase. Like almost all its tribe [flycatchers] it is full of spirit, and exceedingly active.[38]

Wilson's Warbler

This neat and active little species I have never met with in the works of any European naturalist. It is an inhabitant of the swamps of the southern states and has been several times seen in the lower parts of the states of New Jersey and Delaware. Amidst almost unapproachable thickets of deep morasses it commonly spends its time, during summer, and has a sharp squeaking note, no ways musical. It leaves the southern states early in October.[39]

Canada Warbler

This is a solitary, and in the lower parts of Pennsylvania, rather a rare species; being more numerous in the interior, particularly near the mountains, where the only two I ever met with were shot. They are silent birds, as far as I could observe; and were busily darting among the branches after insects. From the specific name given them it is probably that they are more plenty in Canada than in the United States; where it is doubtful whether they be not mere passengers in spring and autumn.[40]

Current name: Yellow Warbler, *Setophaga petechia* (Linnaeus)
Wilson's name: Yellow Red-poll Warbler, *Sylvia petechia*
Original name: *Motacilla petechia* Linnaeus 1766

Current name: Yellow-billed Cuckoo, *Coccyzus americanus* (Linnaeus)
Wilson's name: Yellow-billed Cuckoo, *Cuculus carolinensis*
Original name: *Cuculus americanus* Linnaeus 1758

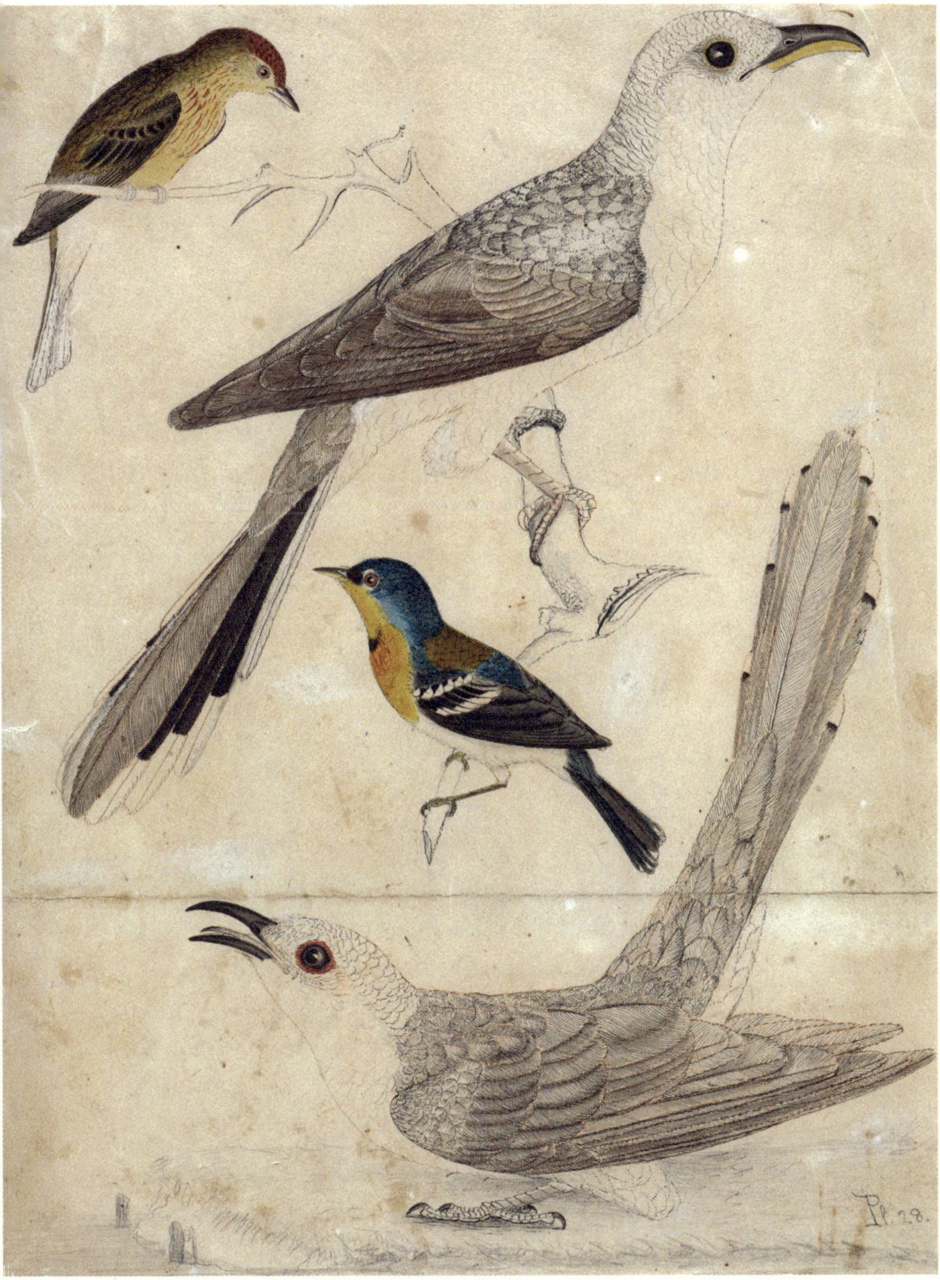

Figure 3.30. Yellow Warbler (*top left*); Yellow-billed Cuckoo (*upper middle*); Northern Parula (*lower middle*); Black-billed Cuckoo (*bottom*). (From the archives of the Ernst Mayr Library, Museum of Comparative Zoology, Harvard University.)

Current name: Northern Parula, *Setophaga americana* (Linnaeus)
Wilson's name: Blue Yellow-back Warbler, *Sylvia pusilla*
Original name: *Parus americanus* Linnaeus 1758

Current name: Black-billed Cuckoo, *Coccyzus erythropthalmus* (Wilson)
Wilson's name: Black-billed Cuckoo, *Cuculus erythropthalmus*
Original name: Black-billed Cuckoo *Cuculus erythropthalmus* Wilson 1811

Commentary

This drawing of the Black-billed Cuckoo, with its erect tail, makes for an artful composition when juxtaposed with that of the Yellow-billed Cuckoo, perched with tail depressed. It also allowed Wilson to include two small warblers on the same plate. Although Wilson often let the engraver fill in the details of the branches, stumps and ground, he never left details of the bills, eyes, and more distinctive plumage characteristics to the engraver. In this particular drawing he did indicate the ground under the Black-billed Cuckoo's feet with some grass-like marks.

Yellow Red-poll Warbler (Yellow Warbler)

This delicate little bird arrives in Pennsylvania early in April, while the maples are yet in blossom, among the branches of which it may generally be found at that season, feeding on the stamina of the flowers, and on small winged insects. Low, swampy thickets are its favorite places of resort.[41]

Yellow-billed Cuckoo

A stranger who visits the United States, for the purpose of examining their natural productions, and passes through our woods in the month of May or June, will sometimes hear, as he traverses the borders of deep, retired, high-timbered hollows, an uncouth, guttural sound or note, resembling the syllables *kowe, kowe, kowe, kowe kowe* beginning slowly, but ending so rapidly, that the notes seem to run into each other; and *vice versa:* he will hear this frequently, without being able to discover the bird or animal from which it proceeds, as it is both shy and solitary, seeking always the thickest foliage for concealment. This is the Yellow-billed Cuckoo.[42]

Blue Yellow-back Warbler (Parula Warbler)

Notwithstanding the respectability of the above authorities [Linnaeus, Catesby, Latham, and Bartram], I must continue to consider this bird as a species of Warbler. Its habits, indeed, partake something of the Titmouse; but the form of its bill is decidedly that of the *Sylvia* genus. It is remarkable for frequenting the tops of the tallest trees, where it feeds on the small winged insects and caterpillars that infest the young leaves and blossoms.[43]

Black-billed Cuckoo

This Cuckoo is nearly as numerous as the former (Yellow-billed Cuckoo), but has hitherto escaped the notice of European naturalists; or, from its general resemblance, has been confounded with the preceding. Its particular markings, however, and some of its habits, sufficiently characterize it as a distinct species.

. . . The Black-billed Cuckoo is particularly fond of the sides of creeks, feeding on small shell fish, snails, &c. I have also often found broken pieces of oyster shells in its gizzard, which, like that of the other [Yellow-billed Cuckoo], is covered with fine downy hair.[44]

Current name: Pileated Woodpecker, *Dryocopus pileatus* (Linnaeus)
Wilson's name: Pileated Woodpecker, *Picus pileatus*
Original name: *Picus pileatus* Linnaeus 1758

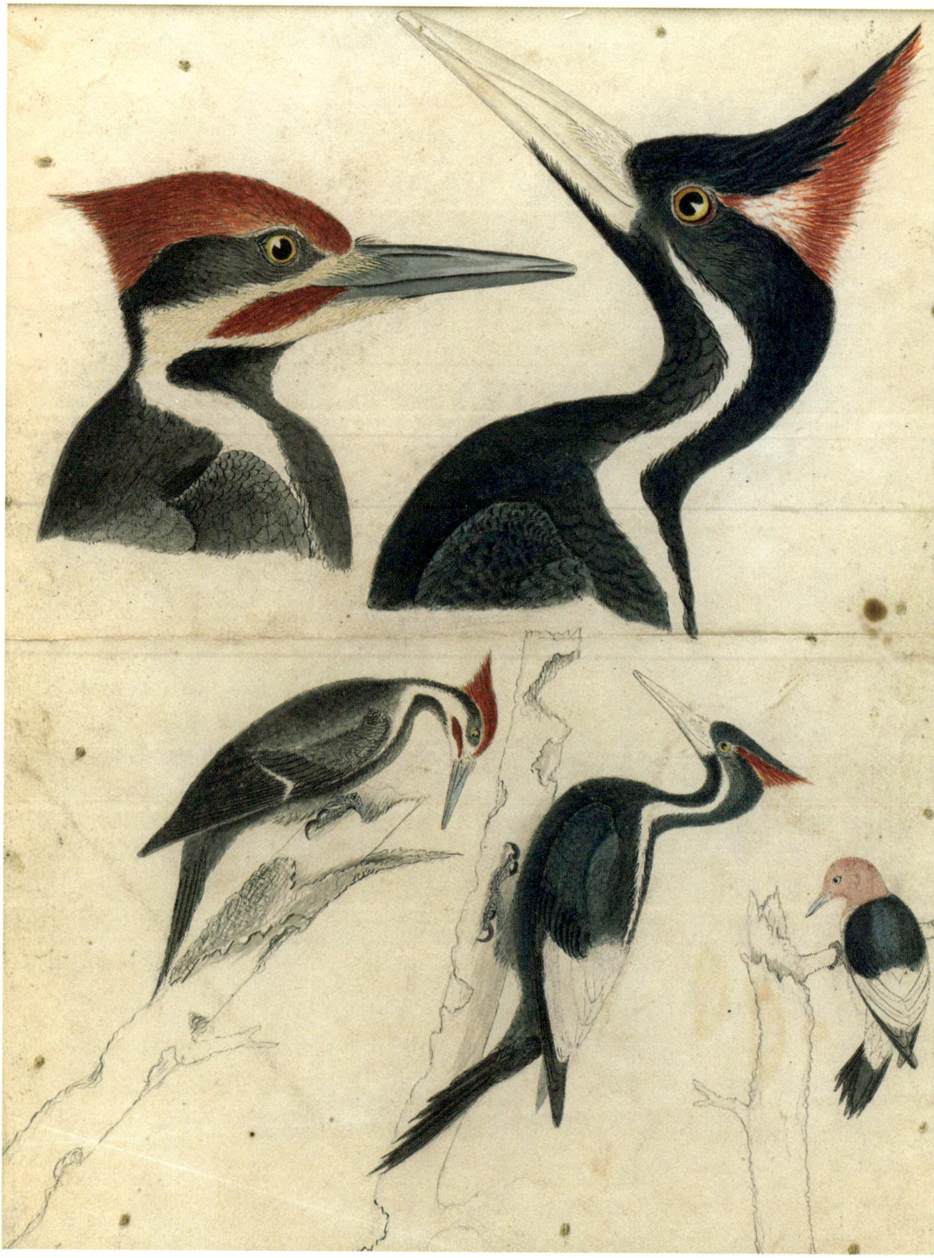

Figure 3.31. Pileated Woodpecker (*upper and lower left*); Ivory-billed Woodpecker (*upper right and lower middle*); Red-headed Woodpecker (*lower right*). (From the archives of the Ernst Mayr Library, Museum of Comparative Zoology, Harvard University.)

Current name: Ivory-billed Woodpecker, *Campephilus principalis* (Linnaeus)
Wilson's name: Ivory-billed Woodpecker, *Picus principalis*
Original name: *Picus principalis* Linnaeus 1758

Current name: Red-headed Woodpecker, *Melanerpes erythrocephalus* (Linnaeus)
Wilson's name: Red-headed Woodpecker, *Picus erythrocephalus*
Original name: *Picus erythrocephalus* Linnaeus 1758

Commentary

In these watercolor washes over pencil, Wilson is careful to portray the lower three birds on the same scale, while the upper portraits are life size. Wilson tried to show the details that would help readers identify birds in the wild. The heads are presented in side view and do not differ substantially from images of the same species presented in modern field guides. This is the final draft for the finished plate shown in Figure 3.14.

Pileated Woodpecker

> Almost every old trunk in the forest where he resides bears the marks of his chisel. Wherever he perceives a tree beginning to decay, he examines it round and round with great skill and dexterity, strips off the bark in sheets of five or six feet in length, to get at the hidden cause of the disease, and labors with a gayety and activity really surprising. I have seen him separate the greatest part of the bark from a large, dead pine tree, for twenty or thirty feet, in less than a quarter of an hour.[45]

Ivory-billed Woodpecker

> In looking over the accounts given of the Ivory-billed Woodpecker by the naturalists of Europe, I find it asserted that it inhabits from New Jersey to Mexico. I believe, however, that few of them are ever seen

to the north of Virginia, and very few of them even in that state. The first place I observed this bird at, when on my way to the south, was about twelve miles north of Wilmington in North Carolina. There I found the bird from which the drawing of the figure in the plate was taken. This bird was only wounded slightly in the wing, and, on being caught, uttered a loudly reiterated and most piteous note, exactly resembling the violent crying of a young child; which terrified my horse so, as nearly to have cost me my life. It was distressing to hear it. I carried it with me in the chair [saddle], under cover, to Wilmington. In passing through the streets, its affecting cries surprised every one within hearing, particularly the females, who hurried to the doors and windows with looks of alarm and anxiety. I drove on, and on arriving at the piazza of the hotel, where I intended to put up, the landlord came forward, and a number of other persons who happened to be there, all equally alarmed at what they heard; this was greatly increased by my asking whether he could furnish me with accommodations for myself and my baby. The man looked blank and foolish, while the others stared with still greater astonishment. After diverting myself for a minute or two at their expense, I drew my Woodpecker from under the cover, and a general laugh took place.[46]

Red-headed Woodpecker

There is perhaps no bird in North America more universally known than this. His tricolored plumage, red, white, and black, glossed with steel blue, is so striking and characteristic, and his predatory habits in the orchards and corn-fields, added to his numbers, and fondness for hovering along the fences, so very notorious, that almost every child is acquainted with the Red-headed Woodpecker.[47]

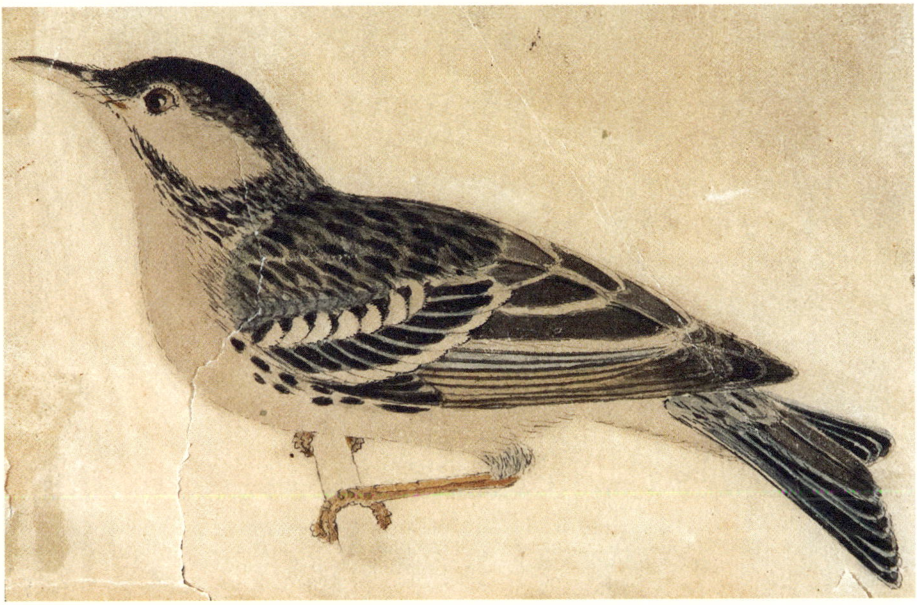

Figure 3.32. Blackpoll Warbler. (From the archives of the Paisley Museum, Renfrewshire Council, Scotland.)

Current name: Blackpoll Warbler, *Setophaga striata* (Forster)
Wilson's name: Black-poll Warbler, *Sylvia striata*
Original name: *Muscicapa striata* Forster 1772

Commentary

This fully colored individual is the same as the one in plate 30 of *American Ornithology*. Presumably it is the one from which the plate was prepared. Alternatively, it may have been prepared by Wilson as a reference for the colorists and then sent to his father and stepmother in Paisley in one of his letters or in a copy of volume 4 when that was sent to them. The painting is in the archives of the Paisley Public Library Museum.

Blackpoll Warbler

This bird may be considered as occupying an intermediate station between the Flycatchers and the Warblers, having the manners of the former, and the bill, partially of the latter. The nice gradations by which Nature passes from one species to another, even in this department of the great chain of beings, will for ever baffle all the artificial rules and systems of man. And this truth every fresh discovery must impress more forcibly on the mind of the observing naturalist.[48]

Current name: Red-winged Blackbird, *Agelaius phoeniceus* (Linnaeus)
Wilson's name: Red-winged Starling, *Sturnus predatorius*
Original name: *Oriolus phoeniceus* Linnaeus 1766

Commentary

This cut fragment of a fully rendered Red-winged Blackbird drawing is yet another instance of Wilson reusing one of his completed drawings. It emphasizes that he was not interested in producing gallery art, but rather was focused on producing images for his *American Ornithology*.

Red-winged Starling (Red-winged Blackbird)

They arrive in Pennsylvania late in March. Their general food at this season, as well as during the early part of the summer, . . . consists of grub-worms, caterpillars, and various other larvæ, the silent but deadly enemies of all vegetation, and whose secret and insidious attacks are more to be dreded by the husbandman than the combined forces of the whole feathered tribes together. For these vermin the Starlings search with great diligence; in the ground, at the roots of plants, in orchards, and meadows, as well as among buds, leaves and blossoms; and from their known voracity the multitudes of these

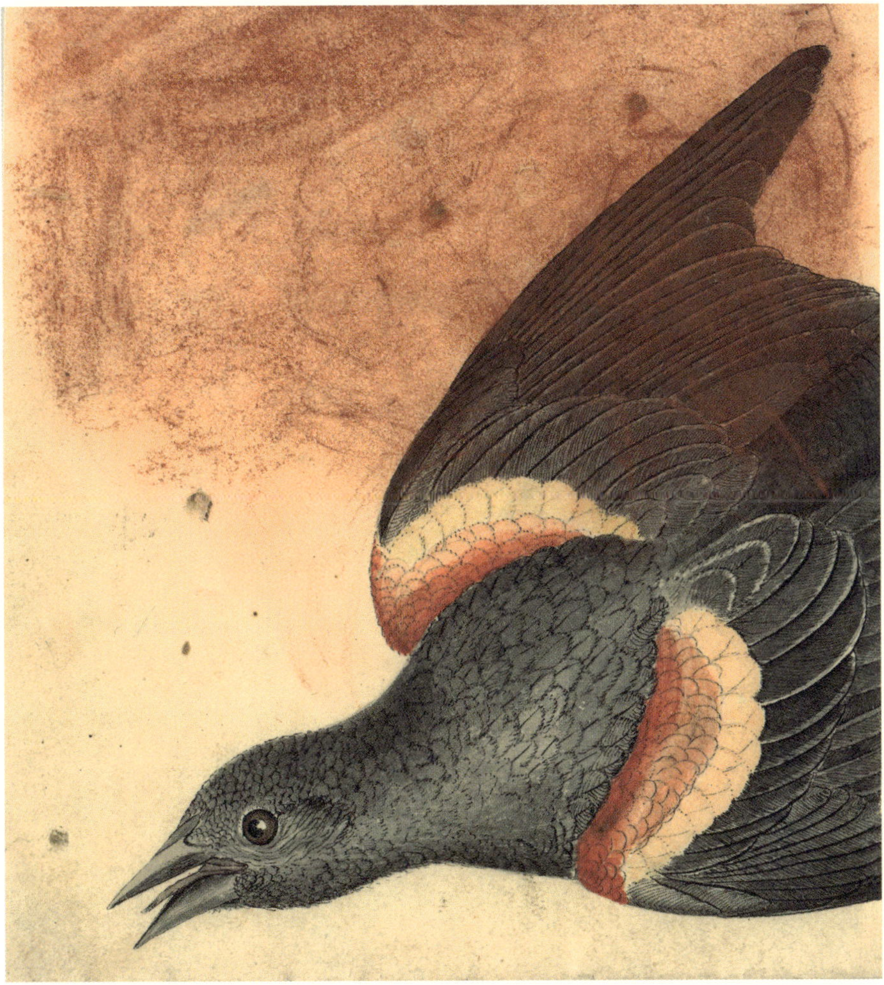

Figure 3.33. The male Red-winged Blackbird cut from the final drawing for the plate (Figure 3.34). (From the archives of the Ernst Mayr Library, Museum of Comparative Zoology, Harvard University.)

insects which they destroy must be immense. Let me illustrate this by a short computation. If we suppose each bird to, on an average to devour fifty of these larvæ in a day, (a very moderate allowance,) a single pair in four months, the usual time such food is sought after, will consume upwards of twelve thousand. It is believed, that not less than a million pair of these birds are distributed over the whole extent of the United States in summer; whose food being nearly the same, would swell the amount of vermin destroyed to twelve thousand millions. But the number of young birds may be fairly estimated at double that of their parents, and as these are constantly fed on larvæ for at least three weeks, making only the same allowance for them as the old ones, their share would amount to four thousand two hundred millions; making a grand total of sixteen thousand two hundred millions of noxious insects destroyed in the space of four months by this single species! The combined ravages of such a hideous host of vermin would be sufficient to spread famine and desolation over a wide extent of the richest and best cultivated country on earth. All this, it may be said, is mere supposition. It is, however, supposition founded on known and acknowledged facts.[49]

Current name: Common Redpoll, *Carduelis flammea* (Linnaeus)
Wilson's name: Lesser Redpoll, *Fringilla linaria*
Original name: *Fringilla flammea* Linnaeus 1758

Commentary

Plate 30 from Wilson's *American Ornithology* has the fragment (Figure 3.33) positioned to show where it was cut out of the final drawing. This plate is also interesting in that Audubon copied both the male and female for his portrait of the Red-winged Blackbird.

Lesser Redpoll (Common Redpoll)

This species extends throughout the whole northern parts of Europe, is likewise found in the remote wilds of Russia, was seen by Steller in Kampsehatlea and probably inhabits corresponding climates round the whole habitable parts of the northern hemisphere.[50]

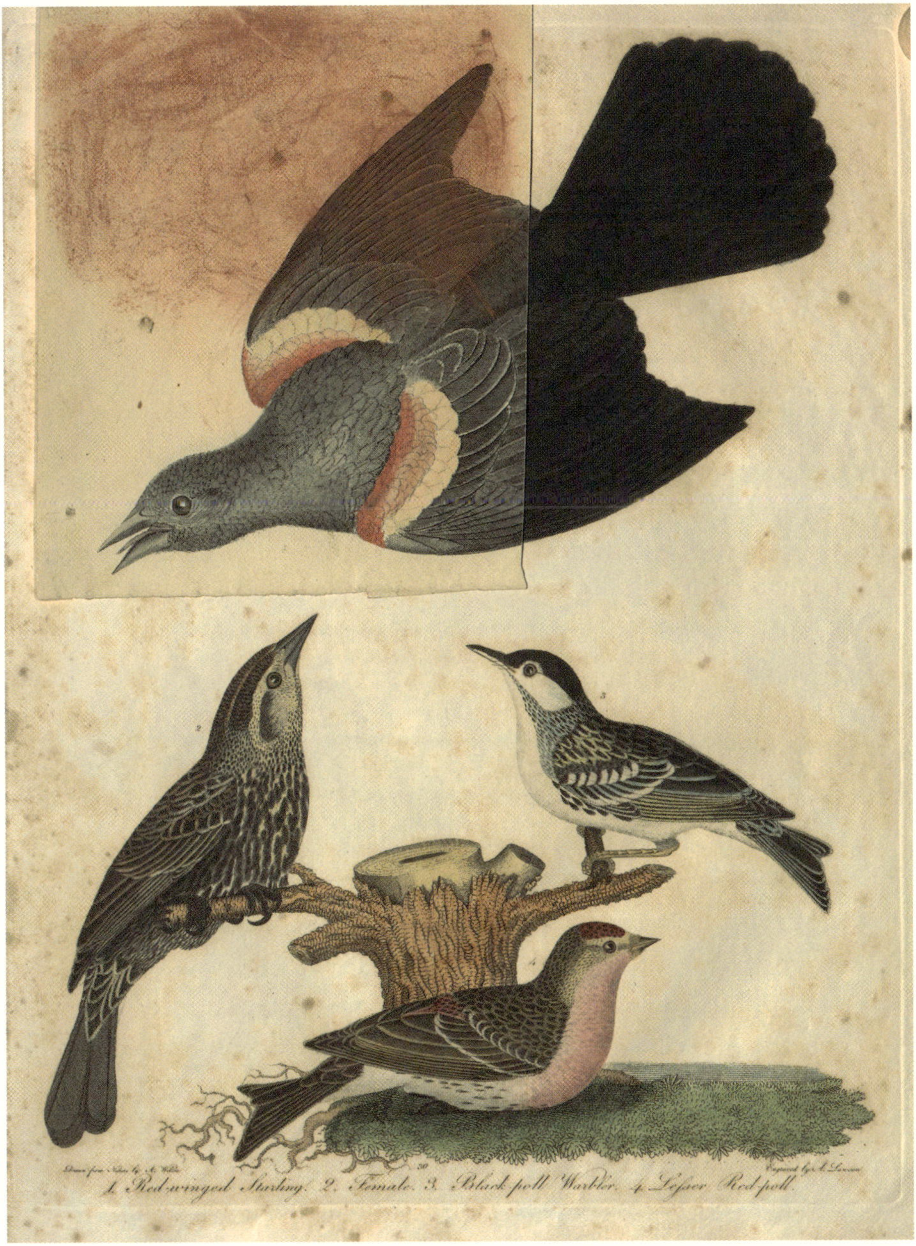

Figure 3.34. Plate 30 of *American Ornithology*. Male Redwinged Blackbird (*top*), female (*middle left*), Blackpoll Warbler (*middle right*), Common Redpoll (*bottom*). (From the archives of the Ernst Mayr Library, Museum of Comparative Zoology, Harvard University.)

Current name: Black-billed Magpie, *Pica hudsonia* (Linnaeus)
Wilson's name: Magpie, *Corvus pica*
Original name: *Corvus pica* Linnaeus 1758

Current name: Northern Saw-whet Owl, *Aegolius acadicus* (Gmelin)
Wilson's name: Little Owl, *Strix passerina*
Original name: *Strix acadica* Gmelin 1788

Commentary

These are two of Wilson's more interesting drawings. The eye of the Magpie is properly positioned in the drawing, but too far forward in the final plate. The owl's pose is different from the usual full-face pose in which the bird returns the observer's gaze. The flank feathers were done with broad brush strokes over pencil and the feathers of the facial disc are particularly accomplished. The back, tail, and head feathers are fully outlined in ink. The fine pencil outline for the background log and the stump on which the owl rests are very Bartramesque.

Magpie (Black-billed Magpie)

> This bird unites in its character, courage and cunning, turbulency and rapacity. Not inelegantly formed, and distinguished by gay as well as splendid plumage, he has long been noted in those countries where he commonly resides, and his habits and manners are there familiarly known. He is particularly pernicious to plantations of young oaks, tearing up the acorns; and also to birds destroying great numbers of their eggs and young, even young chickens, partridges, grouse, and pheasants . . . In 1804, an exploring party under the command of Captains Lewis and Clark, on their route to the Pacific Ocean across the continent, first met with the Magpie somewhere near the great bend of the Missouri, and found that the number of these birds increased as they advanced. Here also the Blue Jay disappeared; as if the territorial boundaries and jurisdiction of these two noisy and voracious families of the same tribe had been mutually

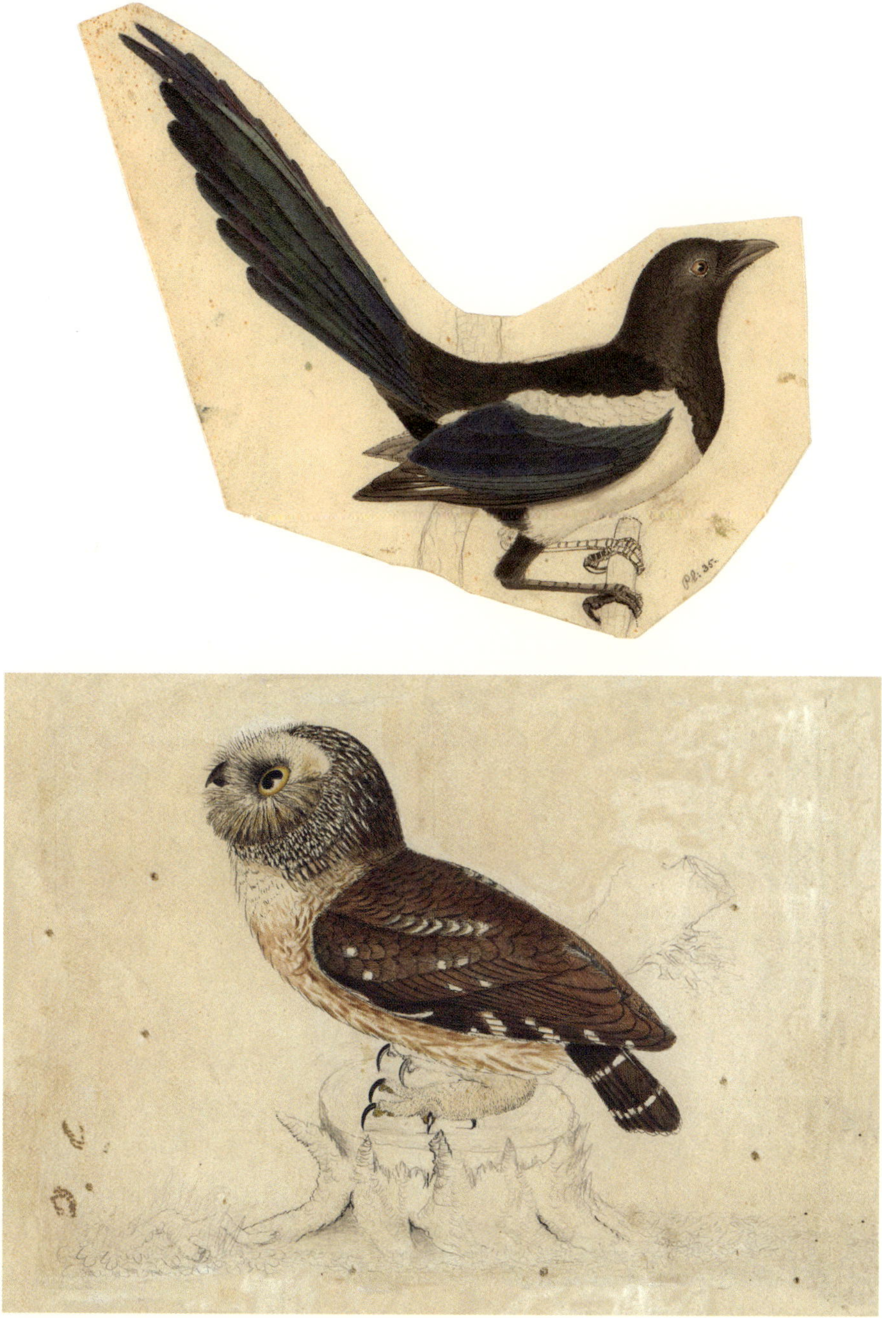

Figure 3.35. Black-billed Magpie (*top*); Northern Saw-whet Owl (*bottom*). (From the archives of the Ernst Mayr Library, Museum of Comparative Zoology, Harvard University.)

agreed on, and distinctly settled. But the Magpie was found to be far more daring than the Jay, dashing into their very tents, and carrying off the meat from the dishes. One of the hunters who accompanied the expedition informed me, that they frequently attended him while he was engaged in skinning and cleaning the carcass of the deer, bear, or buffalo he had killed, often seizing the meat that hung within a foot or two of his head. On the shores of the Koos-koos-ke river, on the west side of the great range of the Rocky Mountains, they were found to be equally numerous.[51]

Little Owl (Northern Saw-whet Owl)

This species is a general and constant inhabitant of the middle and northern states; but is found most numerous in the neighborhood of the sea-shore, and among woods and swamps of pine trees. It rarely rambles much during day; but, if disturbed, flies a short way, and again takes shelter from the light; at the approach of twilight it is all life and activity, being a noted and dexterous mouse-catcher.[52]

Current name: Seaside Sparrow, *Ammodramus maritimus* (Wilson)
Wilson's name: Seaside Finch, *Fringilla maritima*
Original name: *Fringilla maritima* Wilson 1811

Commentary

This fully colored rendering is in the pose of the Seaside Sparrow in the *American Ornithology* plate, but the bird in the plate has a shorter body and larger head, more like the proportions of the live bird. The gray cast and ill-defined markings of the bird in the painting, however, are more similar to the live bird than to the bird in the plate with its clearly defined facial striping and browner coloring. The coloring in the plate may reflect the colorist's interpretation; Wilson's interpretation may be better represented by the colors of this painting.

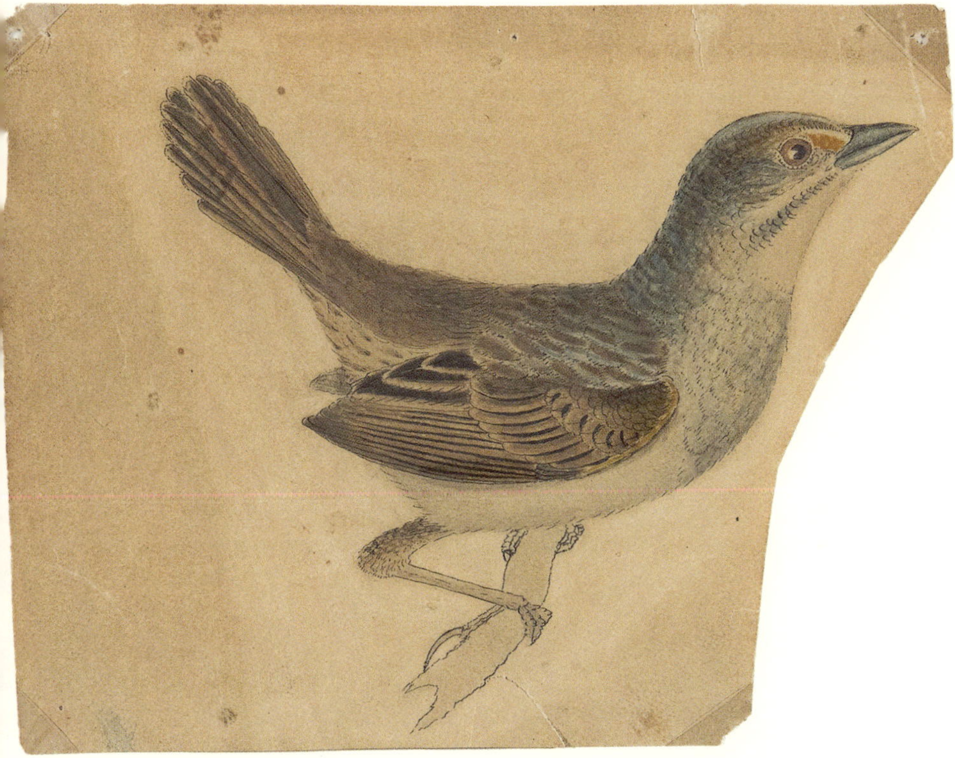

Figure 3.36. Seaside Sparrow. (From the archives of the Paisley Museum, Renfrewshire Council, Scotland.)

Sea-side Finch (Seaside Sparrow)

This species derives its whole subsistence from the sea. I examined a great number of individuals by dissection, and found their stomachs universally filled with fragments of shrimps, minute shell fish, and broken limbs of small sea crabs. Its flesh, also, as was to be expected, tasted of fish, or was what is usually termed *sedgy*. Amidst the recesses of these wet sea marshes it seeks the rankest growth of grass, and sea weed, and climbs along the stalks of the rushes with as much dexterity as it runs along the ground, which is rather a singular circumstance, most of our climbers being rather awkward at running.[53]

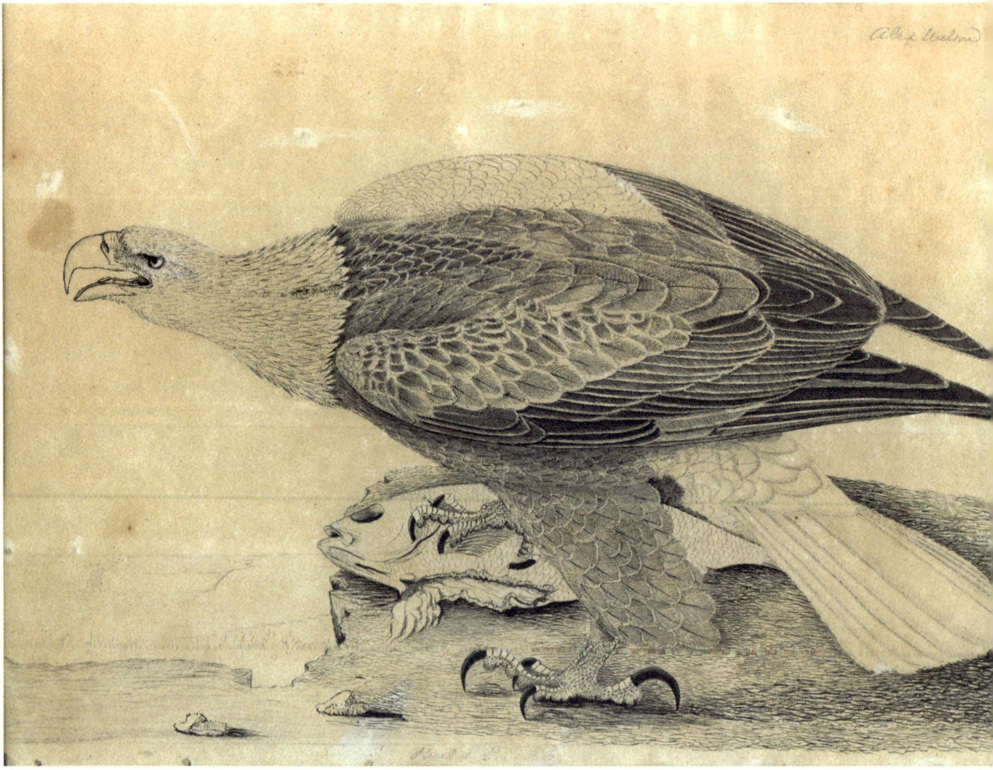

Figure 3.37. Bald Eagle. (From the archives of the Ernst Mayr Library, Museum of Comparative Zoology, Harvard University.)

Current name: Bald Eagle, *Haliaeetus leucocephalus* (Linnaeus)
Wilson's name: White-headed or Bald Eagle, *Falco leucocephalus*
Original name: *Falco leucocephalus* Linnaeus 1766

Commentary

The legs, feet, and feather renderings in ink over pencil are exceptional in this illustration. The background is more complete than in most of Wilson's earlier drawings and includes a scavenged fish with a sunken eye (which suggests that it has been dead for some time). This drawing conveys the important information that Bald Eagles are primarily fish eaters and scavengers. Note too that the talons are *in* the fish. Wilson was a close observer who rendered the details accurately.

White-headed or Bald Eagle

> This distinguished bird, as he is the most beautiful of his tribe in this part of the world, and the adopted emblem of our country, is entitled to particular notice. He was drawn from one of the largest and most perfect specimens I have yet met with. In the back ground is seen a distant view of the celebrated Cataract of Niagara, a noted place of resort for these birds, as well on account of the fish procured there, as for the numerous carcasses of squirrels, deer, bears, and various other animals, that, in their attempts to cross the river above the Falls, have been dragged into the current, and precipitated down that tremendous gulf, where, among the rocks that bound the Rapids below, they furnish a rich repast for the Vulture, the Raven, and the Bald Eagle . . .
>
> The appetite of the Bald Eagle, though habituated to long fasting, is of the most voracious, and often the most indelicate kind. Fish, when he can obtain them, are preferred to all other fare. Young lambs and pigs are dainty morsels, and made free with on all favorable occasions. Ducks, Geese, Gulls, and other sea fowl, are also seized with avidity. The most putrid carrion, when nothing better can be had, is acceptable; and the collected groups of gormandizing Vultures, on the approach of this dignified personage, instantly disperse, and make way for their master, waiting his departure in sullen silence, and at a respectful distance, on the adjacent trees.[54]

Commentary

This appears to be a preliminary sketch of the face of the eagle that is portrayed in full in the previous drawing (Figure 3.37). In the present sketch

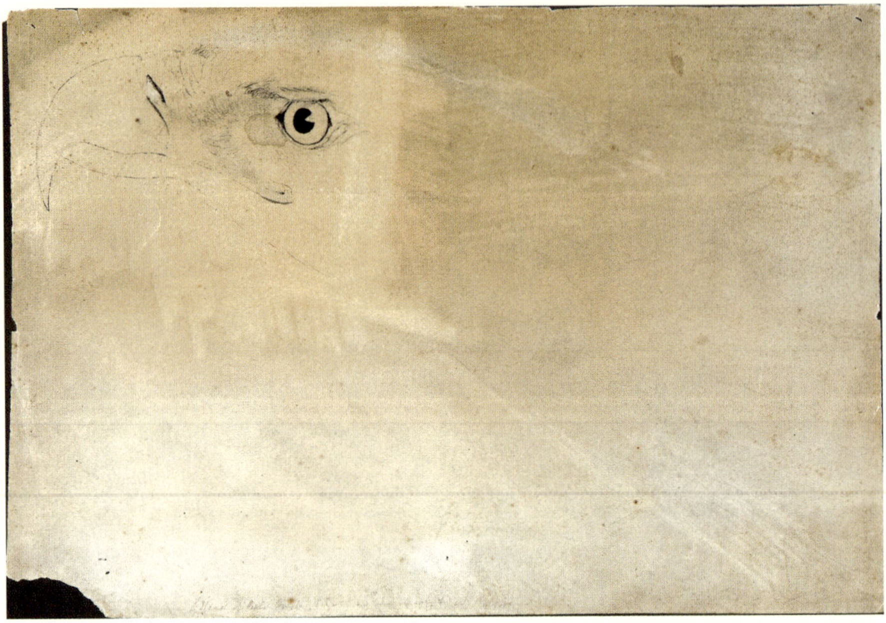

Figure 3.38. Wilson's preliminary sketch of the Bald Eagle seen fully realized in Figure 3.37. (From the archives of the Paisley Museum, Renfrewshire Council, Scotland.)

the bill is closed and the brow not as prominent as in the preceding portrait. These features combine to give the eagle in the sketch an intent look, whereas the eagle in the fully realized drawing and in the plate in *American Ornithology* has a much more ferocious look. With one foot planted on the fish, it seems to dare anything to try to take that fish away.

Current name: Osprey, *Pandion haliaetus* (Linnaeus)
Wilson's name: Fish-hawk, or Osprey, *Falco haliaetus*
Original name: *Falco Haliaetus* Linnaeus 1758

Current name: Fish Crow, *Corvus ossifragus* Wilson
Wilson's name: Fish Crow, *Corvus ossifragus*
Original name: Fish Crow, *Corvus ossifragus* Wilson 1812

Current name: Least Sandpiper, *Calidris minutilla* (Vieillot)
Wilson's name: Little Sandpiper, *Tringa pusilla*
Original name: *Tringa minutilla* Vieillot 1819

Current name: Semipalmated Plover, *Charadrius semipalmatus* Bonaparte
Wilson's name: Ringed or Ring Plover, *Charadrius hiaticula* Linnaeus
Original name: *Charadrius semipalmatus* Bonaparte 1825

Commentary

This drawing is not one of Wilson's best efforts. The composition is poor with a lot of empty space, the Osprey is awkwardly posed, the legs are too short in the Least Sandpiper, and the running pose of the Semipalmated Plover is unrealistic. Still, one can see how Wilson's astute observations of behavior and ecology, as well as his accumulated experience and knowledge, have come to play an important role in how he presents his birds: the Bald Eagle chasing the Osprey introduces action and natural history, the plover is also engaged in a characteristic action, and the Fish Crow has a fish.

Fish-hawk or Osprey (Osprey)

> The nest of the Fish Hawk is usually built on the top of a dead or decaying tree, sometimes not more than fifteen, often upwards of fifty feet, from the ground . . . In my late excursions to the sea shore, I ascended to several of these nests that had been built in from year to year, and found them constructed as follows: Externally, large sticks, from half an inch to an inch and a half in diameter, and two or three feet in length, piled to the height of four or five feet, and from two to three feet in breadth; these were intermixed with corn-stalks, sea-weed, pieces of wet turf, in large quantities, mullein-stalks, and lined with dry sea-grass; the whole forming a mass very observable at half a mile's distance, and large enough to fill a cart, and be no inconsiderable load for a horse.[55]

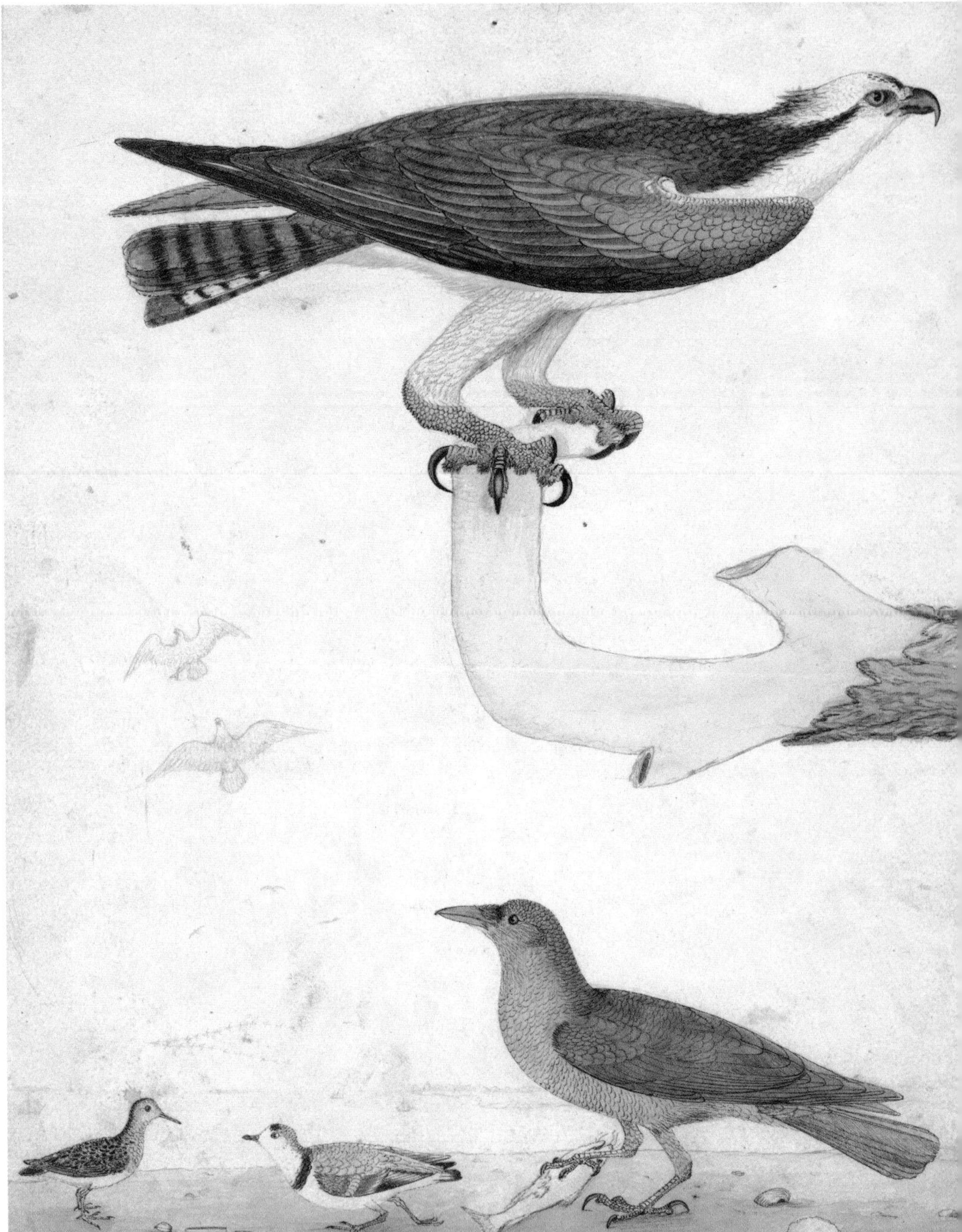

Fish-Crow

I first met with this species on the sea-coast of Georgia, and observed that they regularly retired to the interior as evening approached, and came down to the shores of the River Savannah by the first appearance of the day. Their voice first attracted my notice, being very different from that of the Common Crow, more hoarse and guttural, uttered as if something stuck in their throat, and varied into several modulations as they flew along.[56]

Little Sandpiper (Least Sandpiper)

This is the least of its tribe in this part of the world, and in its mode of flight has much more resemblance to the Snipe than to the Sandpiper. It is migratory, departing early in October for the South. It resides chiefly among the sea marshes, and feeds among the mud at low water; springs with a zigzag irregular flight, and a feeble twit.[57]

Ringed Plover (Semipalmated Plover)

As you approach near their nests, they seem to court your attention, and the moment they think you observe them, they spread out their wings and tail, dragging themselves along, and imitating the squeaking of young birds; if you turn from them they immediately resume their proper posture until they have again caught your eye, when they display the same attempts at deception as before. A flat dry sandy beach, just beyond the reach of the summer tides, is their favorite place for breeding.[58]

Figure 3.39. Osprey (*top*); Fish Crow (*lower right*); Least Sandpiper (*lower left*); Semipalmated Plover (*lower middle*). (From the archives of the Ernst Mayr Library, Museum of Comparative Zoology, Harvard University.)

The present species, or true Ring Plover, and also the former, or light colored bird [Piping Plover], both arrive on the sea-coast of New Jersey late in April. The present kind continues to be seen in flocks until late in May, when they disappear on their way farther north; the light colored bird remains during the summer, forms its nest in the sand, and generally produces two broods in the season . . . The present species is never seen to breed here; and, though I have opened great numbers of them as late as the 20th of May, the eggs which the females contained were never larger than small bird-shot; while, at the same time, the colored kind had every where begun to lay in the little cavities which they had dug in the sand on the beach. These facts being considered, it seems difficult to reconcile such difference of habits in one and the same bird.[59]

Current name: Barn Swallow, *Hirundo rustica* Linnaeus
Wilson's name: Barn Swallow, *Hirundo Americana*
Original name: *Hirundo rustica* Linnaeus 1758

Current name: Scarlet Ibis, *Eudocimus ruber* (Linnaeus)
Wilson's name: Scarlet Ibis, *Tantalus ruber*
Original name: *Scolopax rubra* Linnaeus 1758

Current name: Bank Swallow, *Riparia riparia* (Linnaeus)
Wilson's name: Bank Swallow, or Sand Martin, *Hirundo riparia*
Original name: *Hirundo riparia* Linnaeus 1758

Figure 3.40. Barn Swallow (*top* and *middle*); Scarlet Ibis (*lower left*); Bank Swallow (*lower right*). (From the archives of the Ernst Mayr Library, Museum of Comparative Zoology, Harvard University.)

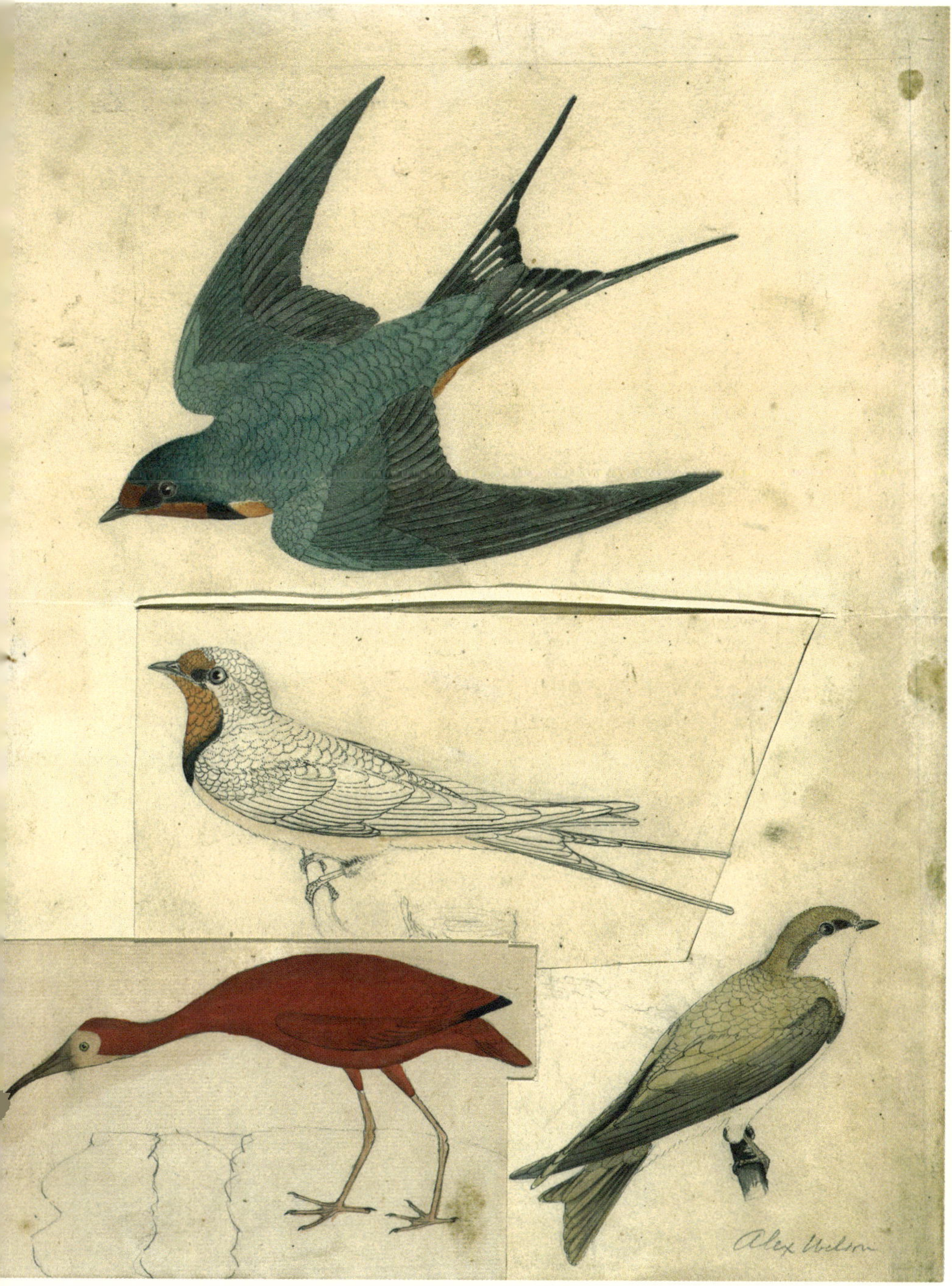
Alex Wilson

Commentary

Wilson frequently formed a plate by arranging cutout drawings of several species and then pasting them into the desired positions to complete the composition. This is the case here with the lower Barn Swallow drawing. The inclusion of the Scarlet Ibis, however, is something of a mystery. It is found in plate 66 of Wilson's *American Ornithology,* while the three swallows are in plate 38. The ibis occupies a position on this layout that is occupied by a Tree Swallow in the printed plate. Furthermore, it may not be Wilson's drawing: note the lack of detail in the legs, feet, and wings compared to the detail of the feet and feathers in the swallows. It may be that this image was not painted by Wilson, but was later copied from the earlier volume of *American Ornithology* and pasted into its current position.

Barn Swallow

There are but few persons in the United States unacquainted with this gay, innocent, and active little bird. Indeed the whole tribe are so distinguished from the rest of small birds by their sweeping rapidity of flight, their peculiar aërial evolutions of wing over our fields and rivers, and through our very streets, from morning to night, that the light of heaven itself, the sky, the trees, or any other common objects of Nature, are not better known than the Swallows. We welcome their first appearance with delight, as the faithful harbingers and companions of flowery spring and ruddy summer; and when, after a long frost-bound, and boisterous winter, we hear it announced, that "the Swallows are come," what a train of charming ideas are associated with the simple tidings![60]

Scarlet Ibis

This beautiful bird is found in the most southern parts of Carolina, also in Georgia and Florida, chiefly about the sea-shore and its vicinity. In most parts of America within the tropics, and in almost all the West India islands, it is said to be common; also in the Bahamas. Of its manners, little more has been collected, than that it frequents the borders of the sea, and shores of the neighboring rivers, feeding on small fry, shell fish, sea worms, and small crabs.[61]

Bank Swallow

This appears to be the most sociable with its kind, and the least intimate with man, of all our Swallows; living together in large communities of sometimes three or four hundred. On the high sandy bank of a river, quarry, or gravel-pit, at a foot or two from the surface, they commonly scratch out holes for their nests, running them in a horizontal direction to the depth of two and sometimes three feet.[62]

Current name: Purple Martin, *Progne subis* (Linnaeus)
Wilson's name: Purple Martin, *Hirundo subis*
Original name: *Hirundo subis* Linnaeus, 1758

Current name: Chimney Swift, *Chaetura pelagica* (Linnaeus)
Wilson's name: Chimney Swallow, *Hirundo pelacgia*
Original name: *Hirundo pelagica* Linnaeus 1758

Current name: Connecticut Warbler, *Oporornis agilis* (Wilson)
Wilson's name: Connecticut Warbler, *Sylvia agilis*
Original name: Connecticut Warbler, *Sylvia agilis* Wilson 1812

Commentary

This composition is yet another example of Wilson's technique of drawing birds on separate pieces of paper and pasting the drawings together to create a layout with the desired balance. The tail feathers of the swift are accurately portrayed and the position of the tail is accurate for a swift perched in a hol-

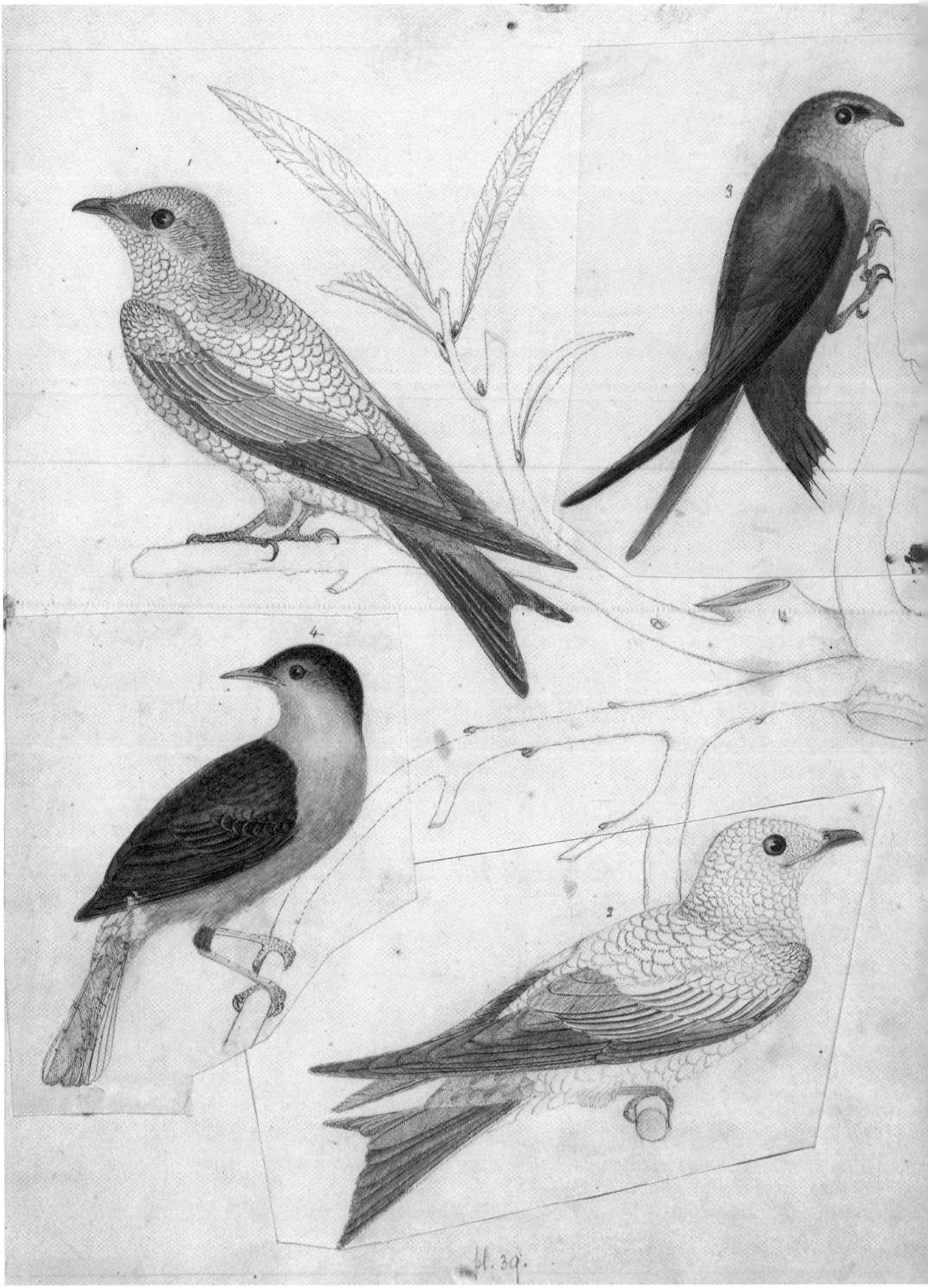

low tree or chimney. The foot, however, is anatomically inaccurate. The four toes should point forward because swifts have no opposable toe.

Purple Martin

This well-known bird is a general inhabitant of the United States, and a particular favorite wherever he takes up his abode. I never met with more than one man who disliked the Martins, and would not permit them to settle about his house. This was a penurious, close-fisted German, who hated them, because, as he said, "they eat his *peas.*" I told him he must certainly be mistaken, as I never knew an instance of Martins eating *peas;* but he replied with coolness, that he had many times seen them himself "blaying near the hife, and going *schnip, schnap,*" by which I understood that it was his *bees* that had been the sufferers; and the charge could not be denied.[63]

Chimney Swallow (Chimney Swift)

This swallow, like all the rest of its tribe in the United States, is migratory, arriving in Pennsylvania late in April or early in May, and dispersing themselves over the whole country wherever there are vacant chimneys in summer sufficiently high and convenient for their accommodation. In no other situation with us are they observed at present to build. This circumstance naturally suggests the query, Where did these birds construct their nests before the arrival of Europeans in this country, when there were no such places for their accommodation? I would answer, Probably in the same situations in which they still continue to build in the remote regions of our western forests, namely in the hollow of a tree, which in some cases has the nearest resemblance to their present choice, of any other.[64]

Figure 3.41. Purple Martin (*upper left* and *lower right*); Chimney Swift (*upper right*); Connecticut Warbler (*lower left*). (From the archives of the Ernst Mayr Library, Museum of Comparative Zoology, Harvard University.)

Connecticut Warbler

This is a new species, first discovered in the state of Connecticut, and twice since met with in the neighborhood of Philadelphia . . . It was found in every case among low thickets, but seemed more than commonly active, not remaining for a moment in the same position. In some of my future rambles I may learn more of this solitary species.[65]

Current name: Common Nighthawk, *Chordeiles minor* (Forster)
Wilson's name: Night Hawk, *Caprimulgus minor*
Original name: *Caprimulgus minor* Forster 1771

Commentary

This is one of Wilson's best efforts. The pose of the upper bird shows fully the distribution of white on the wings and tail that is important in the bird's identification. The very effective texture to this fully painted work was achieved by applying opaque watercolors over wash. Here as with the illustration of the Orchard Oriole, Wilson has included the egg in the drawing, thereby adding to the natural history value of the portrait, especially so in this case since the lack of nest, mentioned in the text, is illustrated in the plate.

Night-Hawk (Common Nighthawk)

The eggs are placed on the bare ground, in all cases on a dry situation, where the color of the leaves, ground, stones, or other circumjacent parts of the surface, may resemble the general tint of the eggs, and thereby render them less easy to be discovered . . . They also sit lengthwise on the branch, fence, or limb, on which they roost, and never across, like most other birds: this seems occasioned by the shortness and slender form of their legs and feet, which are not at all calculated to grasp the branch with sufficient firmness to balance their bodies.

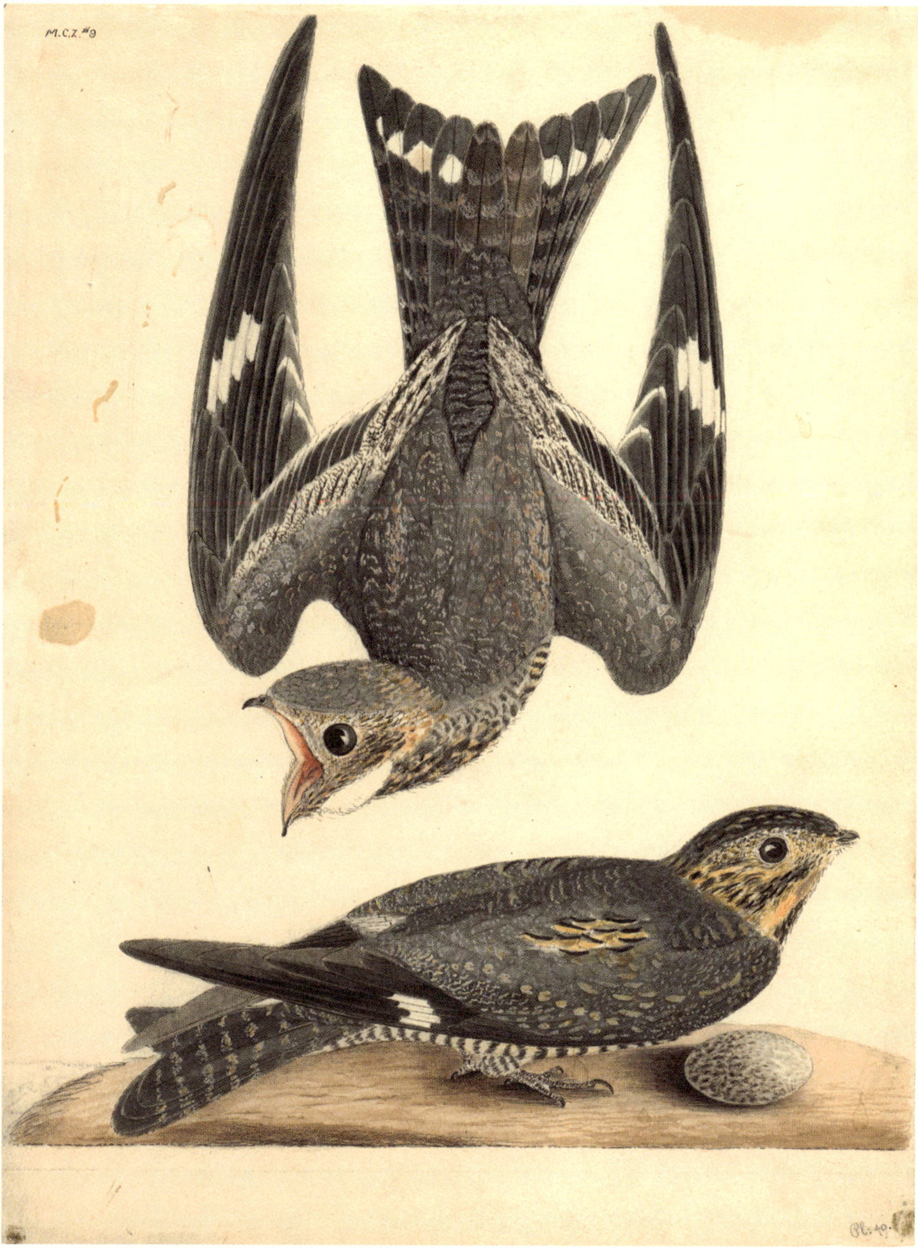

Figure 3.42. Common Nighthawk. (From the archives of the Ernst Mayr Library, Museum of Comparative Zoology, Harvard University.)

The ridiculous name Goatsucker,—which was first bestowed on the European species, from a foolish notion that it sucked the teats of the goats, because, probably, it inhabited the solitary heights where they fed, which nickname has been since applied to the whole genus,—I thought proper to omit. There is something worse than absurd in continuing to brand a whole family of birds with a knavish name, after they are universally known to be innocent of the charge. It is not only unjust, but tends to encourage the belief in an idle fable that is totally destitute of all foundation.[66]

Current name: Eastern Whip-poor-will, *Antrostomus vociferus* (Wilson)
Wilson's name: Whip-poor-will, *Caprimulgus vociferus*
Original name: Whip-poor-will *Caprimulgus vociferous* Wilson 1812

Commentary

This is another fully painted plate that illustrates one of Wilson's composition and drawing techniques described earlier. In this case, the image of the chick is first drawn on a separate piece of paper (Figure 3.43*a*). Next the back of the paper is covered with iron oxide (*b*) in order to transfer an image of the chick onto the final drawing (and eventually, plate) within a larger composition (*c*). The iron oxide emphasizes the lined texture of the paper on which the chick is drawn. Such paper is called laid paper and was made by pouring a slush of wood pulp into a frame and letting it dry, at which time the piece of paper was taken out of the frame. This was the dominant method of making paper until the 1790s. By the early 1800s paper was made by pressing the wood pulp into what was known as wove paper, which had a much smoother surface. Most of Wilson's artwork is on wove paper. The text accompanying these illustrations emphasizes that Wilson drew from life in the field.

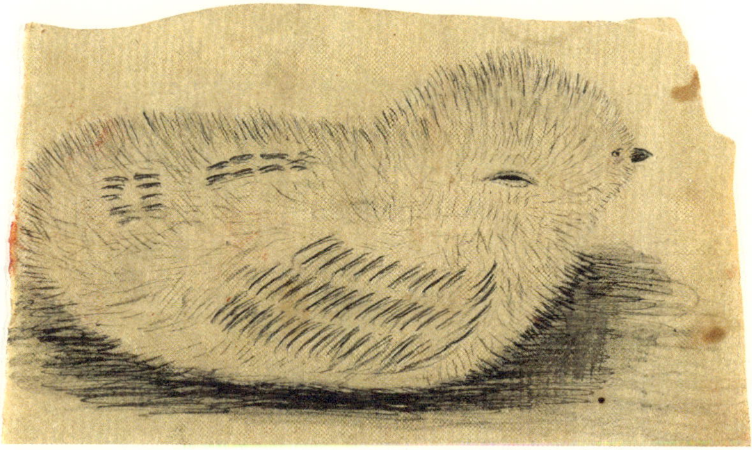

(*a*)

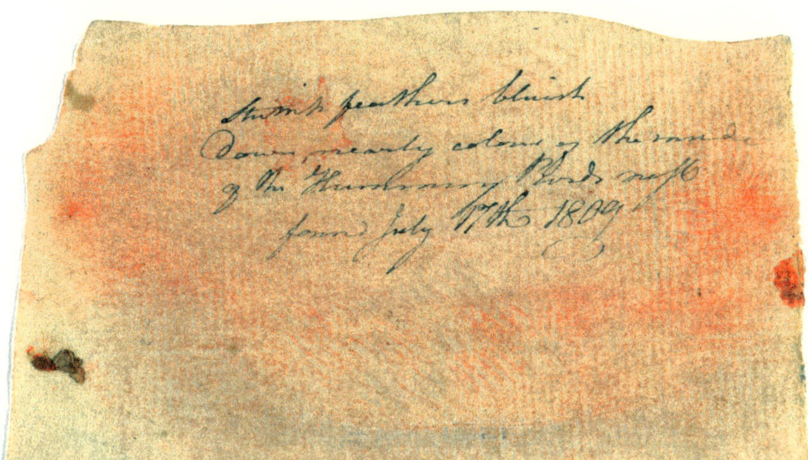

(*b*)

Figure 3.43a–c. Whip-poor-will. Wilson first sketched a chick (*a*), then transferred the sketch to the larger composition (*c on following page*) by rubbing iron oxide on the reverse side of the sketch paper (*b*), thereby allowing it to be retraced onto the plate. (From the archives of the Ernst Mayr Library, Museum of Comparative Zoology, Harvard University.)

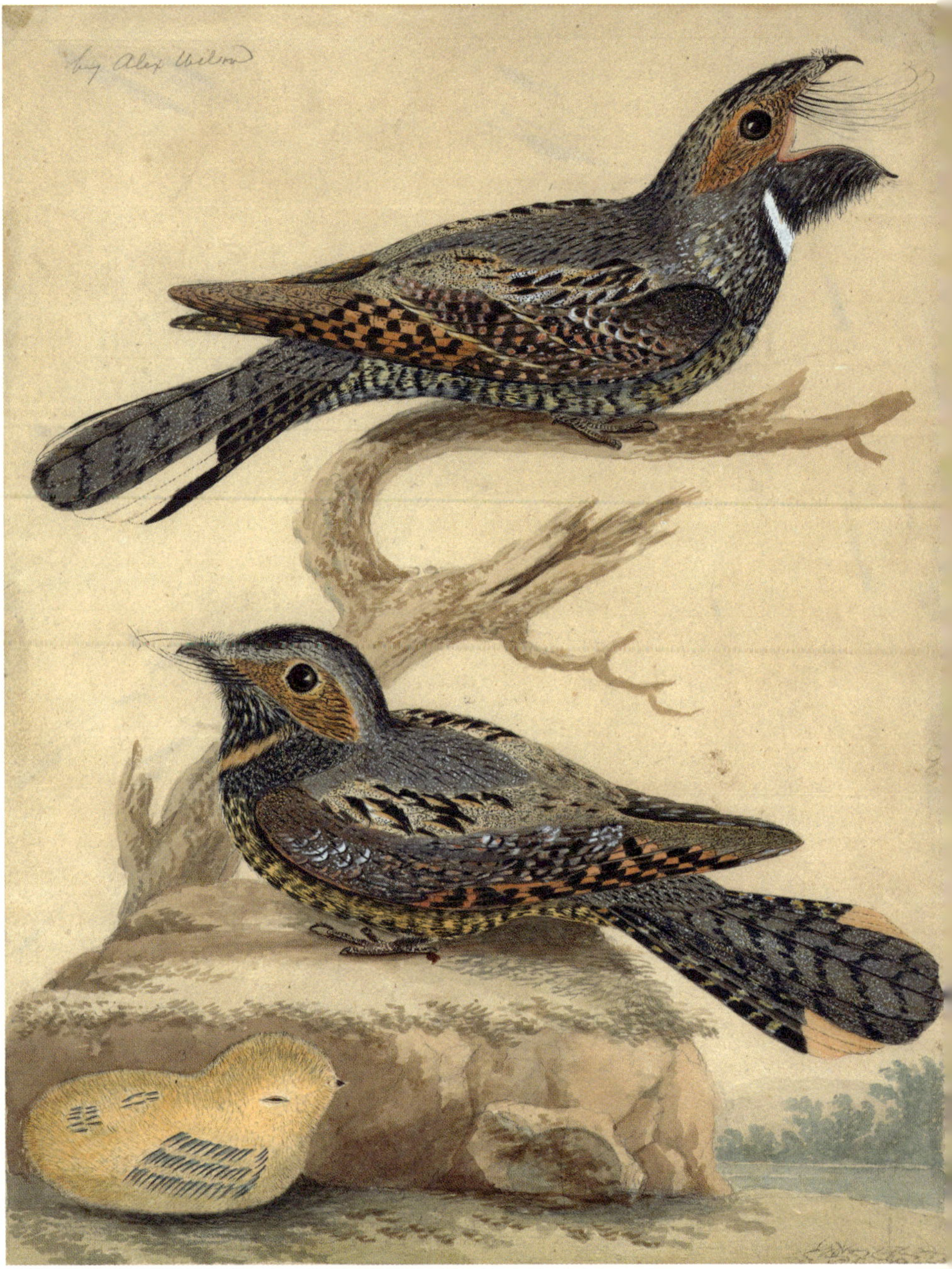

(*c*)

Whip-poor-will

In traversing the woods one day in the early part of June, along the brow of a rocky declivity, a Whip-poor-will rose from my feet, and fluttered along, sometimes prostrating herself, and beating the ground with her wings, as if just expiring. Aware of her purpose, I stood still, and began to examine the space immediately around me for the eggs or young, one or other of which I was certain must be near. After a long search, to my mortification, I could find neither; and was just going to abandon the spot, when I perceived somewhat like a slight mouldiness among the withered leaves, and, on stooping down, discovered it to be a young Whip-poor-will, seemingly asleep, as its eyelids were nearly closed; or perhaps this might only be to protect its tender eyes from the glare of day. I sat down by it on the leaves, and drew it as it then appeared.[67]

Current name: Eastern Screech-Owl, *Megascops asio* (Linnaeus)
Wilson's name: Red Owl, *Strix asio*
Original name: *Strix Asio* Linnaeus 1758

Current name: Hermit Thrush, *Catharus guttatus* (Pallas)
Wilson's name: Hermit Thrush, *Muscicapa solitaries*
Original name: *Muscicapa guttata* Pallas 1811

Current name: Veery, *Catharus fuscescens* (Stephens)
Wilson's name: Tawny Thrush, *Turdus mustelinus*
Original name: *Turdus Fuscescens* Stephens 1817

Current name: Black-throated Blue Warbler, *Setophaga caerulescens* (Gmelin)
Wilson's name: Pine-swamp Warbler, *Sylvia leucoptera*
Original name: *Moticilla caerulescens* Gmelin 1789

Commentary

The owl is watercolor over ink, over pencil. The thrushes are cutouts pasted into position to create the final layout, a variant on the technique in previous plates where blocks of paper with drawings on them were pasted up. The warbler described as a Pine-swamp Warbler is actually a female Black-throated Blue Warbler, as indicated by Wilson's faithful depiction of the small white wing patch that serves to identify this otherwise drab and featureless bird. Wilson had fallen into the trap of describing males and females of a strongly plumaged dimorphic bird as separate species—a mistake that is forgivable since, as he states in his commentary for the Black-throated Blue Warbler (male), "I know little of this bird. It is one of those transient visitors that, in the month of April, pass through Pennsylvania, on its way to the north, to breed." Wilson had no opportunity to observe a mated pair and thus could not have connected two such strikingly different looking birds as the male and female of this species.

Red Owl (Eastern Screech-Owl)

> Throughout the day, it was all stillness and gravity,—its eyelids half shut, its neck contracted, and its head shrunk, seemingly, into its body; but scarcely was the sun set, and the twilight began to approach, when its eyes became full and sparkling, like two living globes of fire; it crouched on its perch, reconnoitred every object around with looks of eager fierceness; alighted and fed; stood on the meat with clenched talons, while it tore it in morsels with its bill; moaned out its melancholy notes, with many lively gesticulations, not at all accordant with the pitiful tone of its ditty, which reminded one of the shivering moanings of a half-frozen puppy.[68]

Figure 3.44a–d. (*a*) Eastern Screech-Owl; (*b*) Hermit Thrush; (*c*) Veery; (*d*) Black-throated Blue Warbler (*c and d on following page*). (From the archives of the Ernst Mayr Library, Museum of Comparative Zoology, Harvard University.)

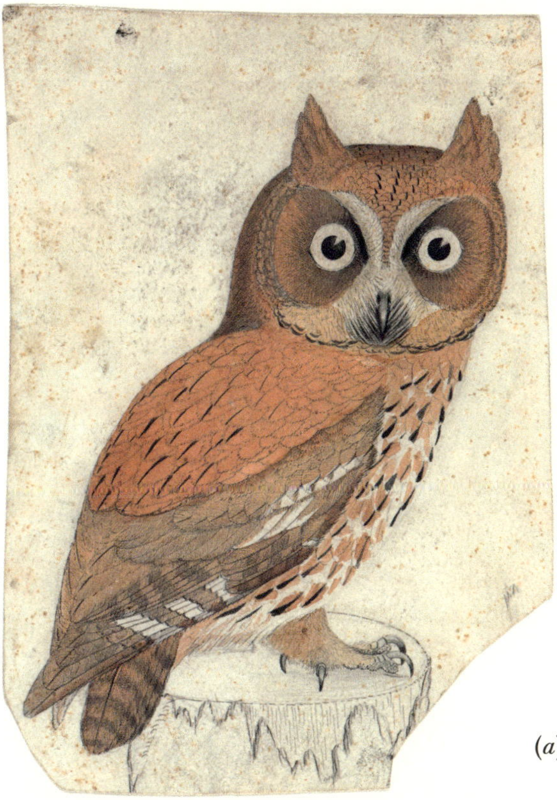

(*a*)

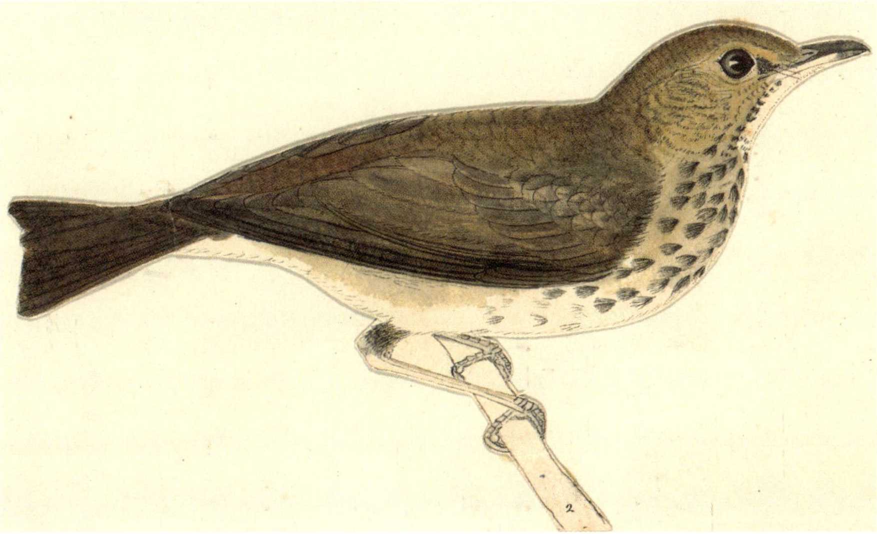

(*b*)

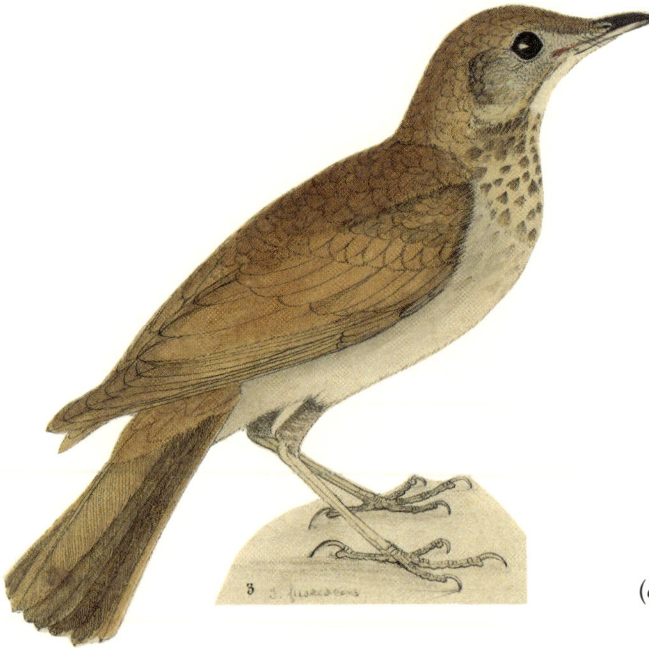

(*c*)

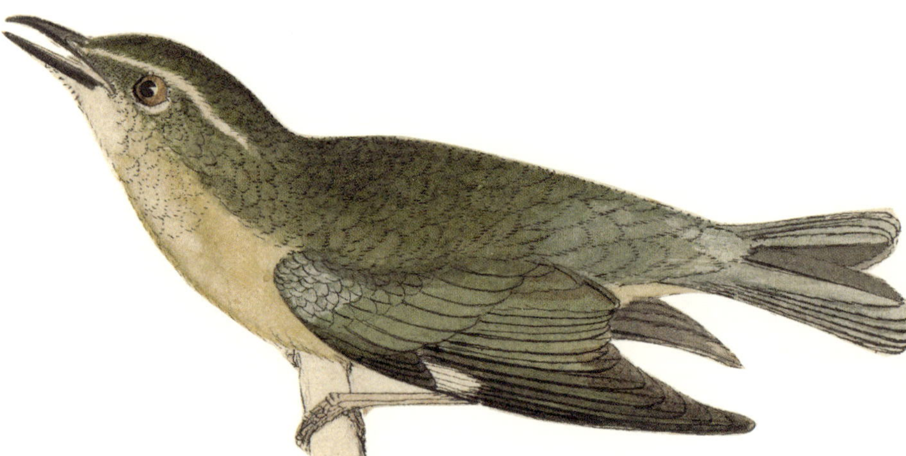

(*d*)

Hermit Thrush

A superficial observer would instantly pronounce this to be only a variety of the Wood Thrush; but, taking into consideration its difference of size, color, manners, want of song, secluded habits, differently formed nest, and spotted eggs, all unlike those of the former, with which it never associates, it is impossible not to conclude it to be a distinct and separate species, however near it may approach to that of the former. Its food, and the country it inhabits, for half the year, being the same, neither could have produced those differences; and we must believe it to be now, what it ever has been, and ever will be, a distinct connecting link in the great chain of this part of animated nature; all the sublime reasoning of certain theoretical closet philosophers to the contrary notwithstanding.[69]

Tawny Thrush (Veery)

I have examined many of these birds in spring, and also on their return in fall, and found very little difference among them between the male and female. In some specimens the wing-coverts were brownish yellow; these appeared to be young birds. I have no doubt but they breed in the northern high districts of the United States; but I have not yet been able to discover their nests.[70]

Pine-swamp Warbler (Black-throated Blue Warbler)

This little bird is, for the first time, figured or described. Its favorite haunts are in the deepest and gloomiest pine and hemlock swamps of our mountainous regions, where every tree, trunk, and fallen log, is covered with a luxuriant coat of moss, that even mantles over the surface of the ground, and prevents the sportsman from avoiding a thousand holes, springs, and swamps, into which he is insensibly plunged.[71]

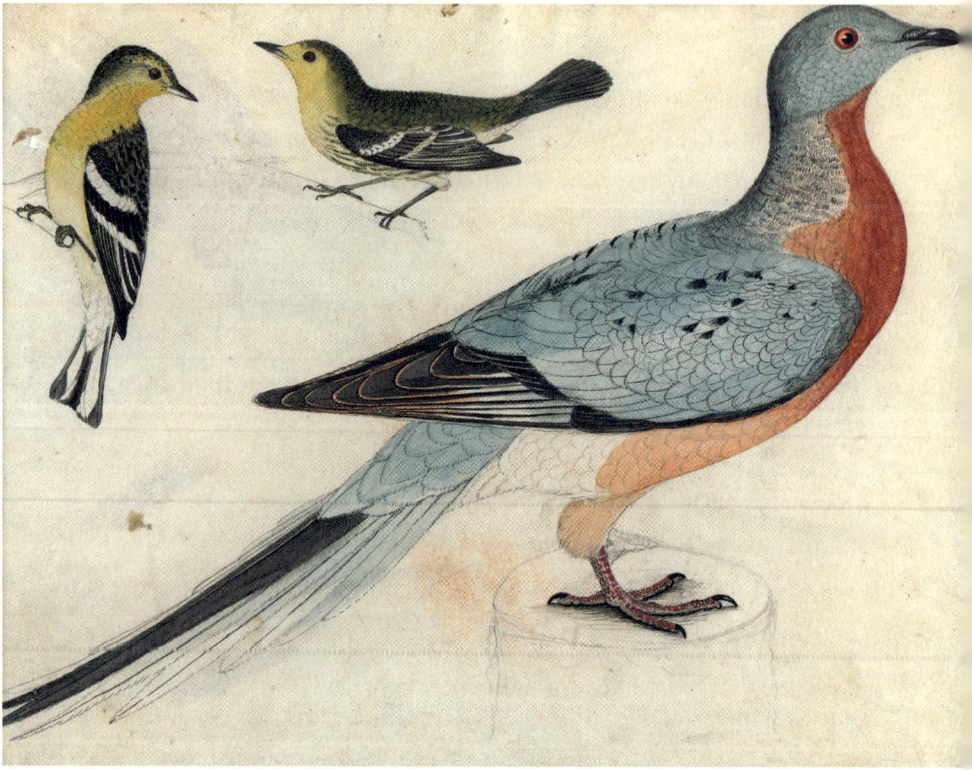

Figure 3.45. Bay-breasted Warbler (*upper left*); Black-throated Green Warbler (*upper middle*); Passenger Pigeon (*bottom diagonal*). (From the archives of the Ernst Mayr Library, Museum of Comparative Zoology, Harvard University.)

Current name: Bay-breasted Warbler, *Setophaga castanea* (Wilson)
Wilson's name: Hemlock Warbler, *Sylvia parus*
Original name: *Sylvia castanea* Wilson 1810

Current name: Black-throated Green Warbler, *Setophaga virens* (Gmelin)
Wilson's name: Blue Mountain Warbler, *Sylvia montana*
Original name: *Motacilla virens* Gmelin 1789

Current name: Passenger Pigeon, *Ectopistes migratorius* (Linnaeus)
Wilson's name: Passenger Pigeon, *Columba migratoria*
Original name: *Columba migratoria* Linnaeus 1766

Commentary

The composition of this figure accommodates the now-extinct Passenger Pigeon painting at life size, using the diagonal for maximum length. The last individual, a female by the name of Martha, aged twenty-nine, died at 1:00 P.M. on 1 September 1914 at the Cincinnati Zoo. She now lies in the Smithsonian Institution, but is not on display. The warbler species identifications are problematic. The mysterious "Hemlock Warbler" that Wilson describes somewhat resembles an immature Bay-breasted Warbler. The Blue Mountain Warbler closely resembles a Black-throated Green Warbler in non-breeding plumage.

Hemlock Warbler (unknown warbler)

It is a most lively and active little bird, climbing among the twigs, and hanging like a Titmouse on the branches; but possessing all the external characters of the Warblers. It has a few low and very sweet notes, at which times it stops and repeats them for a short time, then darts about as before. It shoots after flies to a considerable distance; often begins at the lower branches, and hunts with great regularity and admirable dexterity, upwards to the top, then flies off to the next tree, at the lower branches of which it commences hunting upwards as before.[72]

Blue Mountain Warbler (Black-throated Green Warbler)

This new species was first discovered near that celebrated ridge, or range of mountains, with whose name I have honored it. Several of these solitary Warblers remain yet to be gleaned up from the airy heights of our alpine scenery, as well as from the recesses of our swamps and morasses, whither it is my design to pursue them by every opportunity.[73]

Passenger Pigeon

Not far from Shelbyville, in the state of Kentucky, about five years ago, there was one of these breeding places, which stretched through the woods in nearly a north and south direction; was several miles in breadth and was said to be upwards of forty miles in extent! In this tract, almost every tree was furnished with nests, wherever the branches could accommodate them. The Pigeons made their first appearance there about the 10th of April, and left it altogether, with their young, before the 25th of May.

As soon as the young were fully grown, and before they left the nests, numerous parties of the inhabitants, from all parts of the adjacent country, came with wagons, axes, beds, cooking utensils, many of them accompanied by the greater part of their families, and encamped for several days at this immense nursery. Several of them informed me, that the noise in the woods was so great as to terrify their horses, and that it was difficult for one person to hear another speak, without bawling in his ear. The ground was strewed with broken limbs of trees, eggs, and young Squab Pigeons, which had been precipitated from above, and on which herds of hogs were fattening. Hawks, Buzzards, and Eagles, were sailing about in great numbers, and seizing the squabs from their nests at pleasure; while, from twenty feet upwards to the tops of the trees, the view through the woods presented a perpetual tumult of crowding and fluttering multitudes of Pigeons, their wings roaring like thunder, mingled with the frequent crash of falling timber; for now the axe-men were at work, cutting down those trees that seemed to be most crowded with nests, and contrived to fell them in such a manner, that, in their descent, they might bring down several others; by which means the falling of one large tree sometimes produced two hundred Squabs, little inferior in size to the old ones, and almost one mass of fat.[74]

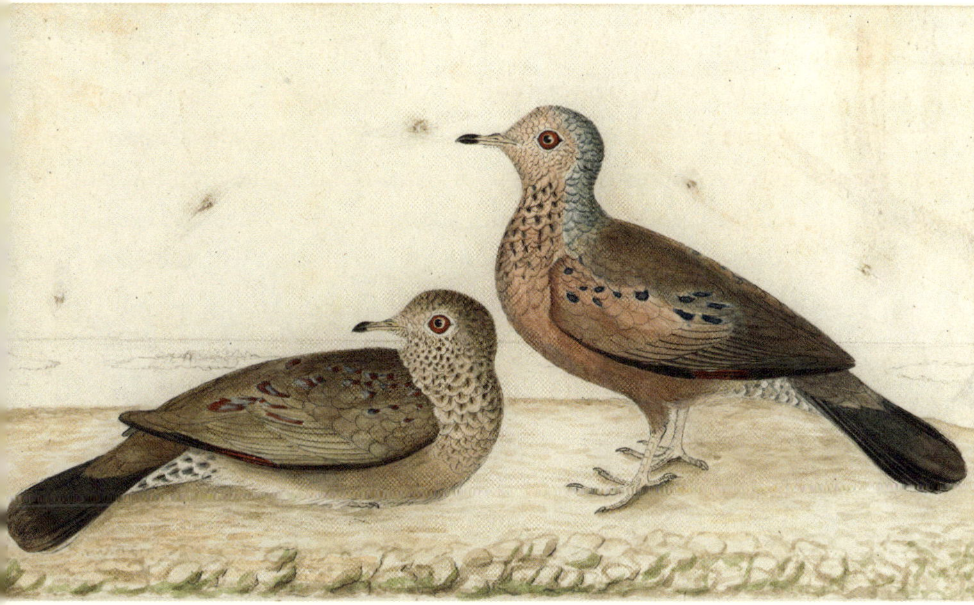

Figure 3.46. Common Ground-Dove. (From the archives of the Ernst Mayr Library, Museum of Comparative Zoology, Harvard University.)

Current name: Common Ground-Dove, *Columbina passerina* (Linnaeus)
Wilson's name: Ground Dove, *Columba passerina*
Original name: *Columba passerina* Linnaeus 1758

Commentary

This fully painted pair of ground-doves is nearly identical to those in plate 46 of *American Ornithology.*

Ground Dove (Common Ground-Dove)

This is one of the least of the Pigeon tribe, whose timid and innocent appearance forms a very striking contrast to the ferocity of the preceding bird [the Sharp-shinned Hawk, first described by Wilson]. Such as they are in nature, such I have endeavored faithfully to represent them. I have been the more particular with this minute species, as no correct figure of it exists in any former work with which I am acquainted.

The Ground Dove is a native of North and South Carolina, Georgia, and the new state of Louisiana, Florida, and the islands of the West Indies. In the latter, it is frequently kept in cages; is esteemed excellent for the table, and honored by the French planters with the name of Ortolan.[75]

Current name: Sora, *Porzana carolina* (Linnaeus)
Wilson's name: Rail, *Rallus carolinus*
Original name: *Rallus carolinus* Linnaeus 1758

Commentary

The Sora in the pencil sketch (*a*) is stepping forward with the foot raised high, a common feature of the Sora's stealthy walk through marsh grass. The tail is up and the bird is looking up. The fully painted Sora is stepping forward, with its tail up (*b*). The bird is looking forward, not up, however, as in the pencil sketch. The third painting of a Sora (*c*) depicts a strongly developed background, which will become a characteristic of drawings for later volumes of *American Ornithology.* The one broken reed adds a touch of realism. These drawings suggest that Wilson did not just draw a bird for the plate, but experimented with posture, behavior, and background before choosing one drawing for the engraving. In this case he chose the second drawing for *American Ornithology.*

Rail (Sora)

The natural history of the Rail, or, as it is called in Virginia, the Sora, and in South Carolina, the Coot, is, to the most of our sportsmen, involved in profound and inexplicable mystery. It comes, they know

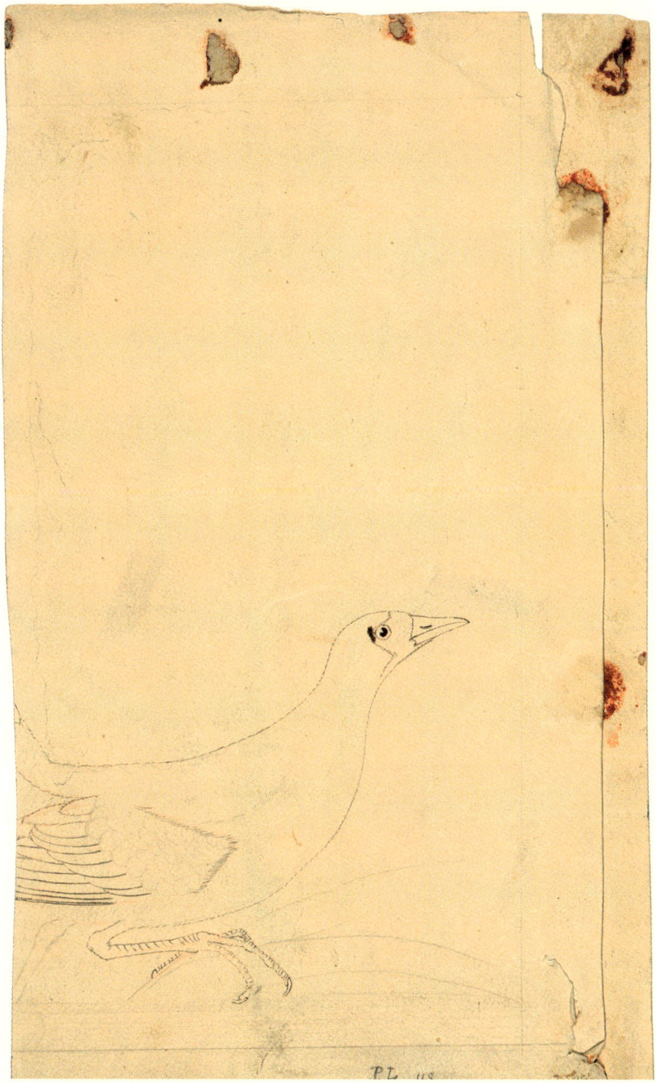

(*a*)

Figure 3.47 a–c. Wilson's pencil sketch of a Sora (*a*) seems to be the one he used to create the illustration for the final plate in *American Ornithology*. The fully painted portrait of the Sora stepping forward (*b*) has the neck more extended and the bill not quite as raised as the one in the final illustration. In illustration (*c*), the Sora is crouching in a posture very unlike that of the bird in *American Ornithology*. (The (*a*) and (*c*) illustrations are from the archives of the Ernst Mayr Library, Museum of Comparative Zoology, Harvard University, the (*b*) Sora is from the archives of the Paisley Museum, Renfrewshire Council, Scotland.)

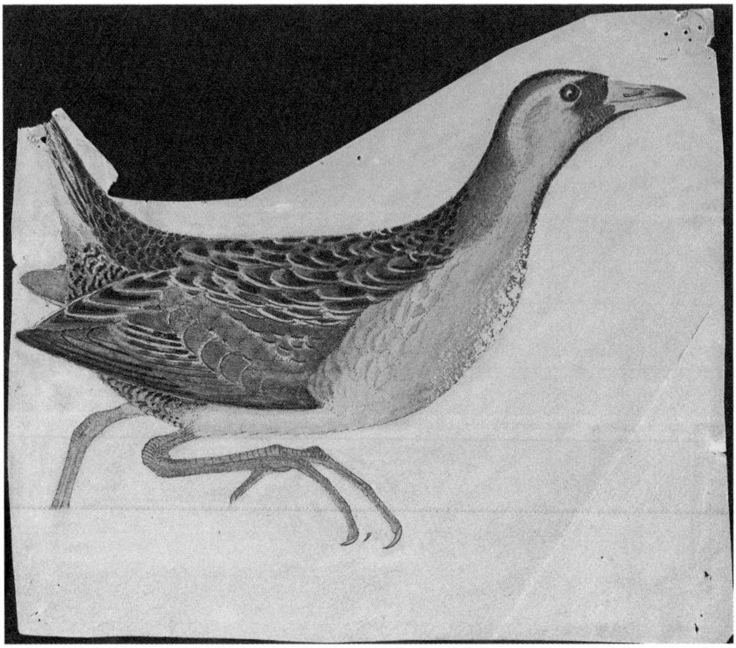

(*b*)

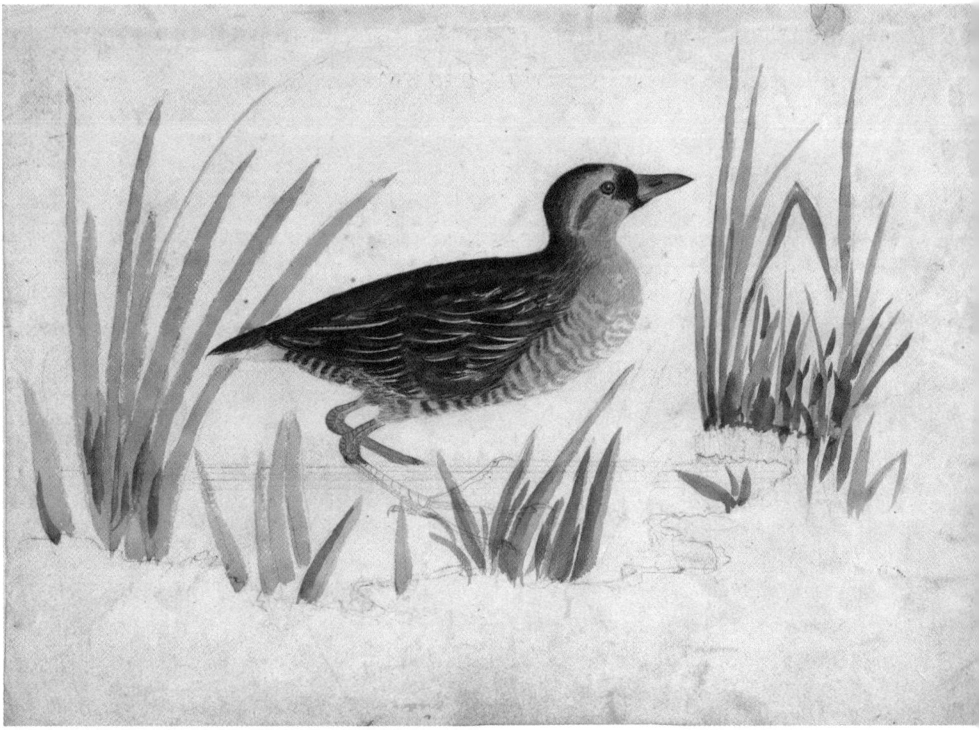

(*c*)

not whence; and goes, they know not where. No one can detect their first moment of arrival; yet all at once the reedy shores and grassy marshes of our large rivers swarm with them, thousands being sometimes found within the space of a few acres. These, when they do venture on wing, seem to fly so feebly, and in such short fluttering flights among the reeds, as to render it highly improbable to most people that they could possibly make their way over an extensive tract of country. Yet, on the first smart frost that occurs, the whole suddenly disappear, as if they had never been.[76]

Current name: Ruffed Grouse, *Bonasa umbellus* (Linnaeus)
Wilson's name: Ruffed Grous, *Tetrao umbellus*
Original name: *Tetrao umbellus* Linnaeus 1766

Commentary

This painting by Wilson is similar to Plate 49 in the *Ornithology* but the subject's head and neck are more extended and there are substantial differences in the plumage patterns and foregrounds.

Ruffed Grous

The drumming, as it is usually called of the Pheasant [Ruffed Grouse] is another singularity of this species. This is performed by the male alone. In walking through solitary woods frequented by these birds a stranger is surprised by suddenly hearing a kind of thumping, very similar to that produced by striking two full blown ox bladders together but much louder; the strokes at first are slow and distinct; but gradually increase in rapidity till they run into each other resembling the rumbling sound of very distant thunder, dying away gradually on the ear. After a few minutes pause this is again repeated; and in a calm day may be heard nearly half a mile off. This drumming is most common in spring and is the call of the cock to

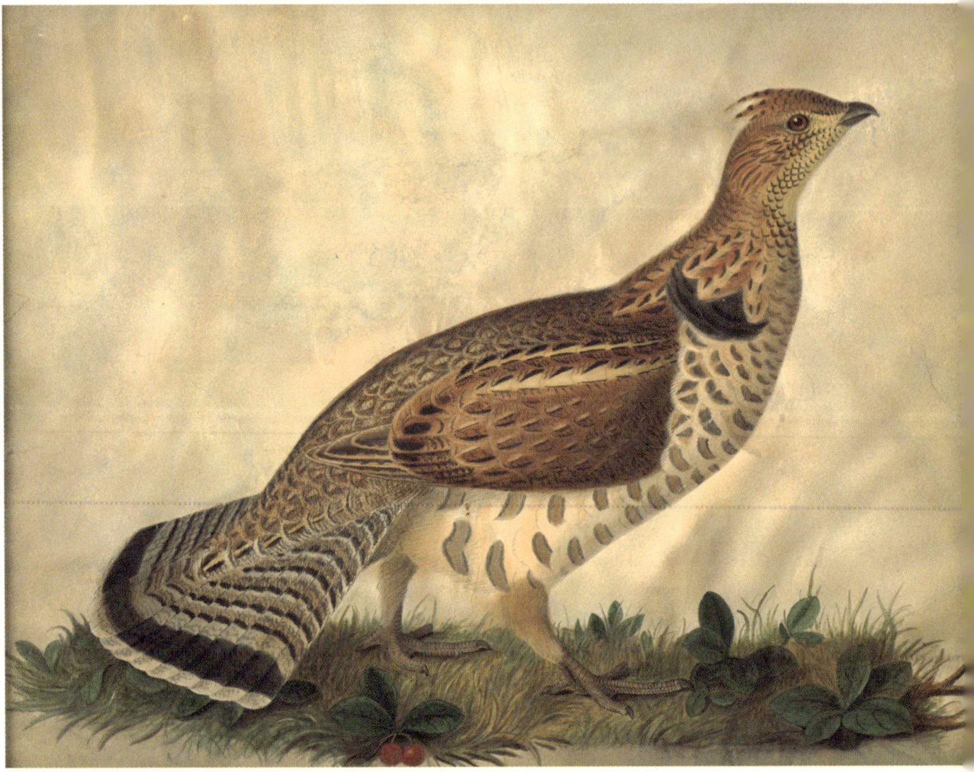

Figure 3.48. Ruffed Grouse (From the archives of the Ernst Mayr Library, Museum of Comparative Zoology, Harvard University.)

his favorite female. It is produced in the following manner. The bird, standing on an old prostrate log, generally in a retired and sheltered situation, lowers his wings, erects his expanded tail, contracts his throat, elevates the two tufts of feathers on the neck and inflates his whole body, something in the manner of a turkey cock, strutting and wheeling about with great stateliness. After a few manœuvres of this kind he begins to strike with his stiffened wings in short quick strokes which become more and more rapid until they run into each other as has already been described.[77]

Current name: Northern Harrier, *Circus cyaneus* (Linnaeus)
Wilson's name: Marsh Hawk, *Falco uliginosus*
Original name: *Falco cyaneus* Linnaeus 1766

Current name: Long-eared Owl, *Asio otus* (Linnaeus)
Wilson's name: Long-eared Owl, *Strix otus*
Original name: *Strix Otus* Linnaeus 1758

Current name: Swallow-tailed Kite, *Elanoides forficatus* (Linnaeus)
Wilson's name: Swallow-tailed Hawk *Falco furcatus*
Original name: *Falco forficatus* Linnaeus 1758

Commentary

It was not until volume 7 of *American Ornithology* that Wilson routinely painted detailed backgrounds. This plate, in which Wilson retains his habit of simply outlining branches and letting the engraver fill in the details, is testament to his respect for Lawson's artistic sense and technical skill. The projecting end of the left wing is uncolored.

Marsh Hawk (Northern Harrier)

This hawk is most numerous where there are extensive meadows and salt marshes, over which it sails very low, making frequent circuitous sweeps over the same ground, in search of a species of mouse . . . and very abundant in such situations. It occasionally flaps the wings, but is most commonly seen sailing about within a few feet of the surface.[78]

Long-eared Owl

This Owl is common to both continents, and is much more numerous in Pennsylvania than the White, or Barn Owl: six or seven were found in a single tree, about fifteen miles from Philadelphia. There

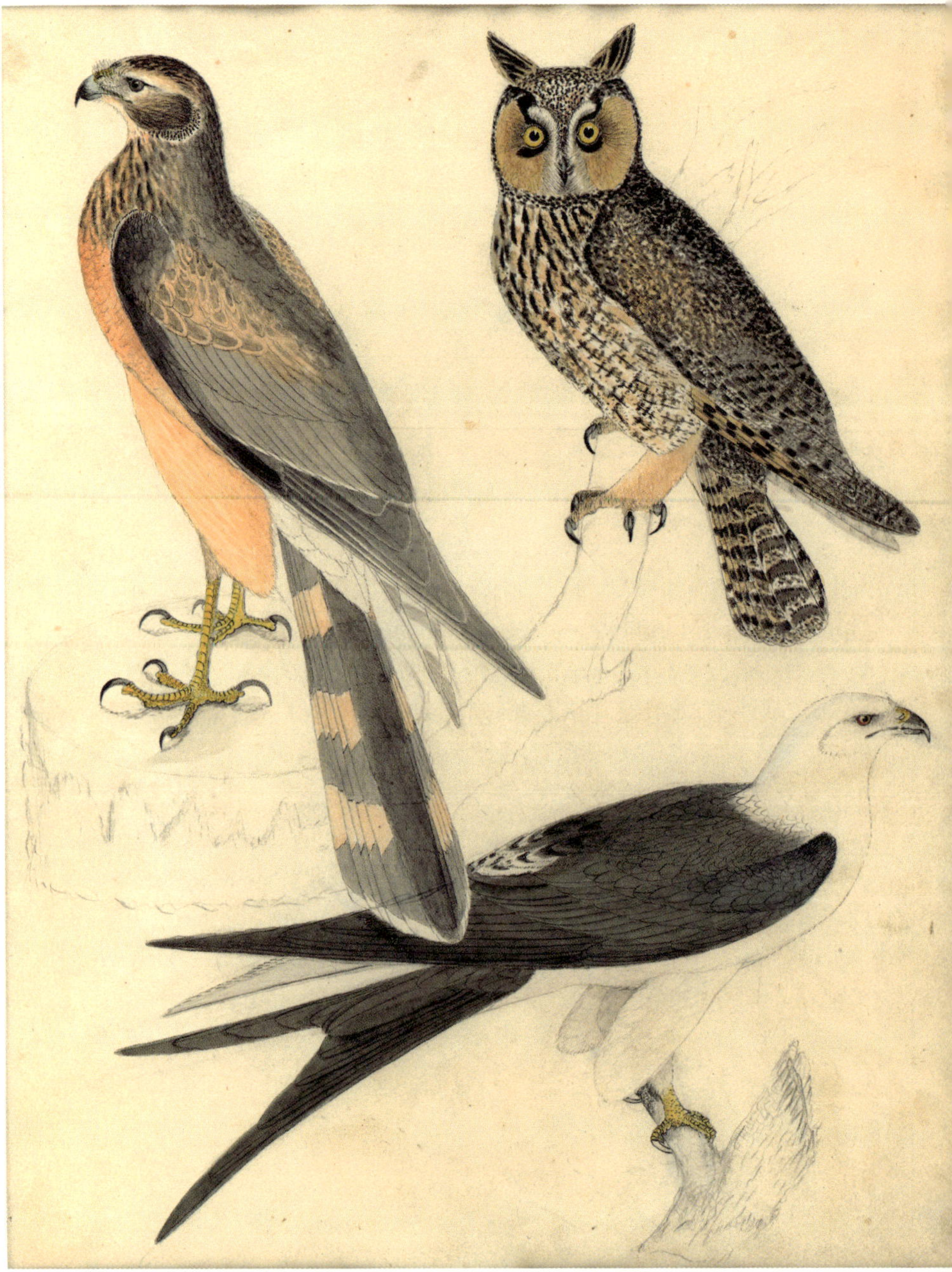

Figure 3.49. Northern Harrier (*upper left*); Long-eared Owl (*upper right*); Swallow-tailed Kite (*bottom*). (From the archives of the Ernst Mayr Library, Museum of Comparative Zoology, Harvard University.)

is little doubt but this species is found inhabiting America to a high latitude; though we have no certain accounts of the fact. Except in size, this species has more resemblance to the Great Horned Owl than any other of its tribe. It resembles it also in breeding among the branches of tall trees; lays four eggs, of nearly a round form, and pure white.[79] The young are grayish white until nearly full grown, and roost during the day close together on a limb, among the thickest of the foliage. This Owl is frequently seen abroad during the day, but is not remarkable for its voice or habits.[80]

Swallow-tailed Hawk (Swallow-tailed Kite)

This very elegant species inhabits the southern districts of the United States in summer; is seldom seen as far north as Pennsylvania, but is very abundant in South Carolina and Georgia, and still more so in West Florida, and the extensive prairies of Ohio and the Indiana Territory. I met with these birds, in the early part of May, at a place called Duck Creek, in Tennessee; and found them sailing about in great numbers near Bayou Manchac, on the Mississippi, twenty or thirty being within view at the same time. At that season, a species of cicada, or locust swarmed among the woods, making a deafening noise, and I could perceive these hawks frequently snatching them from the trees.[81]

Current name: Red-tailed Hawk, *Buteo jamaicensis* (Gmelin)
Wilson's name: Red-tailed Hawk, *Falco borealis*
Original name: *Falco jamaicensis* Gmelin 1788

Commentary

The erect feathers at the back of the head and the prominent ridge above the eye give the hawk a particularly fierce look (*a*). Wilson excelled at painting

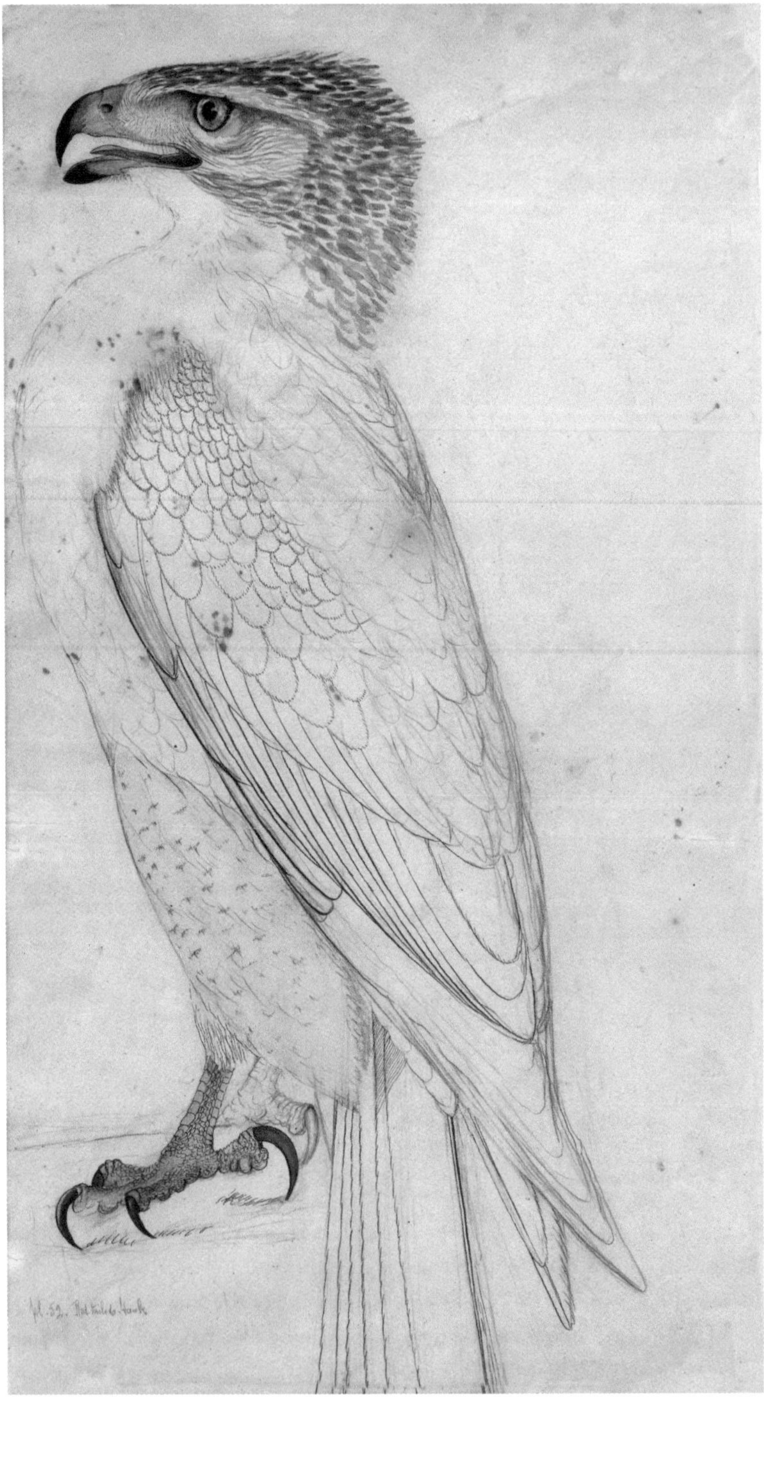

(*a*)

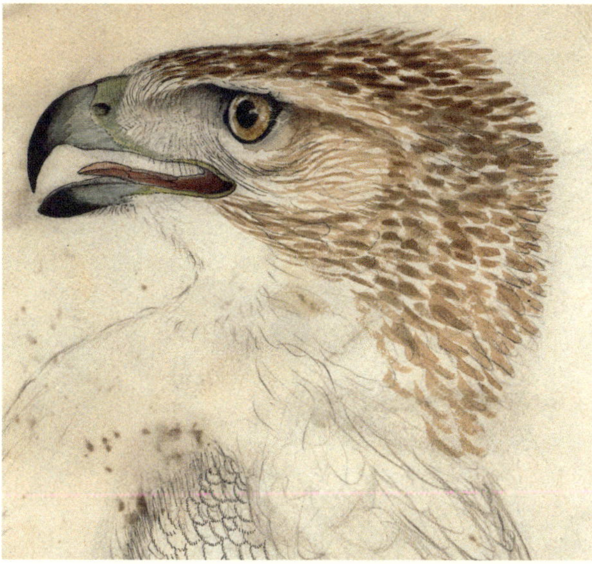

(*b*)

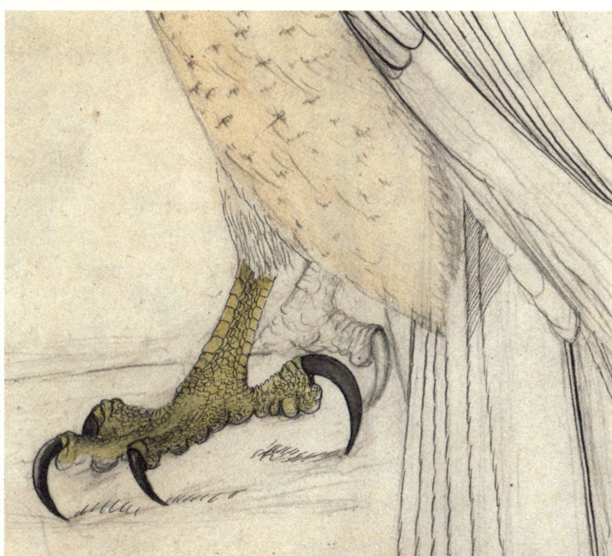

(*c*)

Figure 3.50a–c. Red-tailed Hawk (*a*). The enlargement (*b*) shows how Wilson used short, stiff brush strokes over pencil to show the erect feathers. Wilson is clearly master of his media and artistic technique. The enlargement of the hawk's feet (*c*) shows the precision of Wilson's drawing. (From the archives of the Ernst Mayr Library, Museum of Comparative Zoology, Harvard University.)

eyes that have depth and intensity (*b*). Whatever that hawk is staring at, you can read the concentration in its eye. The legs and feet are anatomically correct and exquisitely done (*c*).

Red-tailed Hawk

> Among the extensive meadows that border the Schuylkill and Delaware, below Philadelphia, where flocks of Larks and where mice and moles are in great abundance, many individuals of this Hawk spend the greater part of the winter. Others prowl around the plantations, looking out for vagrant Chickens; their method of seizing which is, by sweeping swiftly over the spot, and, grappling them with their talons, and bearing them away to the woods.[82]

Current name: Red-shouldered Hawk, *Buteo lineatus* (Gmelin)
Wilson's name: Red-shouldered Hawk, *Falco lineatus*
Original name: *Falco lineatus* Gmelin 1788

Current name: Rough-legged Hawk, *Buteo lagopus* (Pontoppidan)
Wilson's name: Black Hawk, *Falco sancti johannis*
Original name: *Falco lagopus* Pontoppidan 1763

Current name: Baltimore Oriole, *Icterus galbula* (Linnaeus)
Wilson's name: Baltimore Oriole, *Oriolius baltimore*
Original name: *Coracias Galbula* Linnaeus 1758

Current name: Eastern Towhee, *Pipilo erythrophthalmus* (Linnaeus)
Wilson's name: Towhee Bunting, *Emberiza erythropthalma*
Original name: *Fringilla erythrophthalma* Linnaeus 1758

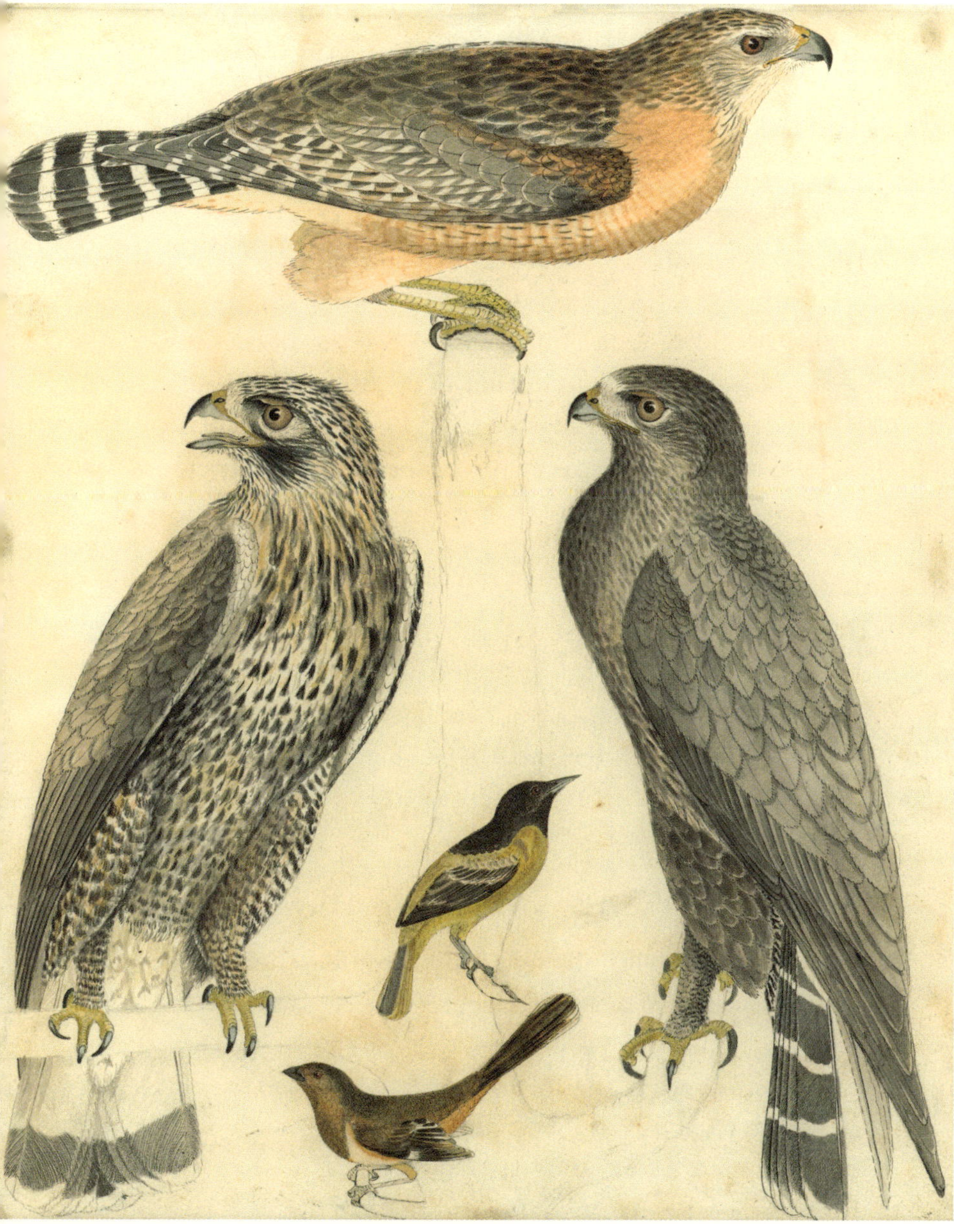

Figure 3.51. Red-shouldered Hawk (*top*); Rough-legged Hawk (*lower left* and *right*); Baltimore Oriole (*middle*); Eastern Towhee (*middle bottom*). (From the archives of the Ernst Mayr Library, Museum of Comparative Zoology, Harvard University.)

Commentary

Wilson's Black Hawk is today's Rough-legged Hawk. This finished painting left only the background to the engraver's imagination. To Wilson's credit he recognized that the Black Hawk (Rough-legged Hawk) had two color morphs, light on the left and dark on the right (see Figure 3.52a for a preliminary drawing of the dark Black Hawk). But how did Wilson decide what was a species and what was a plumage variation related to sex, age, or geography? For example, why were the American and Fish crows considered separate species and the dark and light variants of the Black Hawk deemed one species. Plumage variants of the latter certainly looks more different than the two species of crows, but he correctly placed the crows in separate species and the hawks in the same species.

Red-shouldered Hawk

> This species is more rarely met with than either of the former [Black Hawks (Rough-legged Hawks)]. Its haunts are in the neighborhood of the sea. It preys on Larks, Sandpipers, and the Small Ringed Plover, and frequently on Ducks. It flies high and irregularly, and not in the sailing manner of the Long-winged Hawks.[83]

Black Hawk (Rough-legged Hawk)

> This is a remarkably shy and wary bird, found most frequently along the marshy shores of our large rivers; feeds on mice, frogs, and moles; sails much, and sometimes at a great height; has been seen to kill a Duck on wing; sits, by the side of the marshes, on a stake, for an hour at a time, in an almost perpendicular position, as if dozing; flies with great ease, and occasionally with great swiftness, seldom flapping the wings; seems particularly fond of river shores, swamps, and marshes; is most numerous with us in winter, and but rarely seen in summer; is remarkable for the great size of its eye, length of its wings, and shortness of its toes. The breadth of its head is likewise uncommon.[84]

Baltimore Oriole

These birds are several years in receiving their complete plumage. Sometimes the whole tail of a male individual in spring is yellow, sometimes only the two middle feathers are black, and frequently the black on the back is skirted with orange, and the tail tipped with the same color. Three years, I have reason to believe, are necessary to fix the full tint of the plumage.[85]

In the spring and summer of 1811, a Baltimore took up its abode in Mr. Bartram's garden, whose notes were so singular as particularly to attract my attention; they were as well known to me as the voice of my most intimate friend. On the thirtieth of April, 1812, I was again surprised and pleased at hearing this same Baltimore in the garden, whistling his identical old chant.[86]

Towhee Bunting (Eastern Towhee)

This is a very common, but humble and inoffensive species, frequenting close-sheltered thickets, where it spends most of its time in scratching up the leaves for worms, and the larvae and eggs of insects. It is far from being shy, frequently suffering a person to walk round the bush or thicket, where it is at work, without betraying any marks of alarm, and when disturbed, uttering the notes *tow-he* repeatedly.[87]

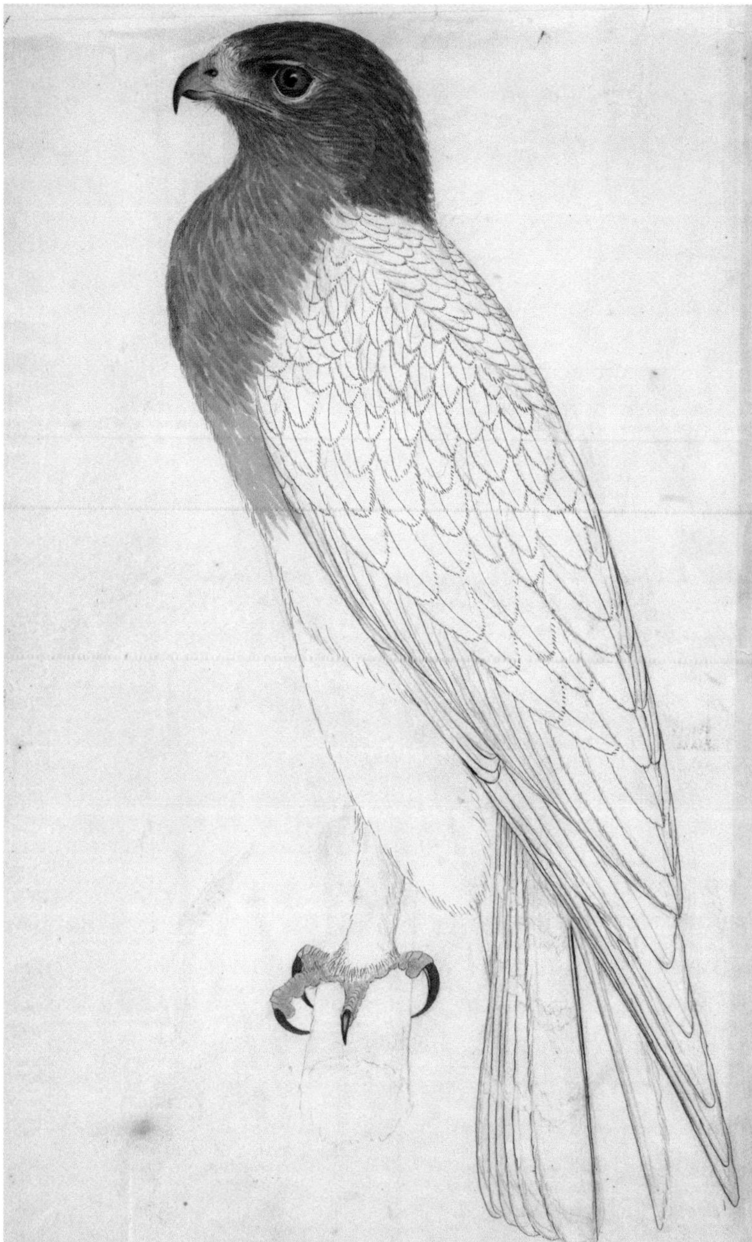

(*a*)

Figure 3.52a, b. A draft (*a*) of the Rough-legged Hawk shown in Figure 3.51. As with the preceding portrait of the Red-tailed Hawk the head and face are superb, the exquisite detail of the eye conveys an intense stare that is positively frightening (*b*). (From the archives of the Ernst Mayr Library, Museum of Comparative Zoology, Harvard University.)

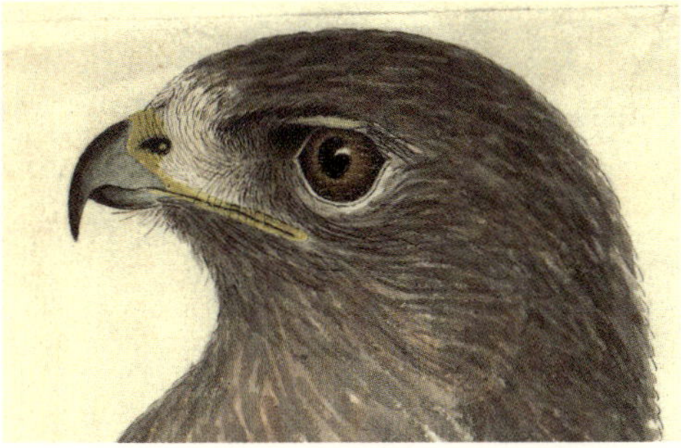

(*b*)

Current name: Rough-legged Hawk, *Buteo lagopus* (Pontoppidan)
Wilson's name: Black Hawk, *Falco sancti johannis*
Original name: *Falco lagopus* Pontoppidan 1763

Commentary

Drawing hawks was Wilson's greatest strength. This magnificent preliminary drawing was full-size with a superbly rendered head, eye, and bill (*b*). As with the Red-tailed Hawk in Figure 3.50b, the intensity of the hawk's gaze borders on frightening.

Black Hawk (Rough-legged Hawk)

> The original color of these birds in their young state may probably be pale brown, as the present individual seemed to be changing to a darker color on the neck and sides of the head. This change, from pale brown to black, is not greater than some of the genus are actually known to undergo. One great advantage of examining living or newly-killed specimens is, that whatever may be the difference of color between any two, the eye, countenance, and form of the head, instantly betray the common family to which they belong; for this family likeness is never lost in the living bird, though in stuffed skins and preserved specimens it is frequently entirely obliterated. I have no hesitation, therefore, in giving it as my opinion, that the present and preceding birds are of the same species, differing only in age, both being males. Of the female I am unable at present to speak.[88]

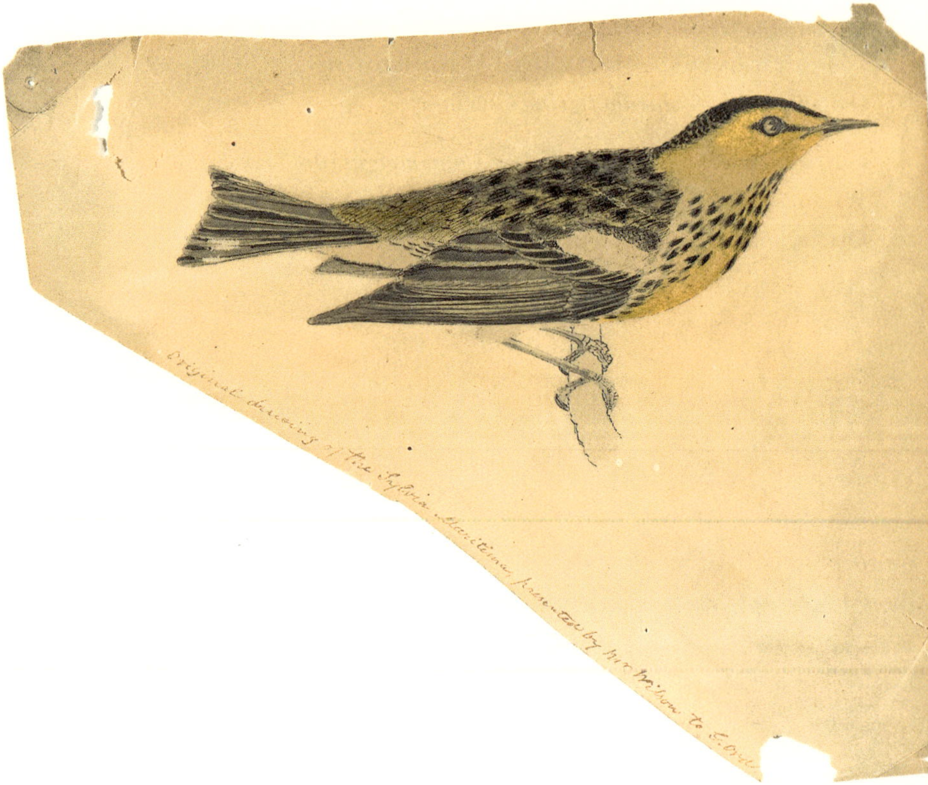
Original drawing of the Sylvia maritima, presented by Mr Wilson to G. Ord

(*a*)

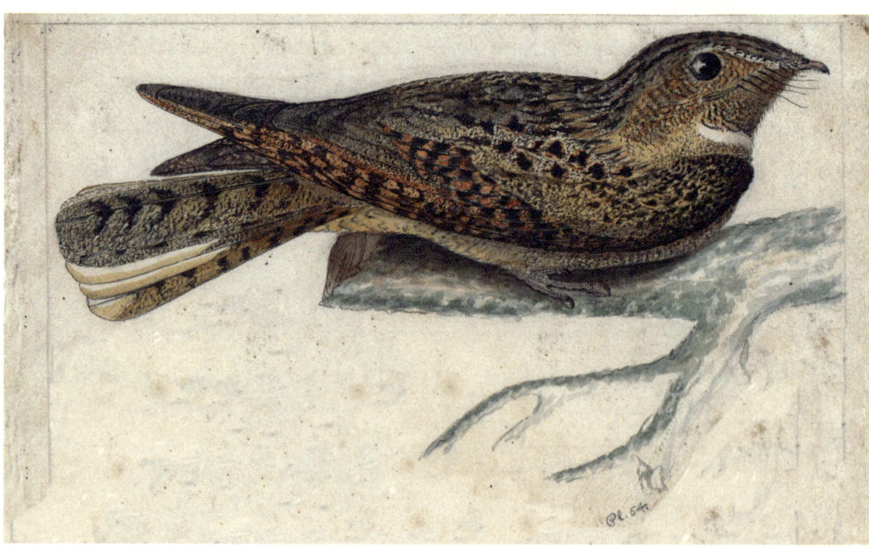
Pl. 54

Current name: Cape May Warbler, *Setophaga tigrina* (Gmelin)
Wilson's name: Cape May Warbler, *Sylvia maritima*
Original name: *Motacilla tigrina* Gmelin 1789

Current name: Chuck-will's-widow, *Caprimulgus carolinensis* (Gmelin)
Wilson's name: Chuck-will's-widow, *Caprimulgus carolinensis*
Original name: *Caprimulgus carolinensis* Gmelin 1789

Commentary

A Cape Way Warbler was observed by Wilson and George Ord on a trip to Cape May, New Jersey, and shot by Ord. This is the only specimen that Wilson ever saw. In the painting, the bird is in the exact position of the warbler in plate 54 of *American Ornithology,* but is substantially thinner and more pointed. Although fully painted and beautifully done, the image of the Chuck-will's-widow lacks the over-painting with opaque watercolors of Wilson's other caprimulgids (see, for example, Figures 3.42 and 3.43).

Cape May Warbler

> The spring had been remarkably cold, with long and violent northeast storms, and many winter birds, as well as passengers from the south, still lingered in the woods as late as the 20th of May, gleaning, in small companies, among the opening buds and infant leaves, and skipping nimbly from twig to twig, which was the case with the bird now before us when it was first observed.[89]

Chuck-will's-widow

> The Chuck-will's-widow, so called from its notes, which seem exactly to articulate those words, arrives on the sea-coast of Georgia about

---

Figure 3.53. Cape May Warbler (*a*). (From the archives of the Paisley Museum, Renfrewshire Council, Scotland.); Chuck-will's-widow (*b*). (From the archives of the Ernst Mayr Library Museum of Comparative Zoology, Harvard University.)

the middle of March, and in Virginia early in April. It commences its singular call generally in the evening, soon after sunset, and continues it, with short, occasional interruptions, for several hours. Towards morning these repetitions are renewed, and continue until dawn has fairly appeared. During the day it is altogether silent. This note, or call, instantly attracts the attention of a stranger, and is strikingly different from that of the Whip-poor-will. In sound and articulation it seems plainly to express the words which have been applied to it, [*Chuck-will's-widow,*] pronouncing each syllable leisurely and distinctly, putting the principal emphasis on the last word. In a still evening it may be heard at the distance of nearly a mile, the tones of its voice being stronger and more full than those of the Whip-poor-will, who utters his with much greater rapidity.[90]

Current name: Whimbrel, *Numenius phaeopus* (Linnaeus)
Wilson's name: Esquimaux Curlew, *Scolopax borealis*
Original name: *Scolopax Phaeopus* Linnaeus 1758

Current name: Marbled Godwit, *Limosa fedoa* (Linnaeus)
Wilson's name: Great Marbled Godwit, *Scolopax fedoa*
Original name: *Scolopax fedoa* Linnaeus 1758

Current name: Willet, *Tringa semipalmatus* (Gmelin)
Wilson's name: Semi-palmated Snipe, *Scolopax semipalmata*
Original name: *Catoptrophorus semipalmatus* Gmelin 1789

Current name: Dunlin, *Calidris alpina* (Linnaeus)
Wilson's name: Red-backed Sandpiper, *Tringa alpina*
Original name: *Tringa alpine* Linnaeus 1758

Commentary

Beginning with volume 7 Wilson added full backgrounds to his drawings, as in this one. The drawing of the Marbled Godwit was apparently pasted in

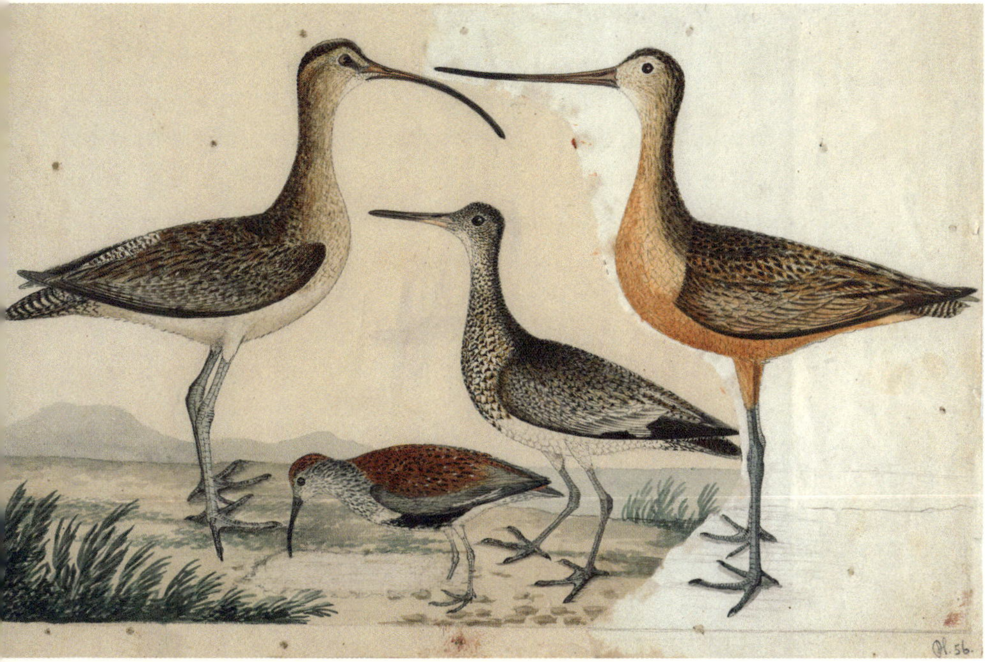

Figure 3.54. Whimbrel (*far left*); Marbled Godwit (*far right*); Willet (*middle right*); Dunlin (*middle left*). (From the archives of the Ernst Mayr Library, Museum of Comparative Zoology, Harvard University.)

after the background had been painted. The legs of the Dunlin were painted in after the background had been finished. The small pool in which the Dunlin stands was added by erasure.

Esquimaux Curlew (Whimbrel)

> The Short-billed Curlew arrives in large flocks on the sea-coast of New Jersey early in May from the south, [and] frequents the salt marshes, muddy shores and inlets, feeding on small worms and minute shell fish. They are most commonly seen on mud flats at low water, in company with various other Waders; and at high water roam along the marshes. They fly high, and with great rapidity. A

few are seen in June and as late as the beginning of July, when they generally move off towards the north. Their appearance on these occasions is very interesting: they collect together from the marshes as if by premeditated design, rise to a great height in the air, usually about an hour before sunset, and forming in one vast line, keep up a constant whistling on their march to the north, as if conversing with one another to render the journey more agreeable.[91]

Great Marbled Godwit (Marbled Godwit)

This is another transient visitant of our sea coasts in spring and autumn, to and from its breeding place in the north. Our gunners call it the *Straight-billed Curlew,* and sometimes the *Red Curlew.* It is a shy, cautious, and watchful bird; yet so strongly are they attached to each other, that on wounding one in a flock, the rest are immediately arrested in their flight, making so many circuits over the spot where it lies fluttering and screaming, that the sportsman often makes great destruction among them. Like the Curlew, they may also be enticed within shot, by imitating their call or whistle; but can seldom be approached without some such manoeuvre. They are much less numerous than the Short-billed Curlews [Whimbrel], with whom, however, they not infrequently associate. They are found among the salt marshes in May, and for some time in June, and also on their return in October and November.[92]

Semi-palmated Snipe (Willet)

The anxiety and affection manifested by these birds for their eggs and young, are truly interesting. A person no sooner enters the marshes than he is beset by the Willets, flying around and skimming over his head, vociferating with great violence their common cry of *pill-will-willet;* and uttering at times a loud, clicking note, as he approaches nearer to their nest. As they occasionally alight and slowly shut their long white wings speckled with black, they have a mournful note, expressive of great tenderness. During the term of incubation, the female often resorts to the sea shore, where, standing up to the belly in water, she washes and dresses her plumage, seeming to enjoy great satisfaction from these frequent immersions. She is also at other times seen to wade more in the water than most of her tribe: and, when wounded in the wing will take to the water without hesitation, and swims tolerably well.[93]

Red-backed Sandpiper (Dunlin)

This bird inhabits both the old and new continents, being known in England by the name of the Dunlin, and in the United States, along the shores of New Jersey, by that of the Red-Back. Its residence here is but transient, chiefly in April and May, while passing to the arctic regions to breed; and in September and October, when on its return southward to winter quarters. During their stay they seldom collect in separate flocks by themselves; but mix with various other species of strand birds, among whom they are rendered conspicuous by the red color of the upper part of their plumage. They frequent the muddy flats and shores of the salt marshes at low water, feeding on small worms and other insects which generally abound in such places. In the month of May, they are extremely fat.[94]

Commentary

The identity of this sandpiper is uncertain. It is not similar to any of the sandpipers that appear in the final plates. "Curlew Sandpiper" is written in pencil on the bottom, but the bird in the sketch lacks the decurved bill characteristic of the Curlew Sandpiper *(Calidris ferruginae).* The red-edged tertials and lesser coverts, the white edge of the greater coverts, the white edging along the primaries, and the slightly dark edges on the breast feathers that give a

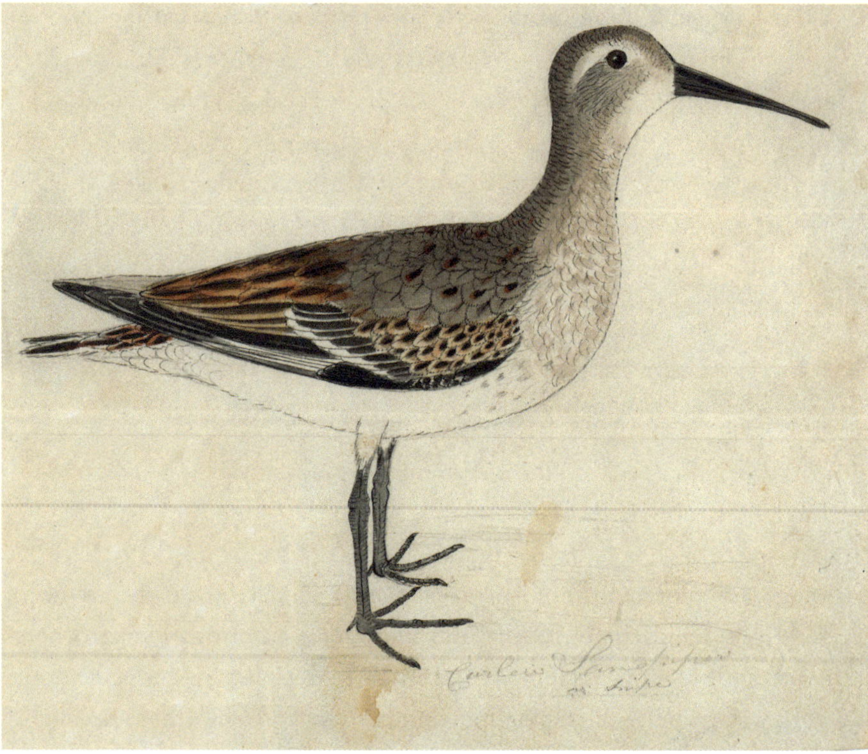

Figure 3.55. A sandpiper, probably a Dunlin, although the species is not labeled. (From the archives of the Ernst Mayr Library, Museum of Comparative Zoology, Harvard University.)

mottled to grayish appearance on the chest are, however, close to the markings of an adult female or juvenile Dunlin migrating south while molting its body feathers. The deep base to the bill is also characteristic of a Dunlin.

Current name: Red Knot (winter plumage), *Calidris canutus* (Linnaeus)
Wilson's name: Ash-colored Sandpiper, *Tringa cinerea*
Original name: *Tringa canutus* Linnaeus 1758

Current name: Black-bellied Plover, *Pluvialis squatarola* (Linnaeus)
Wilson's name: Black-bellied Plover, *Tringa squatarola*
Original name: *Tringa Squatarola* Linnaeus 1758

Current name: Ruddy Turnstone, *Arenaria interpres* (Linnaeus)
Wilson's name: Turn-stone, *Tringa interpres*
Original name: *Tringa Interpres* Linnaeus 1758

Current name: Red Knot (breeding plumage), *Calidris canutus* (Linnaeus)
Wilson's name: Red-breasted Sandpiper, *Tringa rufa*
Original name: *Tringa canutus* Linnaeus 1758

Current name: Sanderling, *Calidris alba* (Pallas)
Wilson's name: Purre (breeding plumage), *Tringa cinclus*
Sanderling Plover (winter plumage), *Charadrius calidris*
Original name: *Trynga alba* Pallas 1764

Commentary

Wilson's Ash-colored and Red-breasted sandpipers are seasonal variants of the Red Knot. Similarly the Purre is the Sanderling in spring plumage whereas the Sanderling Plover is the Sanderling in winter plumage. Such seasonal dimorphism creates confusion that is similar to that created by sexual dimorphism, for example, the very different plumages of the male and female Black-throated Blue Warblers. As with this confusing warbler, Wilson saw the knots only during spring and fall migration. In the drawings for this plate, Wilson has not properly recessed the eyes into the heads, giving them a stuck-on, glass-eyed appearance. The point highlight near the center of the eye, which creates an effect akin to the red-eye phenomenon in flash photographs, emphasizes this look. Wilson eventually recognized that the Sanderling Plover and Purre were the same species.[95]

Ash-colored Sandpiper (Red Knot in winter plumage)

The regularly disposed concentric semicircles of white and dark brown that mark the upper parts of the plumage of this species, distinguish it from all others, and give it a very neat appearance. In activity it . . . traces the flowing and recession of the waves along the sandy beach, with great nimbleness, wading and searching among the loos-

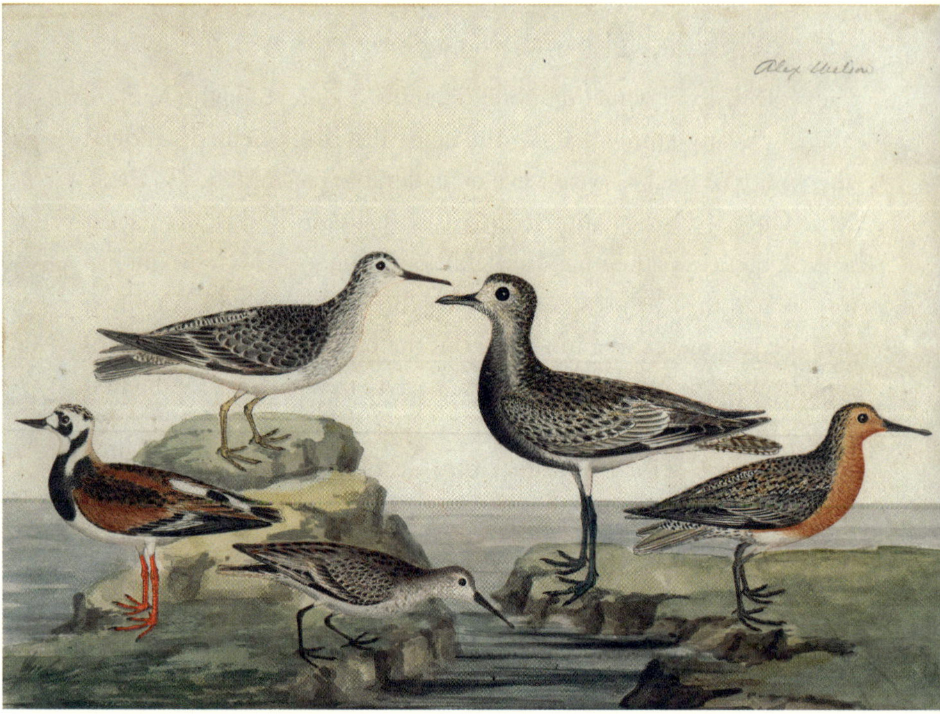

Figure 3.56. Red Knot (winter plumage; *uppermost*); Black-bellied Plover (*middle*); Ruddy Turnstone (*lower left*); Red Knot (breeding plumage; *far right*); Sanderling (*lower middle*). (From the archives of the Ernst Mayr Library, Museum of Comparative Zoology, Harvard University.)

ened particles for its favorite food, which is a small, thin, oval, bivalve shell fish, of a white or pearl color, and not larger than the seed of an apple . . . During the latter part of summer and autumn, these minute shell fish constitute the food of almost all those busy flocks that run with such activity along the sands, among the flowing and retreating waves. They are universally swallowed whole; but the action of the bird's stomach, assisted by the shells themselves, soon reduces them to a pulp. If we may judge from their effects they must be extremely nutritious, for almost all of those tribes that feed on them are at this season mere lumps of fat.[96]

Black-bellied Plover

They have a loud whistling note; often fly at a great height; and are called by many gunners along the coast the Black-bellied Killdeer. The young of the first year have considerable resemblance to those of the Golden Plover; but may be easily distinguished from this last by the largeness of their head and bill, and in being at least two inches more in length. The greater number of those which I have examined have the rudiments of a hind toe; but the character and manners of the Plover are so conspicuous in the bird, as to determine, at the first glance, the tribe it belongs to . . . In every stage the seemingly disproportionate size of the head, and thickness of the bill, will distinguish this species.[97]

Turn-stone (Ruddy Turnstone)

This beautifully-variegated species is common to both Europe and America; consequently [it] extends its migrations far to the north. It arrives from the south on the shores of New Jersey in April; leaves them early in June; is seen on its return to the south in October; and continues to be occasionally seen until the commencement of cold weather, when it disappears for the season. It is rather a scarce species in this part of the world, and of a solitary disposition, seldom mingling among the large flocks of other sandpipers; but either coursing the sands alone, or in company with two or three of its own species. On the coast of Cape May and Egg Harbor this bird is well known by the name of the *Horse-Foot Snipe,* from its living, during the months of May and June, almost wholly on the eggs, or spawn of the great king Crab, called here by the common people the *Horse-Foot.* This animal is the *Monoculus polyphemus* of entomologists.[98]

Red-breasted Sandpiper (Red Knot in breeding plumage)

The common name of this species on our sea coast is the *Gray-back,* and among the gunners it is a particular favorite, being generally a plump, tender and excellent bird for the table; and, consequently, brings a good price in market.

The Gray-Backs do not breed on the shores of the Middle States. Their first appearance is early in May. They remain a few weeks, and again disappear until October. They usually keep in small flocks, alight in a close body together on the sand flats, where they search for the small bivalve shells already described. On the approach of the sportsman they frequently stand fixed and silent for some time; do not appear to be easily alarmed, neither do they run about in the water as much as some others, or with the same rapidity, but appear more tranquil and deliberate. In the month of November they retire to the south.[99]

Purre (Sanderling in breeding plumage)

This is one of the most numerous of our Strand birds, as they are usually called, that frequent the sandy beach on the frontiers of the ocean. In its habits it differs so little from the preceding [Red Knot] . . . except in being still more active and expert in running and searching among the sand, on the reflux of the waves, as it nimbly darts about for food, . . . The Purre continues longer with us, both in spring and autumn than either of the two preceding; many of them remain during the very severest of the winter, tho the greater part retire to the more genial regions of the south; where I have seen them at such seasons, particularly on the sea-coasts of both Carolinas, during the month of February, in great numbers.[100]

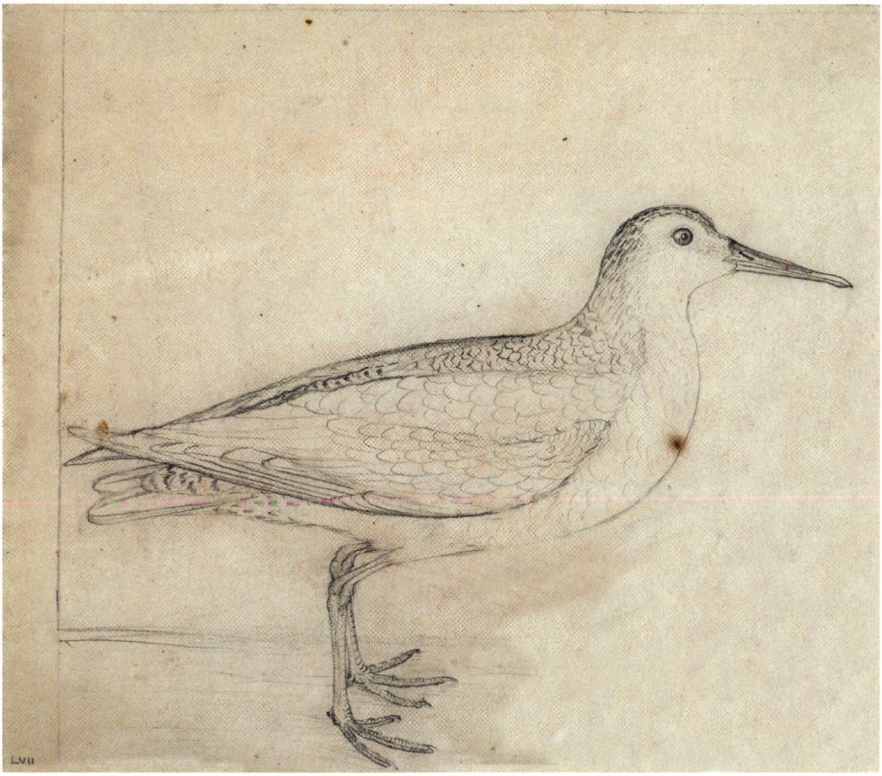

Figure 3.57. Red Knot in summer plumage, perhaps a preliminary sketch for the knot in Figure 3.56. (From the archives of the Ernst Mayr Library, Museum of Comparative Zoology, Harvard University.)

Commentary

The knot in the drawing matches the plate closely, but the eye and bill of the sketch are superior to those in the plate, in which the eye is too large and improperly set in its orbit. As expected, the legs and feet of the knot in the sketch also have the excellent detail characteristic of Wilson's drawings.

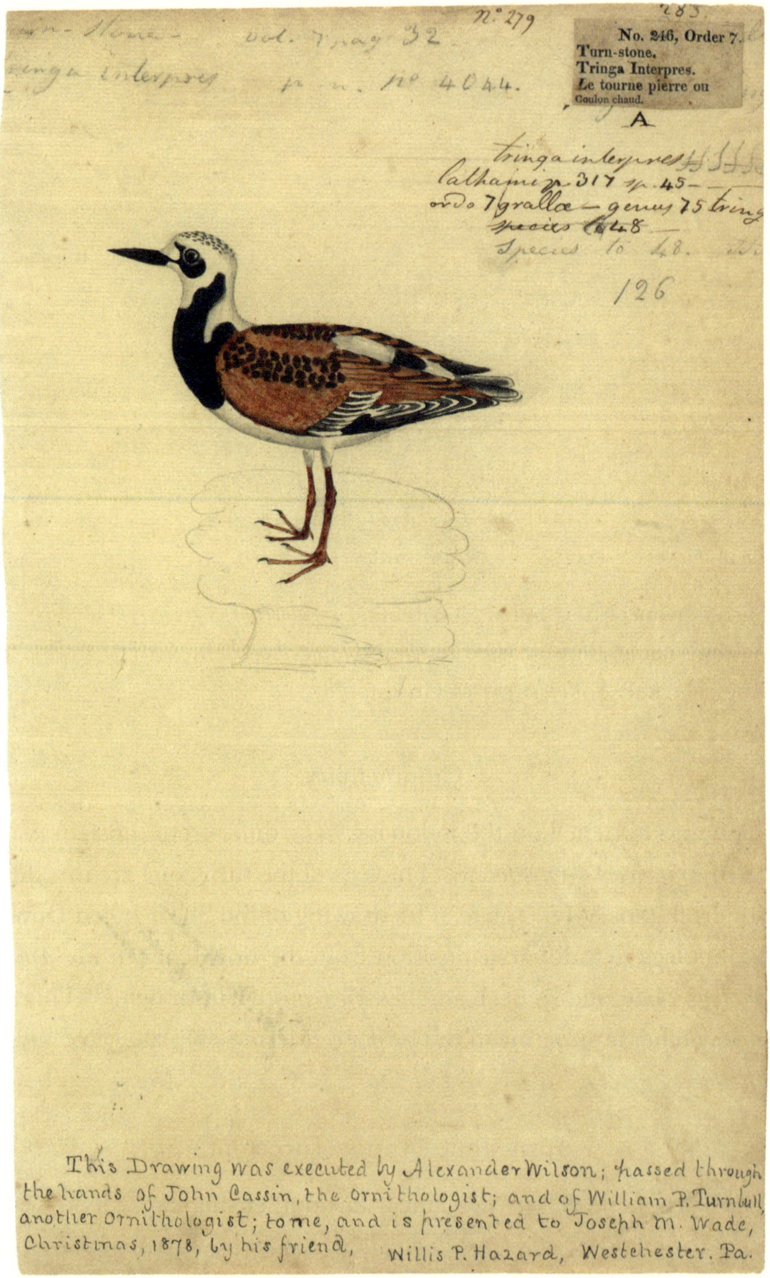

Figure 3.58. Sketch of Ruddy Turnstone perhaps used to indicate coloring for the fully rendered version seen in Figure 3.56. (From the archives of the Ernst Mayr Library, Museum of Comparative Zoology, Harvard University.)

Commentary

This fully rendered turnstone, attributed to Wilson, closely resembles the turnstone of plate 57, although the legs are in a slightly different position. It is atypical of Wilson's usual style in that it lacks the detailed inking of legs, feet, bill, and feathers. Compare this individual with the preceding pencil sketch of a Red Knot (Figure 3.57). Furthermore, the eye is too low in the head. The lack of detail and misplacement of the eye suggest that this sketch may have been drawn hastily as a reference for the colorists.

Current name: Ruddy Turnstone, *Arenaria interpres* (Linnaeus)
Wilson's name: Turn-stone, *Tringa interpres*
Original name: *Tringa Interpres* Linnaeus 1758

Current name: Short-billed Dowitcher, *Limnodromus griseus* (Gmelin)
Wilson's name: Red-breasted Snipe, *Scolopax noveboracensis*
Original name: *Scolopax grisea* Gmelin 1789

Commentary

The drawings in pencil on the following page differ from those in plates 57 and 58 of *American Ornithology.* The legs of the turnstone are in a different position from those of the plate. The drawing of the Short-billed Dowitcher also has its legs in a different position from the dowitcher in the *American Ornithology* plate, and its neck and head are more withdrawn.[101] This sketch of the dowitcher is superior to the bird's rendering on plate 58 of *American Ornithology.*

Turn-stone (Ruddy Turnstone)

> The Turn-stone derives its name from another singularity it possesses, of turning over with its bill small stones and pebbles, in search of various marine worms and insects. At this sort of work it is exceedingly dexterous; and, even when taken and domesticated, is said to retain the same habit. Its bill seems particularly well constructed for

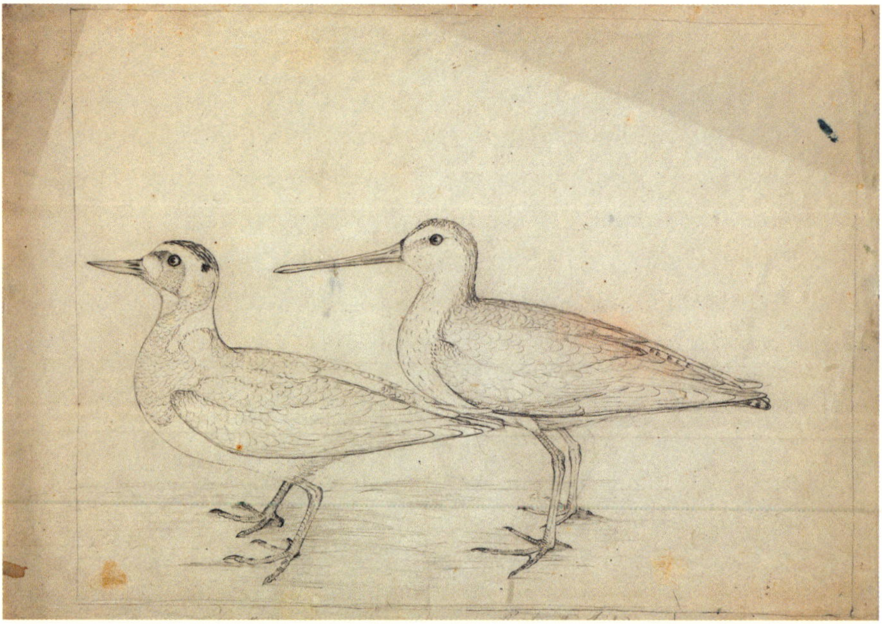

Figure 3.59. Ruddy Turnstone (*left*); Short-billed Dowitcher (*right*). (From the archives of the Ernst Mayr Library, Museum of Comparative Zoology, Harvard University.)

this purpose, differing from all the rest of its tribe, and very much resembling in shape that of the Common Nuthatch.[102]

Red-breasted Snipe (Short-billed Dowitcher)

The Red-breasted Snipe arrives on the sea coast of New Jersey early in April; is seldom or never seen inland: early in May it proceeds to the north to breed, and returns by the latter part of July or beginning of August. During its stay here it flies in flocks, sometimes very high, and has then a loud and shrill whistle, making many evolutions over the marshes; . . . They spring from the marshes with a loud twirling whistle, . . . fly very rapidly, sometimes wheeling, coursing, and doubling along the surface of the marshes; then shooting high in the air, there separating and forming in various bodies, uttering a kind of quivering whistle.[102]

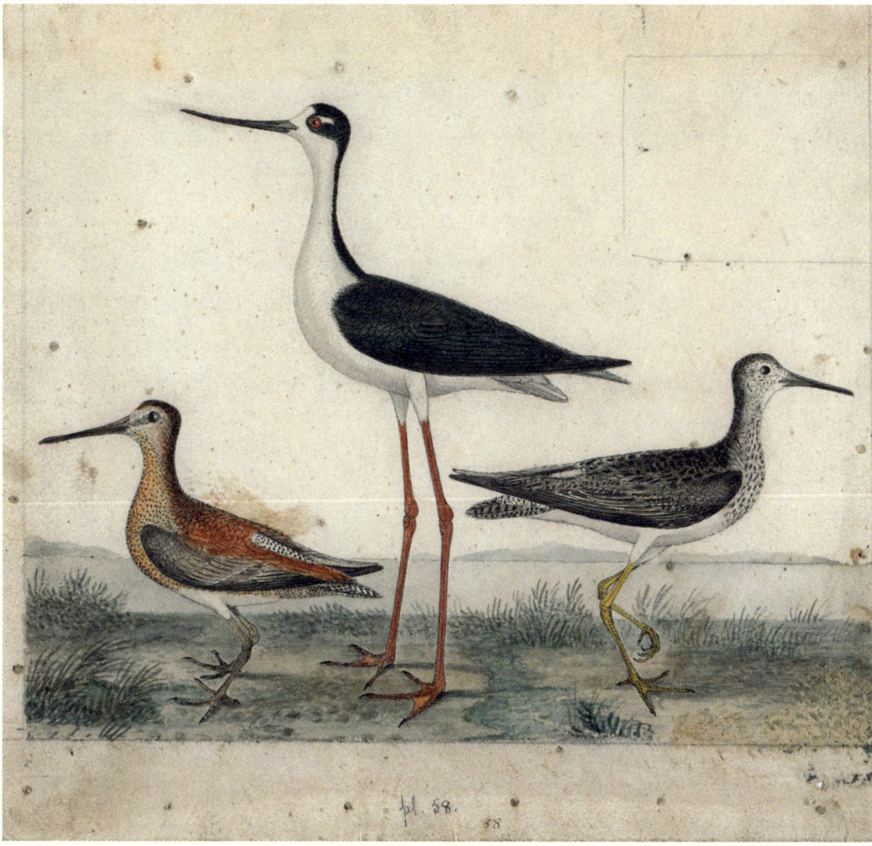

Figure 3.60. Short-billed Dowitcher (*left*); Black-necked Stilt, (*middle*); Lesser Yellowlegs (*right*). (From the archives of the Ernst Mayr Library, Museum of Comparative Zoology, Harvard University.)

Current name: Short-billed Dowitcher, *Limnodromus griseus* (Gmelin)
Wilson's name: Red-breasted Snipe, *Scolopax noveboracensis*
Original name: *Scolopax grisea* Gmelin 1789

Current name: Black-necked Stilt, *Himantopus mexicanus* (Müller)
Wilson's name: Long-legged Avoset, *Recurvirostra himantopus*
Original name: *Charadrius Mexicanus* P. L. S. Müller 1776

Current name: Lesser Yellowlegs, *Tringa flavipes* (Gmelin)
Wilson's name: Yellow-shanks Snipe, *Scolopax flavipes*
Original name: *Scolopax flavipes* Gmelin 1789

Commentary

This is the first of Wilson's fully painted habitats, and so represents an innovation in his presentation of birds. This is not one of Wilson's better paintings, however. The tails of the stilt and yellowlegs are twisted to show the upper surface, a position Wilson uses frequently to emphasize the feather pattern despite the unnatural appearance it creates. The dowitcher too is leaning toward the observer, an awkward position that emphasizes the white stripe along the back. Wilson's stated goal was to promote the identification of species by illustrating critical features as clearly as possible, even if it meant a less artistic representation.

Red-breasted Snipe (Short-billed Dowitcher)

Among many which I opened in May, were several females, that had very little rufous below, and the backs were also much lighter and less marbled with ferruginous. The eggs contained in their ovaries were some of them as large as garden peas . . .

These birds, doubtless breed not far to the northward of the United States, if we may judge from the lateness of the season when they leave us in spring, the largeness of the eggs in the ovaries of the females before they depart, and the short period of time they are absent.[104]

Long-legged Avoset (Black-necked Stilt)

Naturalists have most unaccountably classed this bird with the genus *Charadrius,* or Plover, and yet affect to make the particular conformation of the bill, legs and feet, the rule of their arrangement. In the present subject, however, excepting the trivial circumstance of the want of a hind toe, there is no resemblance whatever of those parts to the bill, legs or feet, of the Plover; on the contrary, they are so entirely different, as to create no small surprise at the adoption and general acceptation of a classification, evidently so absurd and unnatural. This appears the more reprehensible, when we consider the striking affinity there is between this bird and the Common Avoset, not only in the particular form of the bill, nostrils, tongue, legs, feet, wings and tail, but extending to the voice, manners, food, place of breeding, form of the nest, and even the very color of the eggs of both, all of which are strikingly alike, and point out, at once, to the actual observer of Nature, the true relationship of these remarkable birds.[105]

Yellow-shanks Snipe (Lesser Yellowlegs)

Of this species I have but little to say. It inhabits our sea coasts and salt marshes during summer; frequents the flats at low water, and seems particularly fond of walking among the mud, where it doubtless finds its favorite food in abundance. Having never met with its nest, nor with any person acquainted with its particular place or manner of breeding, I must reserve these matters for further observation . . . It has a sharp whistle of three or four notes when about to take wing, and when flying.[106]

Current name: Upland Sandpiper, *Bartramia longicauda* (Bechstein)
Wilson's name: Bartram's Sandpiper, *Tringa bartramia*
Original name: *Tringa longicauda* Bechstein 1812

Current name: Spotted Sandpiper, *Actitis macularius* (Linnaeus)
Wilson's name: Spotted Sandpiper, *Tringa macularia*
Original name: *Tringa macularia* Linnaeus 1766

Commentary

These preliminary drawings are fully painted. The legs of the Spotted Sandpiper are too short, a defect that was corrected in the *American Ornithology*

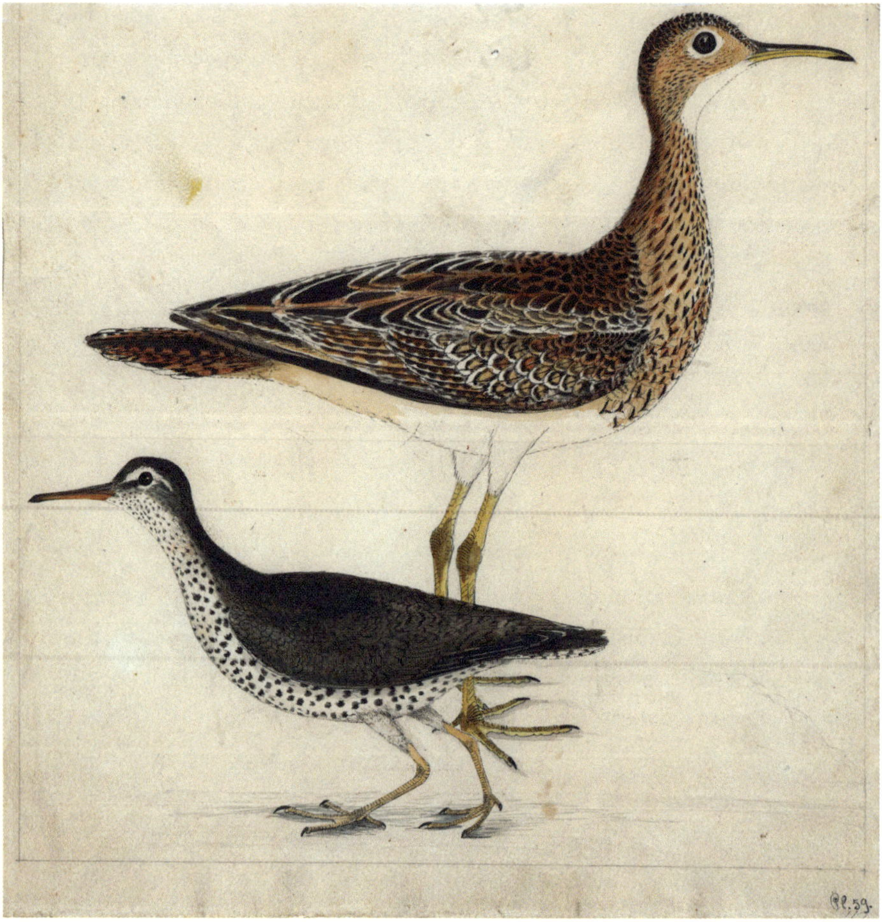

Figure 3.61. Upland Sandpiper (*upper*); Spotted Sandpiper (*lower*). (From the archives of the Ernst Mayr Library, Museum of Comparative Zoology, Harvard University.)

plate. Depicting the sandpiper walking, which is what Spotted Sandpipers are usually doing when seen, adds a dynamic, behavioral quality to this plate. Wilson believed he was describing the Upland Sandpiper for the first time and named it in honor of his friend and mentor William Bartram. Unfortunately, Johann Bechstein had published a description of the same species a few months earlier, although this news had not yet crossed the Atlantic, in part due to the British blockade of the United States at the start of the War of 1812. By international agreement the earlier name has precedence. Lesson revised the taxonomy of sandpipers in 1831, however, and honored Wilson's wish and Bartram's reputation as a naturalist by creating the genus *Bartramia* for this species, which was also known until recently as Bartram's Sandpiper in the European literature.

Bartram's Sandpiper (Upland Sandpiper)

> This bird being as far as I can discover a new species, undescribed by any former author, I have honored it with the name of my very worthy friend, near whose botanic gardens, on the banks of the river Schuylkill, I first found it . . .
>
> Unlike most of their tribe, these birds appear to prefer running about among the grass, feeding on beetles and other winged insects. There were three or four in company; they seemed extremely watchful, silent, and shy, so that it was always with extreme difficulty I could approach them.
>
> . . . They run with great rapidity, sometimes spreading their tail and dropping their wings, as birds do who wish to decoy you from their nest; when they alight, they remain fixed, stand very erect, and have two or three sharp whistling notes as they mount to fly.[107]

Spotted Sandpiper

> This species is as remarkable for perpetually wagging the tail, as some others are for nodding the head; for, whether running on the ground, or on the fences, along the rails, or in the water, this motion seems continual; even the young, as soon as they are freed from the shell, run about, constantly wagging the tail. About the middle of May they resort to the adjoining corn-fields to breed, where I have

frequently found and examined their nests. One of these is now before me, and which was built at the root of a hill of Indian corn, on high ground, is composed wholly of short pieces of dry straw. The eggs are four, of a pale clay or cream color, marked with large, irregular spots of black, and more thinly with others of a paler tint. They are large in proportion to the size of the bird, measuring an inch and a quarter in length, very thick at the great end, and tapering suddenly to the other. The young run about with wonderful speed as soon as they leave the shell, and are then covered with down of a dull drab color, marked with a single streak of black down the middle of the back, and with another behind each ear. They have a weak, plaintive note. On the approach of any person the parents exhibit symptoms of great distress, counterfeiting lameness, and fluttering along the ground with seeming difficulty.[108]

Current name: Black-bellied Plover, *Pluvialis squatarola* (Linnaeus)
Wilson's name: Black-bellied Plover, *Tringa squatarola*
Original name: *Tringa Squatarola* Linnaeus 1758
or
Current name: American Golden-Plover, *Pluvialis dominica* (Müller)
Wilson's name: Golden Plover, Charadrius dominica
Original name: *Charadrius Dominicus* Müller 1776

Current name: Killdeer, *Charadrius vociferus* Linnaeus
Wilson's name: Kildeer Plover, *Charadrius vociferus*
Original name: *Charadrius vociferus* Linnaeus 1758

Commentary

These preliminary drawings differ from the final plates. The pose of the plover is similar to that of the Black-bellied Plover in the *American Ornithology* plate, but the bill on the drawing is too narrow for a Black-bellied Plover, and hence it may have been drawn as a Golden Plover, which has a proportion-

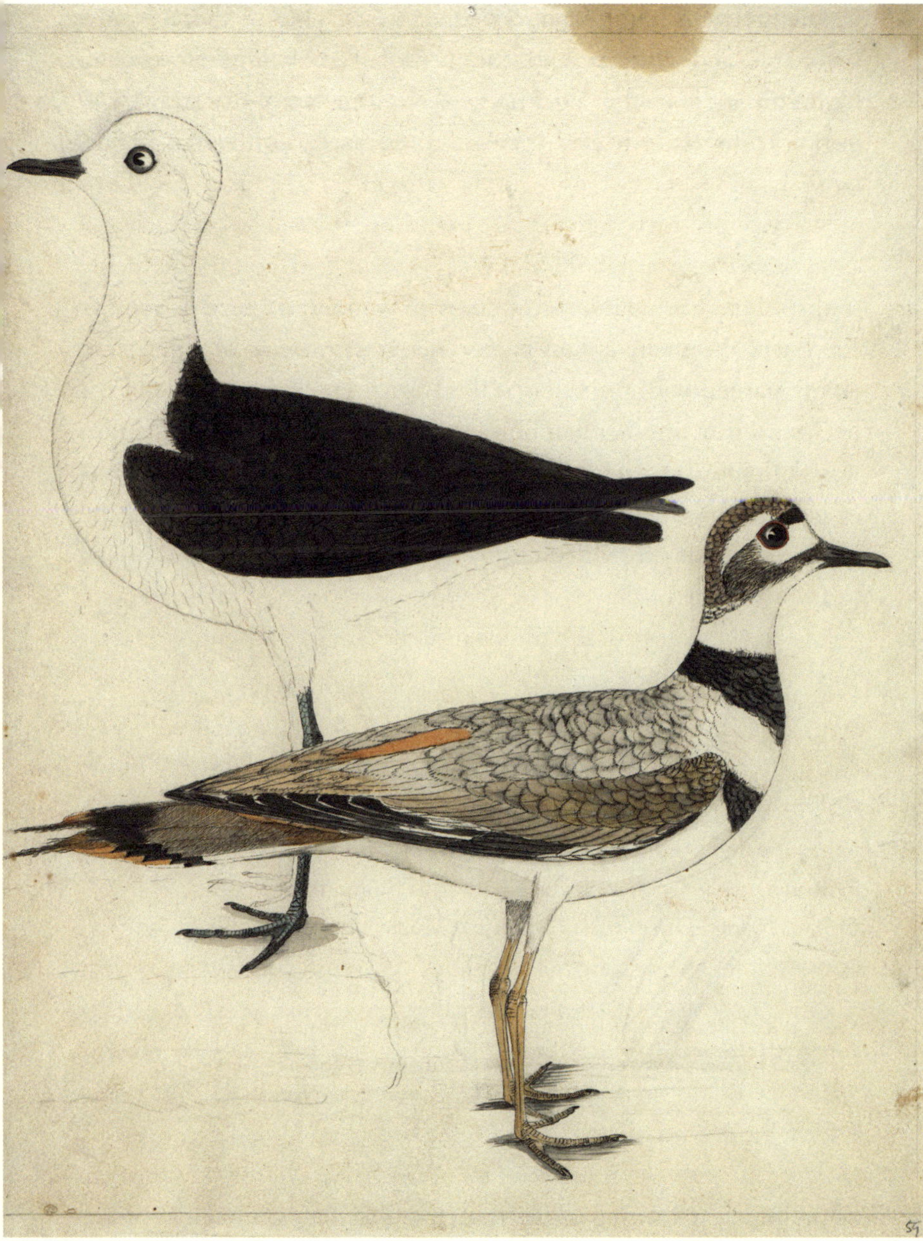

Figure 3.62. Black-bellied or American Golden-Plover (*upper*); Killdeer (*lower*). (From the archives of the Ernst Mayr Library, Museum of Comparative Zoology, Harvard University.)

ally narrower bill. The Golden Plover in the *American Ornithology*'s plate 59, however, has its head looking over its back. The Killdeer of the drawing is slimmer than the bird in the final plate and the head is slightly too small. Wilson probably redrew these species for the final painting that went to the engraver.

Black-bellied Plover

> This bird . . . known in some parts of the country by the name of the Large Whistling Field Plover . . . seems particularly attached to newly ploughed fields, where it forms its nest of a few slight materials, as slightly put together. The female lays four eggs, large for the size of the bird, of a light olive color dashed with black; and has frequently two broods in the same season. It is an extremely shy and watchful bird, though clamorous during breeding time. The young are without the black color on the breast and belly until the second year, and the colors of the plumage above are likewise imperfect till then. They feed on worms, grubs, winged insects, and various kinds of berries . . . About the beginning of September they descend with their young to the sea-coast, and associate with the numerous multitudes then returning from their breeding places in the north.[109]

Golden Plover (American Golden Plover)

> Although these birds are occasionally found along our sea coast from Georgia to Maine, yet they are no where numerous; and I have never met with them in the interior. Our mountains being generally covered with forest, and no species of heath having, as yet, been discovered within the boundaries of the United States, these birds are probably induced to seek the more remote arctic regions of the continent to breed and rear their young in, where the country is more open, and unencumbered with woods.[110]

Kildeer Plover (Killdeer)

Nothing can exceed the alarm and anxiety of these birds during the breeding season. Their cries of *kildeer, kildeer,* as they winnow the air overhead, dive and course around you, or run along the ground counterfeiting lameness, are shrill and incessant. The moment they see a person approach, they fly or run to attack him with their harassing clamor, continuing it over so wide an extent of ground, that they puzzle the pursuer as to the particular spot where the nest or young are concealed; very much resembling, in this respect, the Lapwing of Europe.[111]

Current name: Black Skimmer, *Rynchops niger* Linnaeus
Wilson's name: Black Skimmer or Sheerwater, *Rhynchops nigra*
Original name: *Rynchops niger* Linnaeus 1758

Current name: Sanderling, *Calidris alba* (Pallas)
Wilson's name: Purre (Sanderling in breeding plumage), *Tringa cinclus;* Sanderling Plover (Sanderling in winter plumage), *Charadrius calidris*
Original name: *Trynga alba* Pallas 1764

Commentary

In *American Ornithology* the Sanderling appears in plate 59 in winter plumage and in plate 63 in breeding plumage. As with the Red Knot, Wilson's Red-breasted Sandpiper (spring) and Ash-colored Sandpiper (fall), the Sanderling, a migratory species with very different summer and winter plumages, was mistakenly described as two species (but see the commentary for Figure 3.56). The Sanderling portrayed here differs markedly from that in the two plates and is superior to both. Wilson made numerous sketches of skimmer heads with their unique lower bill. The two figured here were not preliminary drawings of the image in the final plate.

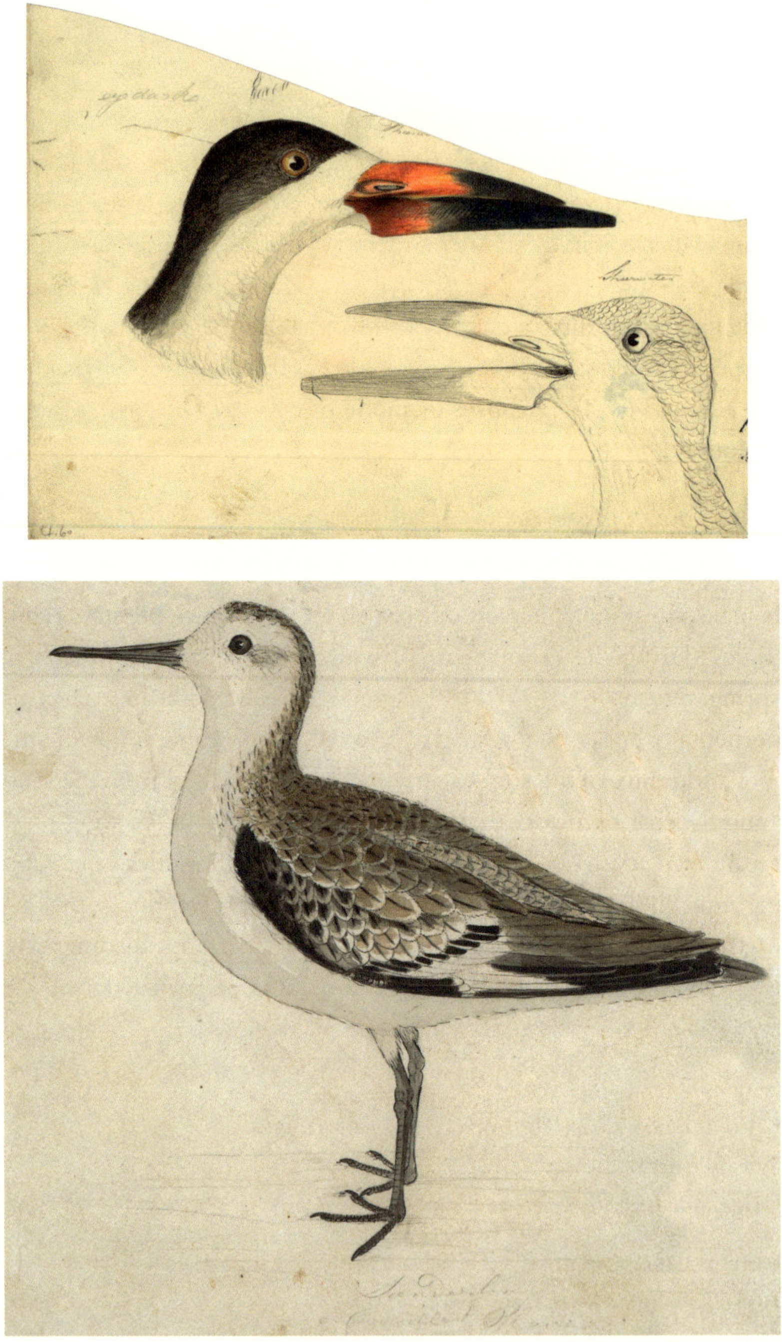

(*a*)

(*b*)

Figure 3.63. Black Skimmer (*a*); Sanderling (*b*). (From the archives of the Ernst Mayr Library, Museum of Comparative Zoology, Harvard University.)

Black Skimmer

The singular conformation of the bill of this bird has excited much surprise; and some writers, measuring the divine proportions of nature by their own contracted standards of conception, in the plenitude of their vanity have pronounced it to be "a lame and defective weapon." Such ignorant presumption, or rather impiety, ought to hide its head in the dust on a calm display of the peculiar construction of this singular bird, and the wisdom by which it is so admirably adapted to the purposes or mode of existence for which it was intended.[112]

Sanderling Plover (Sanderling in fall plumage)

It makes its appearance on our sea coasts early in September; continues during the greater part of winter; and on the approach of spring, returns to the northern regions to breed. While here it seems perpetually busy running along the wave-worn strand, following the flux and reflux of the surf, eagerly picking up its food from the sand amid the roar of the ocean. It flies in numerous flocks, keeping a low meandering course along the ridges of the tumbling surf. On alighting, the whole scatter about after the receding wave, busily picking up those minute bivalves already described. As the succeeding wave returns it bears the whole of them before it in one crowded line.[113]

Current name: Least Tern, *Sternula antillarum* Lesson
Wilson's name: Lesser Tern, *Sterna minuta*
Original name: *Sternula antillarum* Lesson 1847

Current name: Common Tern, *Sterna hirundo* Linnaeus
Wilson's name: Great Tern, *Sterna hirundo*
Original name: *Sterna hirundo* Linnaeus 1758

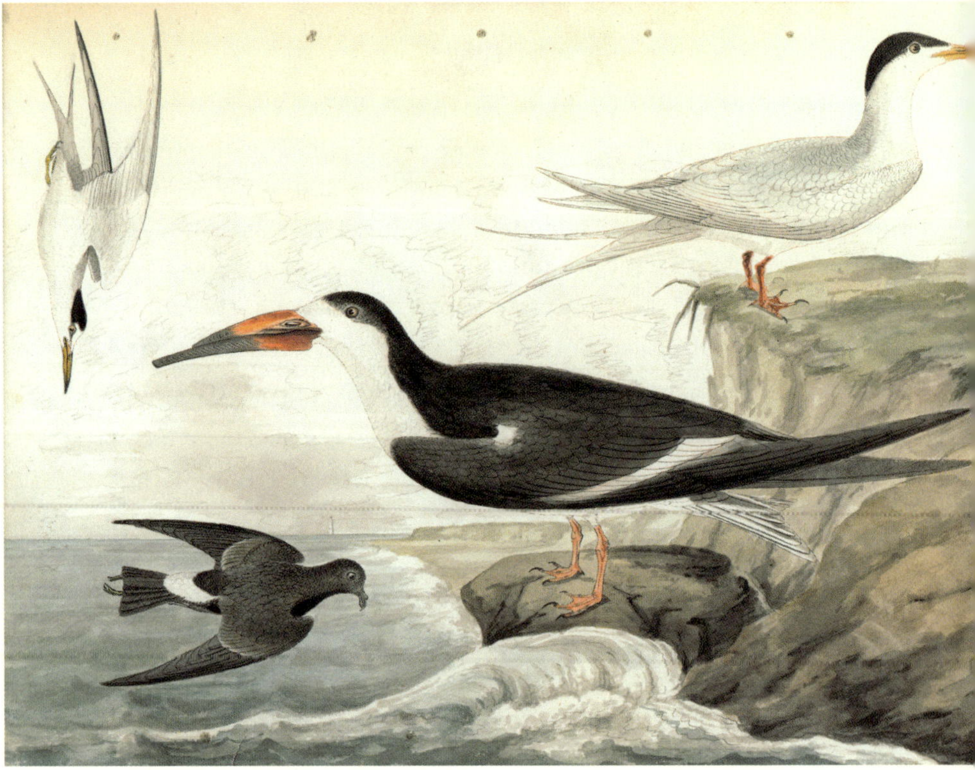

Figure 3.64. Least Tern (*upper left*); Common Tern (*upper right*); Black Skimmer (*middle*); Wilson's Storm-Petrel (*lower left*). (From the archives of the Ernst Mayr Library, Museum of Comparative Zoology, Harvard University.)

Current name: Black Skimmer, *Rynchops niger* Linnaeus
Wilson's name: Black Skimmer or Sheerwater, *Rhynchops nigra*
Original name: *Rynchops niger* Linnaeus 1758

Current name: Wilson's Storm-Petrel, *Oceanites oceanicus* (Kuhl)
Wilson's name: Stormy Petrel, *Procelaria pelagica*
Original name: *Procellaria oceanica* Kuhl 1820

Commentary

This version of the plate has a dynamic quality with a diving tern, flying storm-petrel, and breaking waves. Unfortunately the final plate omitted the breaking waves. The head of the skimmer in this figure lacks the dynamic qualities of the preceding sketches (Figure 3.63). On the basis of Linnaeus's published description, Wilson identified the petrel as the European Storm-Petrel *(Hydrobates pelagicus),* but in 1820 Kuhl would identify important differences and designate it as a new species. Charles Lucien Bonaparte, a nephew of Napoleon, an Italian prince, and an excellent ornithologist, edited and updated the 1828 edition of Wilson's *American Ornithology.* He agreed with Kuhl's decision and gave the bird its common name in Wilson's honor.

Lesser Tern (Least Tern)

> It arrives on the coast somewhat later than the other [Great Tern], but in equal and perhaps greater numbers; coasts along the shores and also over the pools in the salt marshes in search of prawns, of which it is particularly fond; hovers, suspended in the air, for a few moments above its prey, exactly in the manner of some of our small Hawks, and dashes headlong down into the water after it, generally seizing it with its bill; mounts instantly again at the same height, and moves slowly along as before, eagerly examining the surface below.[114]

Great Tern (Common Tern)

> The young are generally produced at intervals of a day or so from each other, and are regularly and abundantly fed for several weeks, before their wings are sufficiently grown to enable them to fly. At first the parents alight with the fish, which they have brought in their mouth or in their bill, and, tearing it in pieces distribute it in such portions as their young are able to swallow. Afterwards they frequently feed them without alighting, as they skim over the spot;

and as the young become nearly ready to fly, they drop the fish among them where the strongest and most active has the best chance to gobble it up. In the mean time, the young themselves frequently search about the marshes, generally not far apart, for insects of various kinds; but so well acquainted are they with the peculiar language of their parents that warn them of the approach of an enemy, that, on hearing their cries they instantly squat, and remain motionless until the danger be over.[115]

Black Skimmer

The Sheerwater is formed for skimming, while on the wing, the surface of the sea for its food, which consists of small fish, shrimps, young fry, &c., whose usual haunts are near the shore, and towards the surface. That the lower mandible, when dipped into and cleaving the water, might not retard the bird's way, it is thinned and sharpened like the blade of a knife; the upper mandible being at such times elevated above water is curtailed in its length, as being less necessary, but tapering gradually to a point, that, on shutting, it may offer less opposition. To prevent inconvenience from the rushing of the water the mouth is confined to the mere opening of the gullet, which indeed prevents mastication taking place there; but the stomach, or gizzard, to which this business is solely allotted, is of uncommon hardness, strength and muscularity, far surpassing in these respects any other water-bird with which I am acquainted. To all these is added a vast expansion of wing, to enable the bird to sail with sufficient celerity while dipping in the water.[116]

Stormy Petrel (Wilson's Storm-Petrel)

When any greasy matter is thrown overboard, these birds instantly collect around it, and facing to windward, with their long wings expanded, and their webbed feet patting the water, the lightness of their bodies and the action of the wind on their wings enable them to do this with ease . . . According to Buffon, it is from this singular habit that the whole genus have obtained the name Petrel, from the apostle Peter, who, as Scripture informs us, also walked on the water . . . One circumstance is worthy of being noticed, and shows the vast range they take over the ocean. In firing at these birds, a quill feather was broken in each wing of an individual, and hung fluttering in the wind, which rendered it so conspicuous among the rest as to be known to all on board. This bird, notwithstanding its inconvenience, continued with us for nearly a week, during which we sailed a distance of more than four hundred miles to the north.[117]

Current name: Black Tern, *Chlidonias niger* (Linnaeus)
Wilson's name: Short-tailed Tern, *Sterna plumbea*
Original name: *Sterna nigra* Linnaeus 1758

Commentary

This drawing is missing the bill and front face of the bird but shows enough of the black head pattern and short tail to identify the species as a Black Tern. Wilson, once again, has written colors on various parts of the sketch. The preceding illustration, Figure 3.64, is similar to the final plate 60 in *American Ornithology,* but it lacks the image of the Black Tern that is included on the final plate. This drawing of the tern is very different from that of the final plate and was probably preliminary.

Short-tailed Tern (Black Tern)

On the sixth of September, 1812, after a violent north east storm, which inundated the meadows of Schuylkill in many places, numerous flocks of this Tern all at once made their appearance, flying over those watery spaces, picking up grasshoppers, beetles, spiders, and

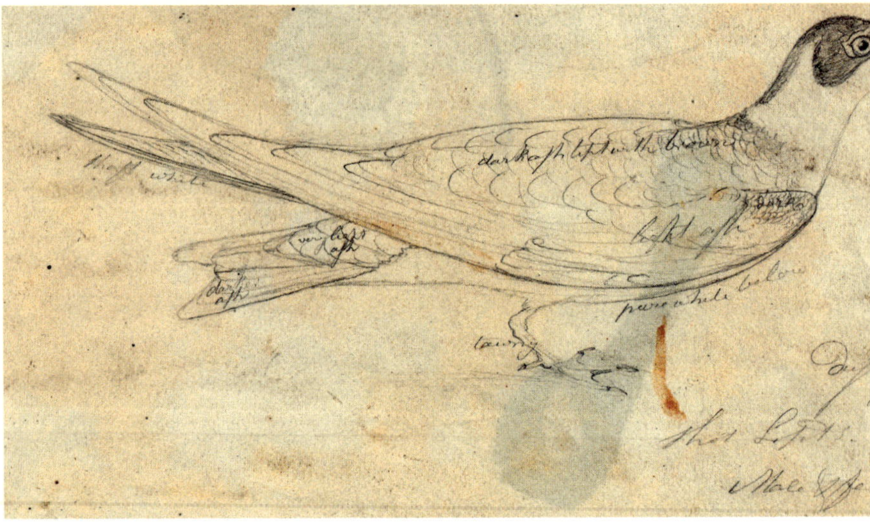

Figure 3.65. Black Tern. (From the archives of the Ernst Mayr Library, Museum of Comparative Zoology, Harvard University.)

> other insects, that were floating on the surface. Some hundreds of them might be seen at the same time, and all seemingly of one sort . . . I examined upwards of thirty individuals of this species, by dissection, and found both sexes alike in colour. Their stomachs contained grasshoppers, crickets, spiders, &c. but no fish.[118]

Current name: Green Heron, *Butorides virescens* (Linnaeus)
Wilson's name: Green Bittern, *Ardea virescens*
Original name: *Ardea virescens* Linnaeus 1758

Commentary

This preliminary sketch for the Green Heron was drawn in pencil, then much of it was overmarked with ink. As in Figures 3.65 and 3.67, Wilson's notes are just visible. He has penciled in coloring instructions for different parts of the bird—for example, "yellow" on the lower mandible and "blue" between the

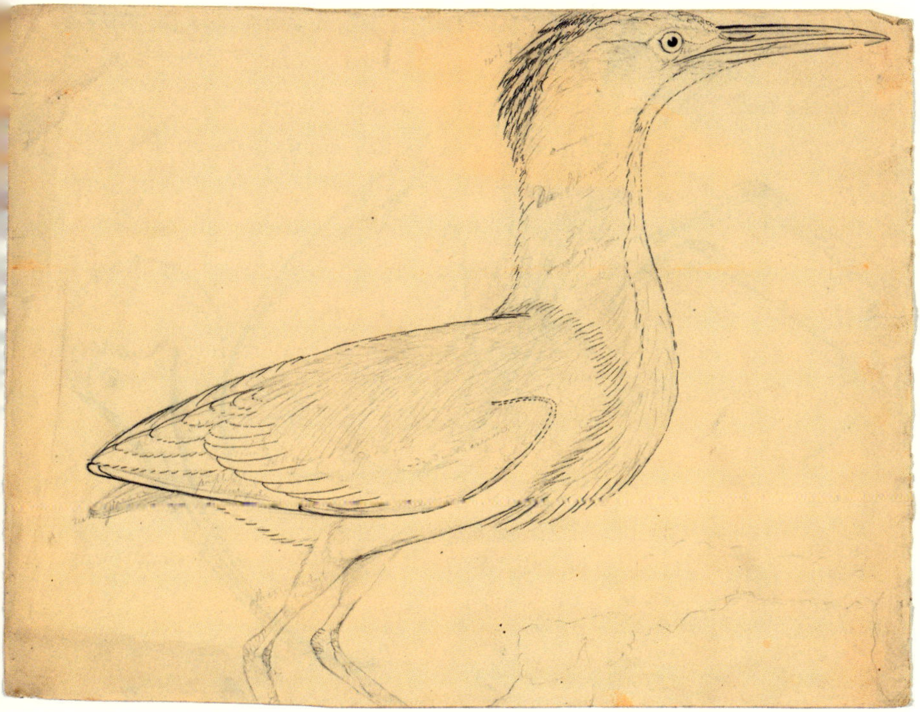

Figure 3.66. Green Heron. (From the archives of the Ernst Mayr Library, Museum of Comparative Zoology, Harvard University.)

eyes and beak. Never one to waste paper, Wilson drew a White Ibis on the back of this sheet (see Figure 3.75).

Green Bittern (Green Heron)

> The capturing of frogs requires much nicer management . . . The Bittern, fixing his penetrating eye on the spot where they disappeared, approaches with slow, stealing step, laying his feet so gently and silently on the ground as not to be heard or felt; and when arrived within reach stands fixed, and bending forwards, until the first glimpse of the frog's head makes its appearance, when, with a stroke instantaneous as lightning, he seizes it in his bill, beats it to death, and feasts on it at his leisure.

This mode of life, requiring little fatigue where game is so plenty, as is generally the case in all our marshes, must be particularly pleasing to the bird; and also very interesting, from the continual exercise of cunning and ingenuity necessary to circumvent its prey. Some of the naturalists of Europe, however, in their superior wisdom, think very differently; and one can scarcely refrain from smiling at the absurdity of those writers, who declare, that the lives of this whole class of birds are rendered miserable by toil and hunger; their very appearance, according to Buffon, presenting the image of suffering anxiety and indigence.[119]

Current name: Great Egret, *Ardea alba* Linnaeus
Wilson's name: Great White Heron, *Ardea egretta*
Original name: *Ardea alba* Linnaeus 1758

Commentary

This is a working sketch as can be seen by the quick zigzag strokes on the head that provide the outline and suggest feathers, but lack detail. Coloration of the bill, facial skin, and eye provides important information on these soft tissues, whose colors fade soon after death. The egret sketched here was shot August 1, 1808, but the plate and account of the Great Egret were not published until volume 7, which appeared in 1813. That suggests that for some species, at least, Wilson accumulated notes for a substantial period of time. The notes, which are in Wilson's handwriting, emphasize Wilson's detailed, quantitative approach to the study of birds. Because they are difficult to read we provide the following transcription:

Ardea alba Great white Crane, shot August 1, 1808 on the Banks of Schuylkill near Mr. Bartrams place The whole plumage is of the purist white—skin of legs black
the latter seamed with lines of dirty white—the bill is of a rich
persimmon yellow with rather a stronger tint of Reddish—from

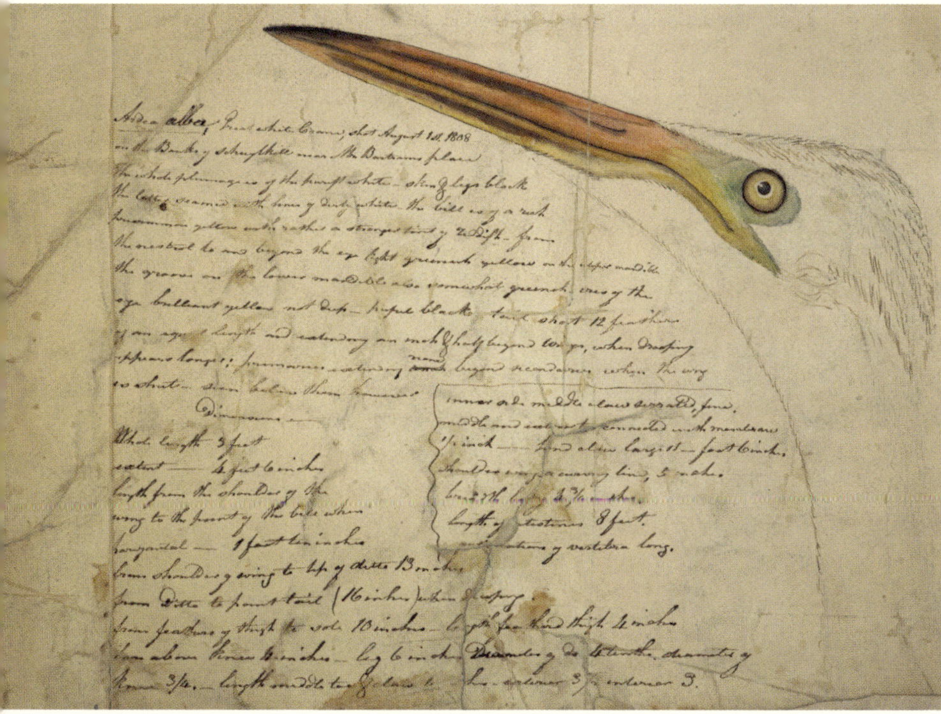

Figure 3.67. Great Egret. (From the archives of the Department of Ornithology, the American Museum of Natural History, New York, Miscellaneous Art no. 252.)

the nostril [*sic*] to and beyond the eye light greenish yellow on the upper mandible
the groove on the lower mandible also somewhat greenish. iris of the
eye brilliant yellow not deep—pupil black tail short—12 feathers
of an equal length and extending an inch & half beyond secondaries when drooping
appears longer: much larger ——— when the wing is shut—seen below them however

inner side middle claw serrated, fine.
middle and ex[terior] t[oe] connected with membrane
½ inch—hind claw largest—foot 6 inches

shoulder wing—a curving line, 5 inches
breadth body 4 ¾ inches
length of intestines 8 feet.
[ar?]t[icu?]lations of vertebra long.

Dimensions
whole length 3 feet
extent—4 feet 6 inches
length from the shoulder of the
wing to the point of the bill when
horizontal—1 foot ten inches
from shoulder of wing to tip of ditto [wing] 13 inches
from Ditto to point [of] tail (16 inches) when drooping.
from feathers of thigh to sole 10 inches—length feathered thigh 4 inches
from above knee 4 inches—leg 6 inches Diameter of do[ditto] 4 tenths.
diameter of
knee ¾.—length middle toe & claw 4 [inc]hes—exterior 3 ½ interior 3.

Great White Heron (Great Egret)

The appearance of this bird, during the first season, when it is entirely destitute of the long flowing plumes of the back, is so different from the same bird in its perfect plumage, which it obtains in the third year, that naturalists and others very generally consider them as two distinct species. The opportunities which I have fortunately had, of observing them with the train in various stages of its progress, from its first appearance to it[s] full growth, satisfy me that the Great White Heron with, and that without the long plumes, are one and the same species, in different periods of age.[120]

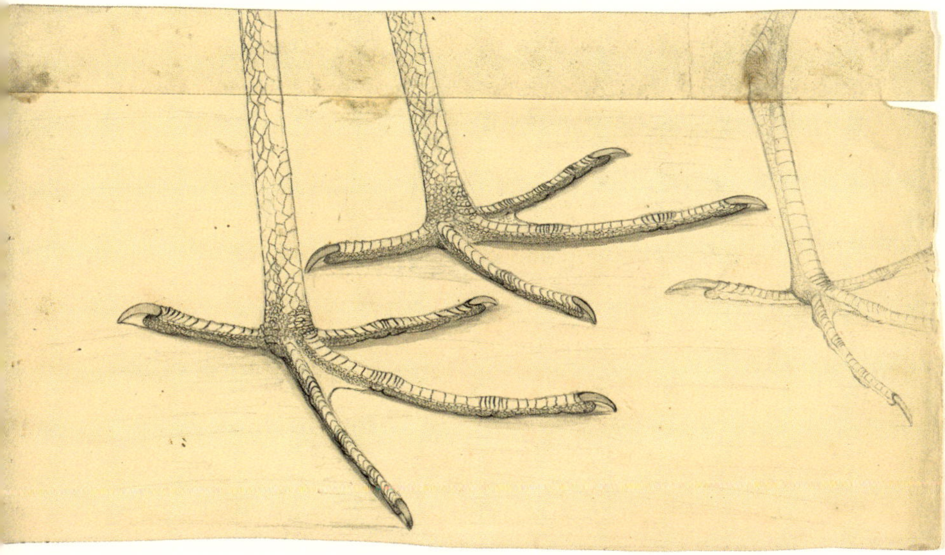

Figure 3.68. Detailed drawing of heron feet. (From the archives of the Ernst Mayr Library, Museum of Comparative Zoology, Harvard University.)

Commentary

This detailed drawing of heron feet exemplifies the exquisite detail that Wilson brought to his drawings. He also describes the feet and serrated middle claw, which is not illustrated in this sketch. Based on his description of the webbing between the "exterior and middle toes" these are the feet of a Great Egret. Note that the right foot is closest to you and the left foot farther away. Is the third foot a right or a left? Note too that the "knee" that Wilson mentions in his description of the Great Egret is actually the ankle, because birds walk on their toes. That is, what we think of as the leg of a bird is a single bone formed by the fusion of those bones that in a human would be called the arch of the foot and some of the bones in the ankle.

Green Bittern (Green Heron)

Legs and feet yellow, tinged before with green, the skin of these thick and movable; . . .

There is one circumstance attending this bird which, I recollect, at first surprised me. On shooting and wounding one, I carried it some distance by the legs, which were at first yellow; but on reaching home, I perceived to my surprise that they were red. On letting the bird remain some time undisturbed, they again became yellow, and I then discovered that the action of the hand had brought a flow of blood into them and produced the change of color. I have remarked the same in those of the Night Heron.[121]

Night Heron (Black-crowned Night-Heron)

Legs and feet a pale yellow cream color; inside of the middle claw serrated.[122]

Great Egret Heron (Great Egret)

The legs are long, stout, and of a black color, as is the bare space of four inches above the knee; the span of the foot measures upwards of six inches; the inner edge of the middle claw is pectinated; the exterior and middle toes are united at the base for about half an inch, by a membrane.[123]

Blue Heron (Little Blue Heron)

Legs blackish green; inner side of the middle claw pectinated.[124]

Snowy Heron (Snowy Egret)

The legs and naked part of the thighs are black; the feet, bright yellow; claws, black, the middle one pectinated. [125]

Great Heron (Great Blue Heron)

Naked thighs, brownish yellow; legs, brownish black, tinctured with yellow, and netted with seams of whitish; in some the legs are nearly black.[126]

American Bittern

Legs and feet, yellow, tinged with pale green; middle claw, pectinated.[127]

Least Bittern

Middle claw pectinated; toes, tinged with pale green; feet, large, the span of the foot measuring two inches and three quarters.[128]

Demi-egret Heron (Tricolored Heron)

The legs and naked thighs, greenish yellow; middle claw, pectinated.[129]

Yellow-crowned Heron (Yellow-crowned Night Heron)

Legs and feet, yellow; middle claw, pectinated.[130]

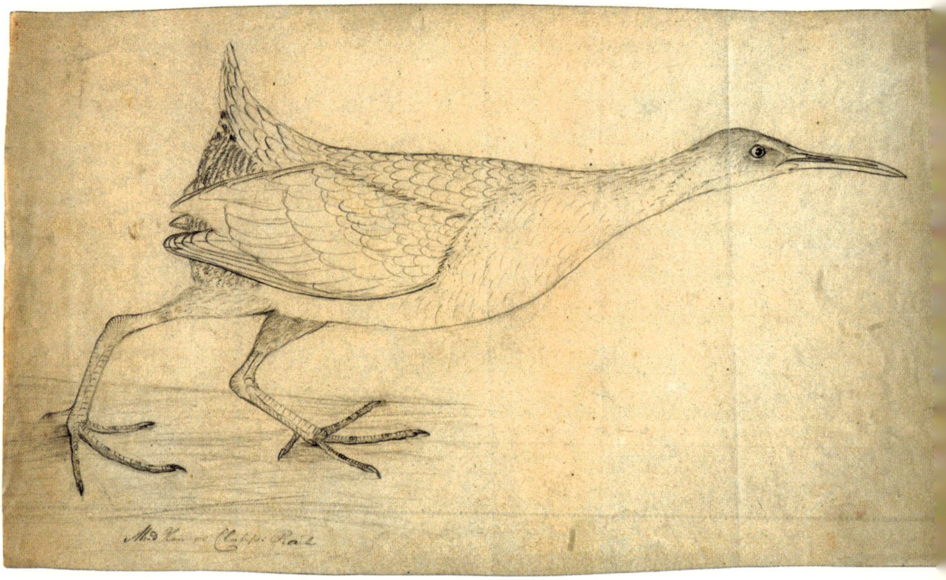

Figure 3.69. Clapper Rail. (From the archives of the Ernst Mayr Library, Museum of Comparative Zoology, Harvard University.)

Current name: Clapper Rail, *Rallus longirostris* Boddaert
Wilson's name: Clapper Rail, *Rallus crepitans*
Original name: *Rallus longirostris* Boddaert 1783

Commentary

This is a preliminary drawing of the Clapper Rail, but the final plate follows it closely in outline. The eye and its inset and placement are superior in the drawing, which is the case for many of Wilson's birds. Lawson apparently had trouble engraving the subtleties of Wilson's meticulous rendering of eyes and bills.

Clapper Rail

The whole defence of this species seems to be in the nervous vigor of its limbs, and thin compressed form of its body, by which it is enabled to pass between the stalks of grass and reeds with great rapidity. There are also every where among the salt marshes covered ways under the flat and matted grass, through which the Rail makes its way like a rat, without a possibility of being seen. There is generally one or more of these from its nest to the water edge, by which it may escape unseen; and sometimes, if closely pressed, it will dive to the other side of the pond, gut, or inlet, rising and disappearing again with the silence and celerity of thought. In smooth water it swims tolerably well, but not fast; sitting high in the water, with its neck erect, and striking with great rapidity. When on shore, it runs with the neck extended, the tail erect, and frequently flirted up. On fair ground they run nearly as fast as a man; having myself, with great difficulty, caught some that were wing broken. They have also the faculty of remaining under water for several minutes, clinging close, head downwards, by the roots of the grass.[131]

Current name: Roseate Spoonbill, *Platalea ajaja* (Linnaeus)
Wilson's name: Roseate Spoon-Bill, *Platalea ajaja*
Original name: *Platalea Ajaia* Linnaeus 1758

Commentary

The drawing is atypical for Wilson. Pencil is used little in the outline or elsewhere and the feathers are drawn in brown rather than black ink. The legs are too short and without Wilson's usual detail. The leg length was corrected in the final plate. Wilson never saw a spoonbill in the wild. The specimen on which the drawing is based (Figure 3.71) was sent to him by the family of William Dunbar, a wealthy planter and naturalist with whom Wilson stayed during his journey along the Mississippi River south of Natchez, Mississippi, in 1810. In fact, it is possible that the illustration shown in Figure 3.70 was painted by William Dunbar, not Alexander Wilson. The specimen was prepared as a mount and placed in the Peale Museum. In 1846 the Peale Museum closed, and in 1850 the specimen was purchased by Moses Kimball and given

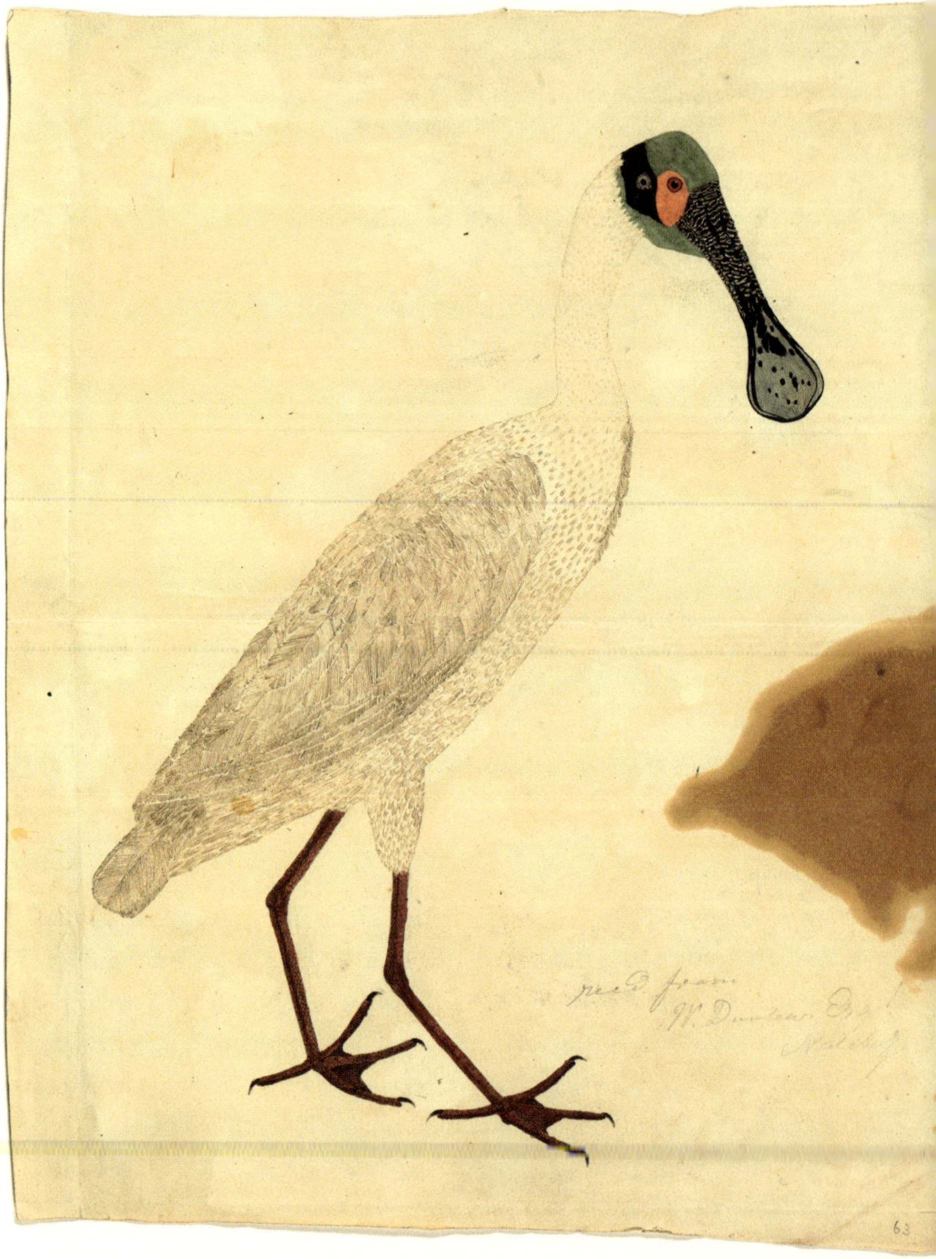

Figure 3.70. Roseate Spoonbill. (From the archives of the Ernst Mayr Library, Museum of Comparative Zoology, Harvard University).

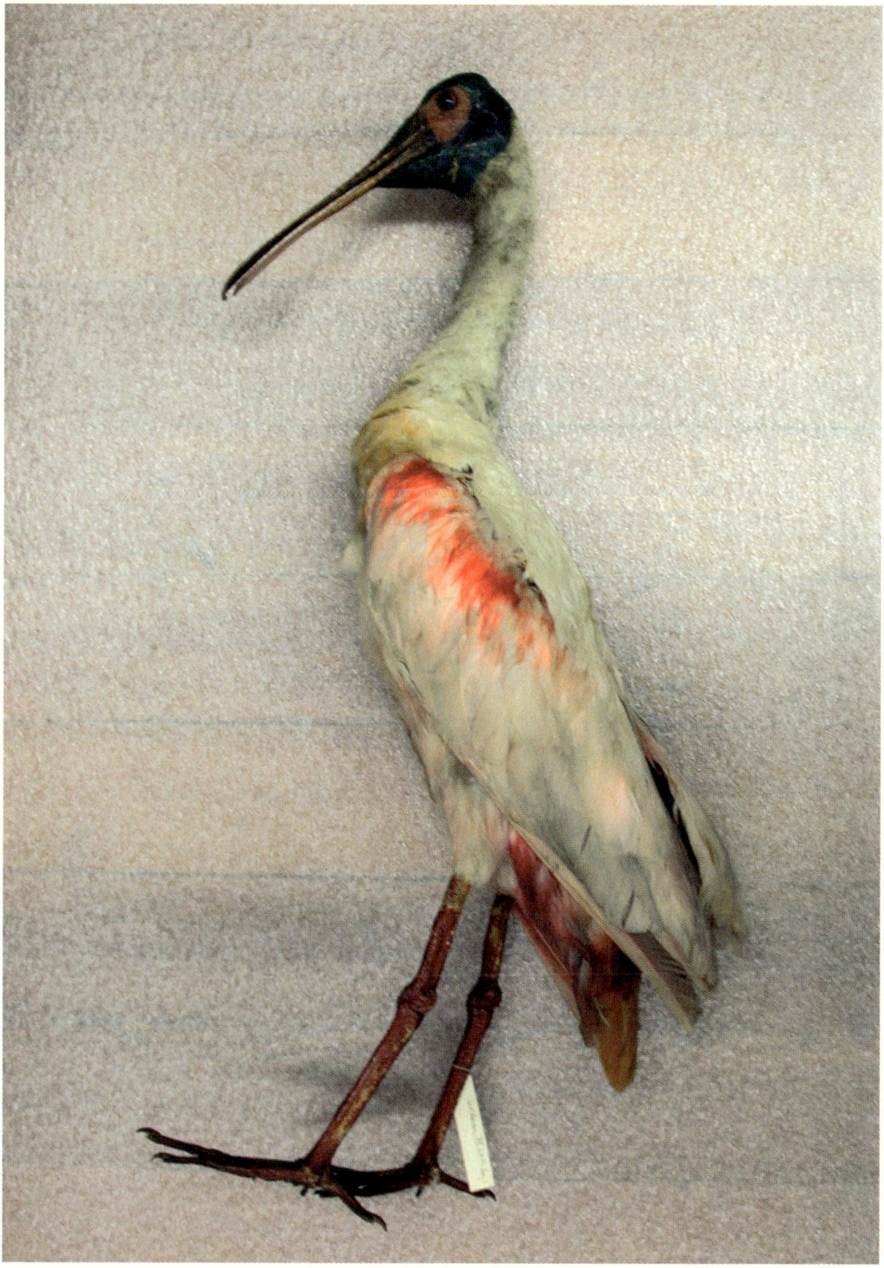

Figure 3.71. Photograph of the specimen of the Roseate Spoonbill from which the bird in Figure 3.70 was drawn. (From the Museum of Comparative Zoology, Harvard University.)

to the Boston Museum. In 1893 it passed to the Boston Society of Natural History and was transferred in 1914 to the Harvard Museum of Comparative Zoology, where it now rests. The colors, markings, and posture are exactly those of the Roseate Spoonbill in this sketch and in Wilson's plate 63.

Roseate Spoon-Bill (Roseate Spoonbill)

This stately and elegant bird inhabits the sea shores of America, from Brazil to Georgia . . . There are few facts on record relative to this very singular bird. It is said that the young are of a blackish chestnut the first year; of the roseate color of the present the second year; and of a deep scarlet the third.

Having never been so fortunate as to meet with them in their native wilds, I regret my present inability to throw any further light on their history and manners. These, it is probable, may resemble in many respects those of the European species, the *White Spoon bill,* once so common in Holland.[132]

Current name: Whooping Crane, *Grus americana* (Linnaeus)
Wilson's name: Hooping Crane, *Ardea americana*
Original name: *Ardea americana* Linnaeus 1758

Current name: Tricolored Heron, *Egretta tricolor* (Müller)
Wilson's name: Louisiana Heron, *Ardea tricolor*
Original name: *Ardea tricolor* Müller 1776

Current name: American Oystercatcher, *Haematopus palliatus* Temminck
Wilson's name: Pied Oyster-catcher, *Haematopus palliatus*
Original name: *Haematopus palliatus* Temminck 1820

Current name: Long-billed Curlew, *Numenius americanus* Bechstein
Wilson's name: Long-billed Curlew, *Numenius americanus*
Original name: *Numenius americanus* Bechstein 1812

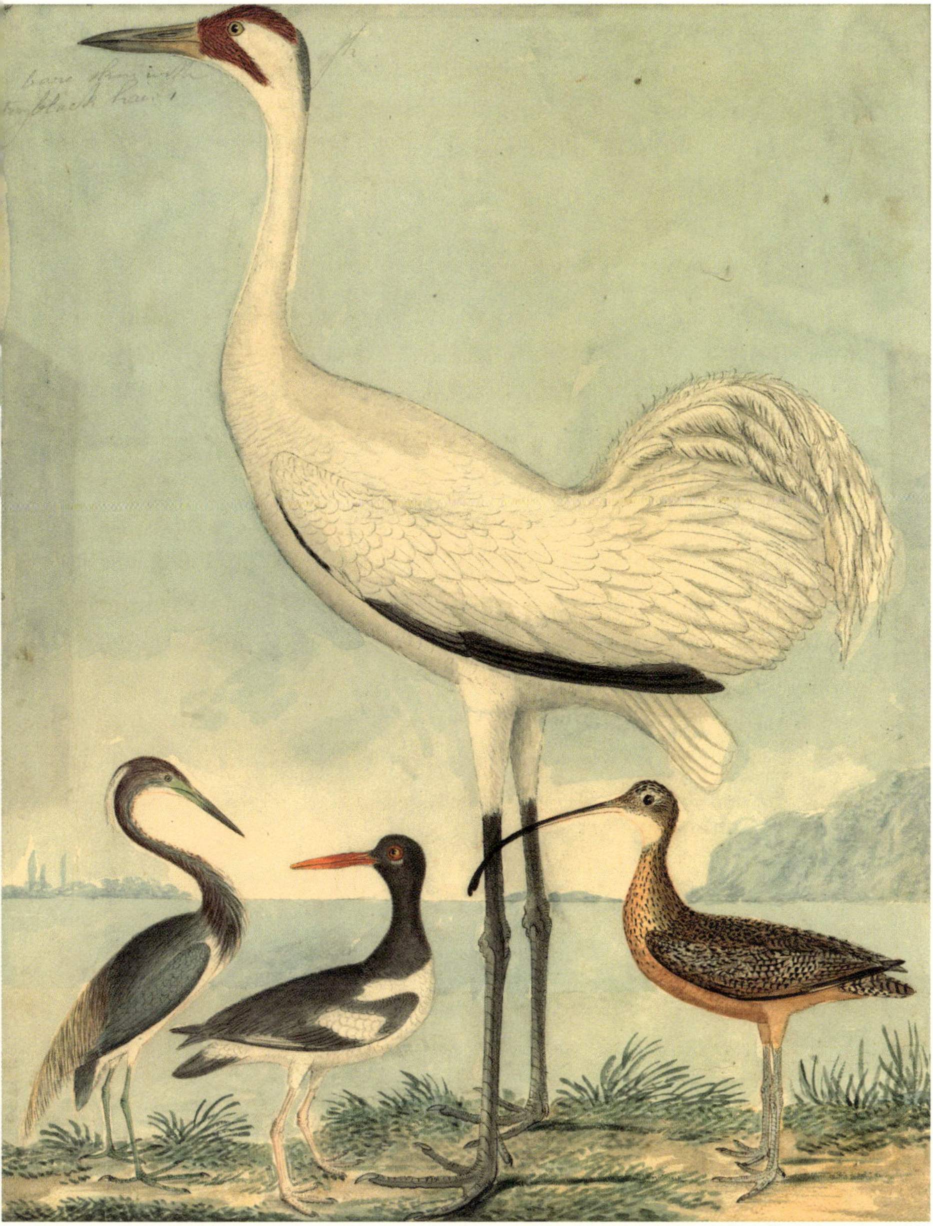

Figure 3.72. Whooping Crane (largest bird); Tricolored Heron (*bottom left*); American Oystercatcher (*bottom middle*); Long-billed Curlew (*bottom right*). (From the archives of the Ernst Mayr Library, Museum of Comparative Zoology, Harvard University.)

Commentary

The sky is washed blue thereby providing the engraver and colorists with a complete conception of the plate. No longer is Wilson supplying only sketches and notes and leaving the composition of the plate to Lawson, his engraver. And no longer are the birds in *American Ornithology* presented on dead sticks or stumps; they are shown in their native habitats. The dominant bird, the Whooping Crane, is figured in full upright stance. Note that the scale of the smaller birds is not in proportion to the crane, nor to the other smaller specimens. Furthermore, the reversed head of the Oystercatcher is a throwback to the earlier style of Edwards, Bartram, and Wilson himself.

Hooping Crane (Whooping Crane)

> This is the tallest and most stately species of all the feathered tribes of the United States; the watchful inhabitant of extensive salt marshes, desolate swamps, and open morasses, in the neighborhood of the sea. Its migrations are regular, and of the most extensive kind, reaching from the shores and inundated tracts of South America to the arctic circle. In these immense periodical journeys, they pass at such a prodigious height in the air as to be seldom observed. They have, however, their resting stages on the route to and from their usual breeding places, the regions of the north.[133]

Louisiana Heron (Tricolored Heron)

> This is a rare and delicately formed species; occasionally found on the swampy river shores of South Carolina, but more frequently along the borders of the Mississippi, particularly below New Orleans. In each of these places it is migratory; and in the latter, as I have been informed, builds its nest on trees, amidst the inundated woods. Its manners correspond very much with those of the [Little] Blue Heron. It is quick in all its motions, darting about after its prey with surprising agility. Small fish, frogs, lizards, tadpoles, and various aquatic insects, constitute its principal food.[134]

Pied Oyster-catcher (American Oystercatcher)

This singular species, although nowhere numerous, inhabits almost every sea shore, both on the new and old continent, but is never found inland. It is the only one of its genus hitherto discovered, and from the conformation of some of its parts one might almost be led by fancy to suppose, that it had borrowed the eye of the Pheasant, the legs and feet of the Bustard, and bill of the Woodpecker . . . The Oyster-Catcher frequents the sandy seabeach of New Jersey, and other parts of our Atlantic coast in summer, in small parties of two or three pairs together. They are extremely shy, and, except about the season of breeding, will seldom permit a person to approach within gun shot. They walk along the shore in a watchful, stately manner, at times probing it with their long, wedge-like bills, in search of small shell fish. This appears evident on examining the hard sands where they usually resort, which are found thickly perforated with oblong holes two or three inches in depth. The small crabs called *fiddlers,* that burrow in the mud at the bottom of inlets, are frequently the prey of the Oyster-Catcher; as are muscles, spout fish, and a variety of other shell-fish and sea insects with which those shores abound.[135]

Long-billed Curlew

This American species has been considered by the naturalists of Europe to be a mere *variety* of their own, notwithstanding its difference of color, and superior length of bill. These differences not being accidental, or found in a few individuals, but common to all, and none being found in America corresponding with that of Europe, we do not hesitate to consider the present as a distinct species, peculiar to this country . . . Like the preceding [Whooping Crane], this bird is an inhabitant of marshes in the vicinity of the sea. It is also found in the interior, where, from its long bill, and loud, whistling note, it is generally known.

The Curlews appear in the salt marshes of New Jersey about the middle of May, on their way to the north; and in September, on their return from their breeding places. Their food consists chiefly of small crabs, which they are very dexterous at probing for, and pulling out of the holes with their long bills; they also feed on those small sea snails so abundant in the marshes, and on various worms and insects. They are likewise fond of bramble berries, frequenting the fields and uplands in search of this fruit, on which they get very fat.[135]

Current name: Great Blue Heron, *Ardea herodias* Linnaeus
Wilson's name: Great Heron, *Ardea herodias*
Original name: *Ardea Herodias* Linnaeus 1758

Current name: Yellow-crowned Night-Heron, *Nyctanassa violacea* (Linnaeus)
Wilson's name: Yellow-crowned Heron, *Ardea violacea*
Original name: *Ardea violacea* Linnaeus 1758

Current name: American Bittern, *Botaurus lentiginosus* (Rackett)
Wilson's name: American Bittern, *Ardea minor*
Original name: *Ardea lentiginosa* Rackett 1813

Current name: Least Bittern, *Ixobrychus exilis* (Gmelin)
Wilson's name: Least Bittern, *Ardea exilis*
Original name: *Ardea exilis* Gmelin 1789

Commentary

Here, as with the Whooping Crane in the previous sketch, a long-necked species is illustrated in full upright stance. The background reverts to a blank sky, but the sketch retains the rather stylized vegetation. The two bitterns and the night-heron are in awkward, unrealistic poses, suggesting that Wilson may have been unfamiliar with their natural stances in the wild. The relative sizes of the four species are, however, approximately correct.

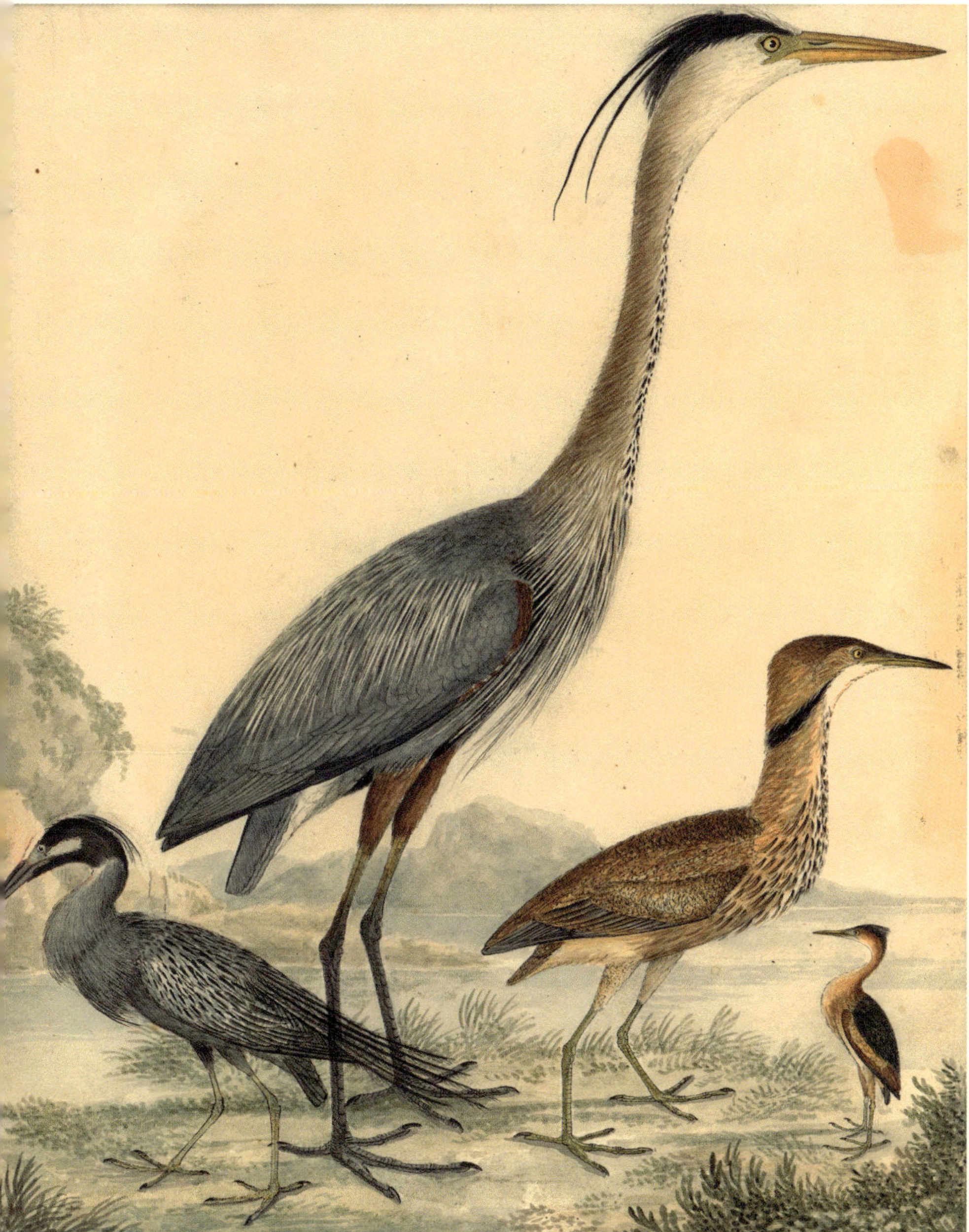

Figure 3.73. Great Blue Heron (*largest bird*); Yellow-crowned Night-Heron (*lower left*); American Bittern (*middle*); Least Bittern (*lower right*). (From the archives of the Ernst Mayr Library, Museum of Comparative Zoology, Harvard University.)

Great Heron (Great Blue Heron)

The principal food of the Great Heron is fish, for which he watches with the most unwearied patience, and seizes them with surprising dexterity. At the edge of the river, pond or sea shore he stands fixed and motionless, sometimes for hours together. But his stroke is quick as thought, and sure as fate to the first luckless fish that approaches within his reach; these he sometimes beats to death, and always swallows head foremost, such being their uniform position in the stomach. He is also an excellent mouser, and of great service to our meadows in destroying the short-tailed or meadow mouse, so injurious to the banks. He also feeds eagerly on grasshoppers, various winged insects, particularly dragon flies, which he is very expert at striking, and also eats the seeds of that species of nymphæ usually called spatter docks, so abundant along our fresh water ponds and rivers.[137]

Yellow-crowned Heron (Yellow-crowned Night-Heron)

This bird inhabits the lower parts of South Carolina, Georgia, and Louisiana, in the summer season; reposing during the day among low, swampy woods, and feeding only in the night. It builds in societies, making its nest with sticks among the branches of low trees, and lays four pale blue eggs. This species is not numerous in Carolina, which, with its solitary mode of life, makes this bird but little known there . . .

The food of this species consists of small fish, crabs, and lizards, particularly the former; it also appears to have a strong attachment to the neighborhood of the ocean.[138]

American Bittern

This is another nocturnal species, common to all our sea and river marshes, tho no-where numerous; it rests all day among the reeds and rushes, and unless disturbed, flies and feeds only during the

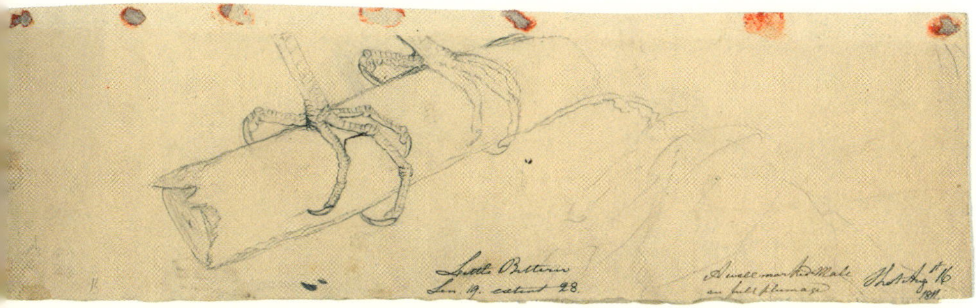

Figure 3.74. Wilson's detailed drawing of the feet of a Least Bittern. (From the archives of the Ernst Mayr Library, Museum of Comparative Zoology, Harvard University.)

> night. In some places it is called the Indian Hen[;] on the sea coast of New Jersey it is known by the name of *Dunkadoo,* a word probably imitative of its common note.[139]

Least Bittern

> This is the smallest known species of the whole tribe. It is commonly found in fresh-water meadows, and rarely visits the salt marshes . . . In the meadows of Schuykill and Delaware below Philadelphia, a few of these birds breed every year; making their nests in the thick tussocks of grass, in swampy places. When alarmed they seldom fly far, but take shelter among the reeds or long grass. They are scarcely ever seen exposed, but skulk during the day; and, like the preceding species, feed chiefly in the night.[140]

Commentary

The detailed drawing of the feet of a Least Bittern appears to have been pasted to a drawing of the rest of the bird, as indicated by the glue spots evident across the top of the drawing. Neither this drawing nor the drawing of the missing body was used in the final plate. That drawing is quite small.

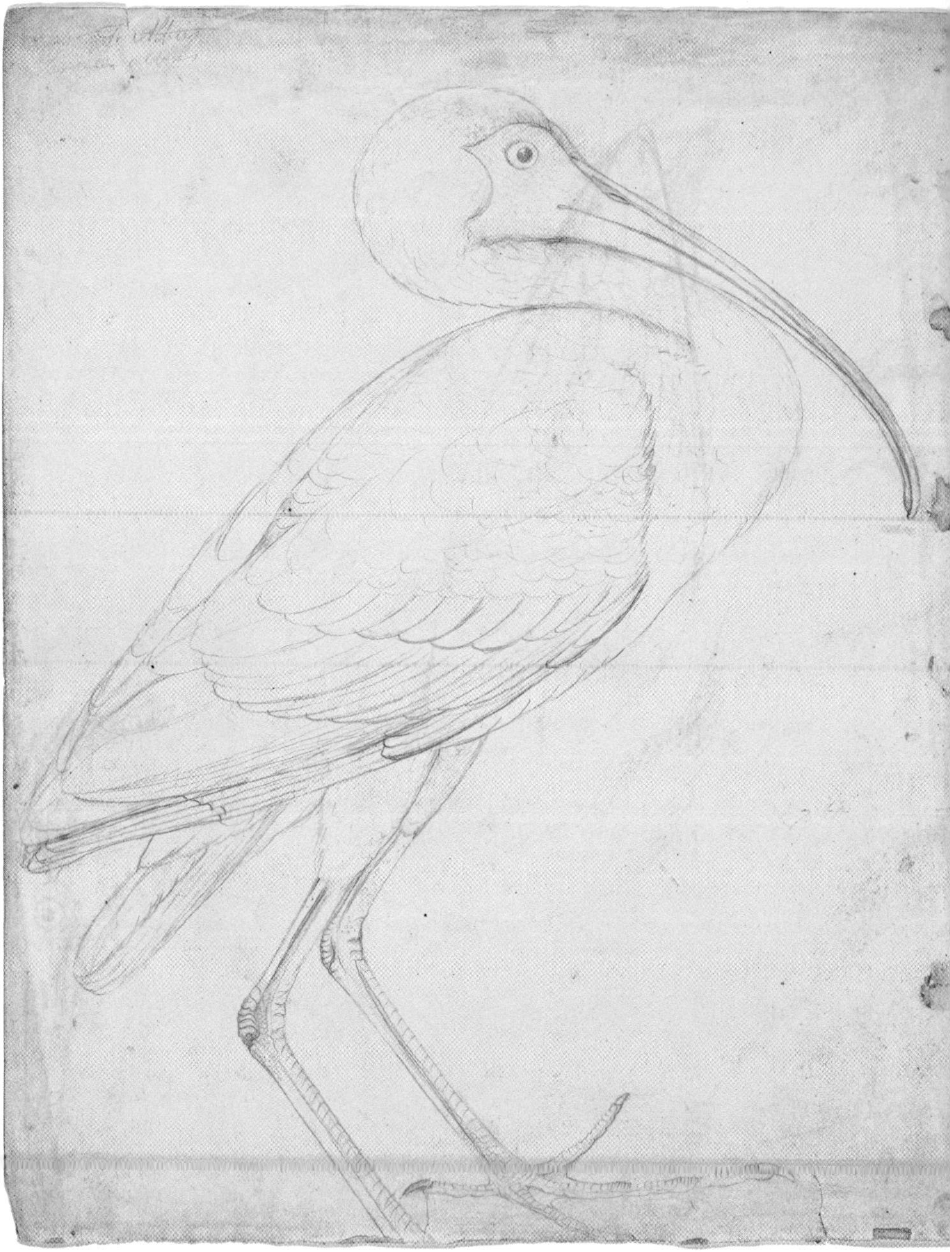

Figure 3.75. Pencil sketch of a White Ibis. (From the archives of the Ernst Mayr Library, Museum of Comparative Zoology, Harvard University.)

Current name: White Ibis, *Eudocimus albus* (Linnaeus)
Wilson's name: White Ibis, *Scolopax alba*
Original name: *Scolopax alba* Linnaeus 1758

Commentary

This graceful drawing of a White Ibis was on the back of the same sheet of paper as the drawing of the Green Heron (Figure 3.66). Paper was expensive and Wilson had limited means. Note that Wilson correctly shows how the coverts overlap the separation between adjacent underlying feathers, much like shingles on a roof overlap the gaps between underlying shingles. Previous illustrators had indicated incorrectly that the shaft of each feather was aligned with the shaft of the underlying feather, which, if true, would have allowed water to penetrate easily.

White Ibis

> This species bears in every respect except that of color, so strong a resemblance to the preceding [Scarlet Ibis], that I have been almost induced to believe it the same, in its white or imperfect stage of color. The length and form of the bill, the size, conformation, as well as color of the legs, the general length and breadth, and even the steel blue on the four outer quill feathers, are exactly alike in both. These suggestions, however, are not made with any certainty of its being the same; but as circumstances which may lead to a more precise examination of the subject hereafter.[141]

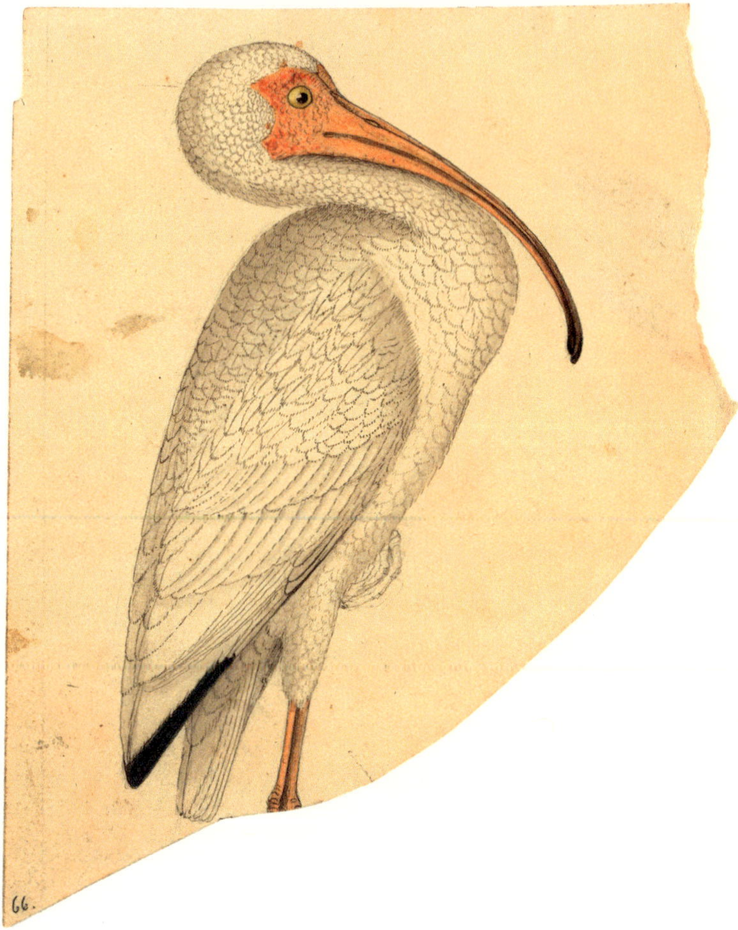

Figure 3.76. A small drawing of a White Ibis by Alexander Wilson, fully colored except for the foot tucked into the lower belly feathers. (From the archives of the Ernst Mayr Library, Museum of Comparative Zoology, Harvard University.)

Commentary

The small, fully rendered White Ibis drawing is similar to the life-sized sketch illustrated in Figure 3.75. This smaller drawing of the ibis has one foot withdrawn into its belly feathers and is the stance depicted in plate 66 of *American Ornithology*. Wilson noticed the feather wear and had the knowledge and insight to relate the wear to the roosting posture of the bird.

Current name: Common Goldeneye, *Bucephala clangula* (Linnaeus)
Wilson's name: Golden-Eye, *Anas clangula*
Original name: *Anas clangula* Linnaeus 1758

Current name: Northern Shoveler, *Anas clypeata* Linnaeus
Wilson's name: Shoveller, *Anas clypeata*
Original name: *Anas clypeata* Linnaeus 1758

Current name: Surf Scoter, *Melanitta perspicillata* (Linnaeus)
Wilson's name: Black or Surf Duck, *Anas perspicillata*
Original name: *Anas perspicillata* Linnaeus 1758

Current name: Bufflehead, *Bucephala albeola* (Linnaeus)
Wilson's name: Buffel-headed Duck, *Anas Albeola*
Original name: *Anas Albeola* Linnaeus 1758

Current name: Ring-necked Duck, *Aythya collaris* (Donovan)
Wilson's name: Tufted Duck, *Anas fuligula*
Original name: *Anas collaris* Donovan 1809

Current name: Canada Goose, *Branta canadensis* (Linnaeus)
Wilson's name: Canada Goose, *Anas canadensis*
Original name: *Anas canadensis* Linnaeus 1758

Commentary

The legs and feet of all the pictured birds, except the Ring-necked Duck, are visible and in use by the swimming birds. Perhaps in an effort to illustrate the legs and feet, Wilson has the ducks riding too high in the water. The open bill of the Northern Shoveler emphasizes its unusual structure, and its flight position enables Wilson to show its distinctive wing pattern. All birds are either flying or swimming, making for a lively composition. The background is complete except for the clouds, which were added by the engraver.

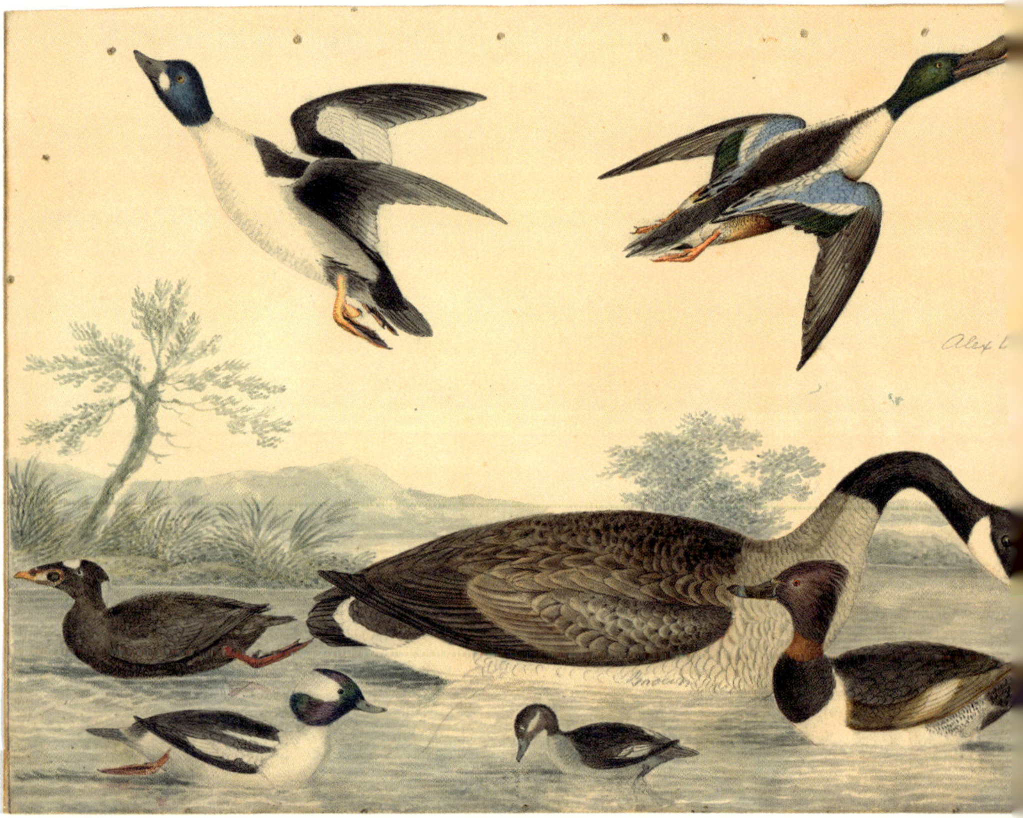

Figure 3.77. Common Goldeneye (*upper left*); Northern Shoveler (*upper right*); Surf Scoter (*middle left*); Bufflehead (male, *lower left;* female, *lower middle*); Ring-necked Duck (*lower right*); Canada Goose (*center to right*). (From the archives of the Ernst Mayr Library, Museum of Comparative Zoology, Harvard University.)

Golden-Eye (Common Goldeneye)

This Duck is well known in Europe, and in various regions of the United States, both along the sea coast and about the lakes and rivers of the interior. It associates in small parties, and may easily be known by the vigorous whistling of its wings as it passes through the air. It swims and dives well; but seldom walks on shore, and then in a waddling awkward manner. Feeding chiefly on shell fish, small fry, &c. their flesh is less esteemed than that of the preceding [Ring-necked Duck]. In the United States they are only winter visitors, leaving us again in the month of April, being then on their passage to the north to breed. They are said to build, like the Wood Duck, in hollow trees.[142]

Shoveller (Northern Shoveler)

If we except the singularly formed and disproportionate size of the bill, there are few Ducks more beautiful, or more elegantly marked than this . . . It occasionally visits the sea-coast; but is more commonly found on our lakes and rivers, particularly along their muddy shores, where it spends the great part of its time in searching for small worms, and the larvæ of insects, sifting the watery mud through the long and finely set teeth of its curious bill, which is admirably constructed for the purpose; being large, to receive a considerable quantity of matter, each mandible bordered with close-set, pectinated rows, exactly resembling those of a weaver's reed, which fitting into each other, form a kind of sieve, capable of retaining very minute worms, seeds, or insects, which constitute the principal food of the bird.[143]

Black or Surf Duck (Surf Scoter)

This Duck is peculiar to America, and altogether confined to the shores and bays of the sea, particularly where the waves roll over the sandy beach. Their food consists principally of those small bivalve shell fish already described, spout fish, and others that lie in the sand near its surface. For these they dive almost constantly, both in the sandy bays and amidst the tumbling surf. They seldom or never visit the salt marshes. They continue on our shores during the winter; and leave us early in May, for their breeding places in the north. Their skins are remarkably strong, and their flesh coarse, tasting of fish.[144]

Buffel-headed Duck (Bufflehead)

This pretty little species, usually known by the name of the *Butter-Box,* or *Butter-Ball,* is common to the sea shores, rivers and lakes of the United States, in every quarter of the country, during autumn and winter. About the middle of April, or early in May, they retire to the north to breed. They are dexterous divers, and fly with extraordinary velocity. So early as the latter part of February the males are observed to have violent disputes for the females. At this time they are more commonly seen in flocks, but, during the preceding part of winter, they usually fly in pairs. Their note is a short *quak* . . .

The *Buffel-headed* Duck, or rather as it has originally been, the *Buffaloe-headed* Duck, from the disproportionate size of its head, is fourteen inches long.[145]

Tufted Duck (Ring-necked Duck)

This is an inhabitant of both continents; it frequents fresh-water rivers, and seldom visits the sea shore. It is a plump, short-bodied Duck; its flesh generally tender and well tasted. They are much rarer than most of our other species, and are seldom seen in market. They are most common about the beginning of winter, and early in the spring. Being birds of passage they leave us entirely during the summer.[146]

Canada Goose

I have never yet visited any quarter of the country where the inhabitants are not familiarly acquainted with the regular passing and repassing of the Wild Geese . . .

The flight of the Wild Geese is heavy and laborious, generally in a straight line, or in two lines approximating to a point, thus >; in both cases the van is led by an old gander, who every now and then pipes his well known *honk,* as if to ask how they come on, and then honk of "All's well" is generally returned by some of the party. Their course

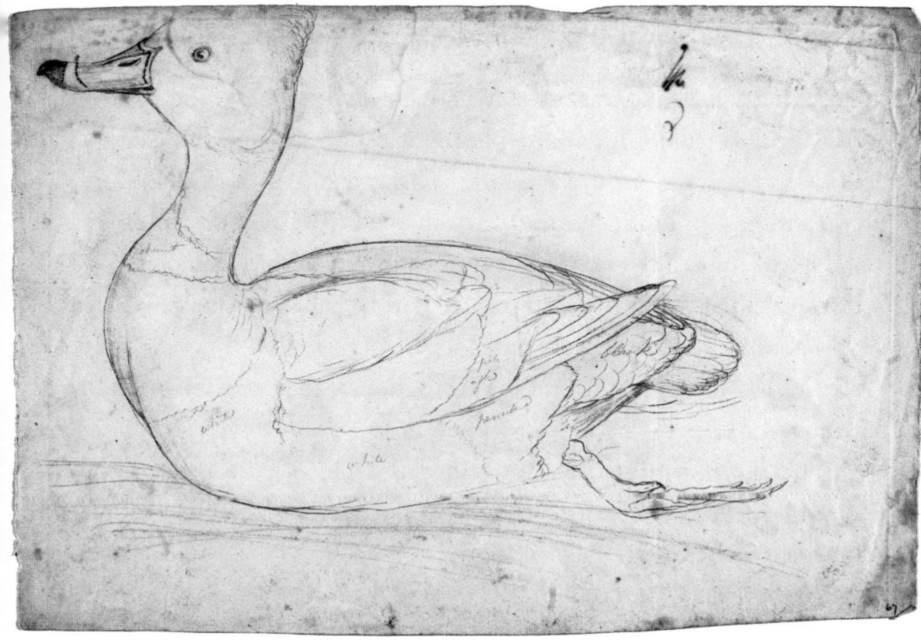

Figure 3.78. Pencil drawing of a Ring-necked Duck by Alexander Wilson, with notes on colors. (From the archives of the Ernst Mayr Library, Museum of Comparative Zoology, Harvard University.)

> is in a straight line, with the exception of the undulations of their flight. When bewildered in foggy weather, they appear sometimes to be in great distress, flying about in an irregular manner, and for a considerable time over the same quarter, making a great clamor.[147]

Commentary

Drawn in pencil with one foot paddling, this lively Ring-necked duck has a more accurate head shape and eye placement than in the preceding figure or plate 67 of *American Ornithology*. Color instructions written lightly in pencil can be seen in several, but not all, areas of the drawing, for example, "chestnut" on the lower neck and "white" on the belly and lower breast.

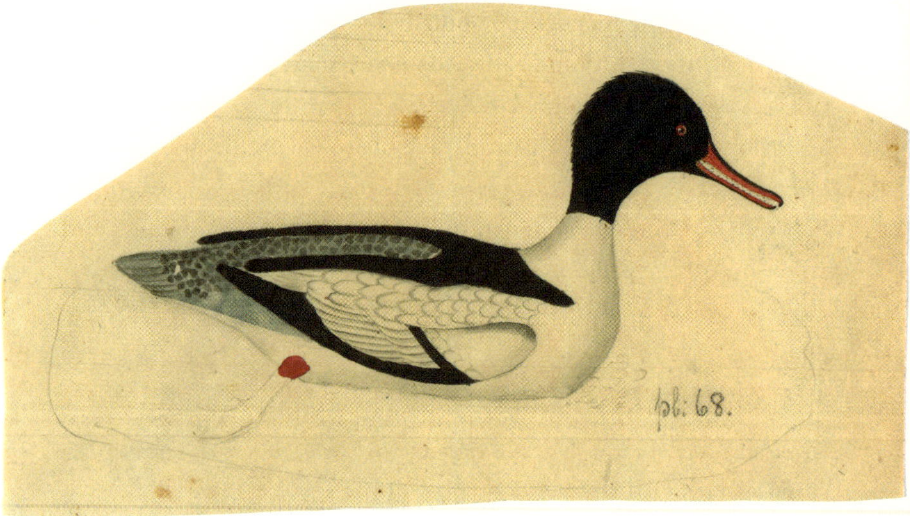

Figure 3.79. Drawing of a Common Merganser. (From the archives of the Ernst Mayr Library, Museum of Comparative Zoology, Harvard University.)

Current name: Common Merganser, *Mergus merganser* Linnaeus
Wilson's name: Goosander, *Mergus merganser*
Original name: *Mergus merganser* Linnaeus 1758

Commentary

This beautifully proportioned and colored drawing is very similar to the final image in plate 68 of *American Ornithology*.

Goosander (Common Merganser)

The young males, which are generally much more numberous than the old ones, so exactly resemble the females in their plumage for at least the first, and part of the second year, as scarcely to be distinguished from them; and, what is somewhat singular, the crests of these and of the females are actually longer than those of the full grown male, though thinner towards its extremities. These circumstances have induced some late Ornithologists to consider them as two different species, the young, or female, having been called the *Dun Diver.* By this arrangement they have entirely deprived the Goosander of his female; for in the whole of my examinations and dissections of the present species, I have never yet found the female in *his* dress. [148]

Current name: Labrador Duck, *Camptorhynchus labradorius* (Gmelin)
Wilson's name: Pied Duck, *Anas labradora*
Original name: *Anas labradoria* Gmelin 1789

Current name: Snow Goose, *Chen caerulescens* (Linnaeus)
Wilson's name: Snow Goose, *Anas hyperborea*
Original name: *Anas caerulescens* Linnaeus 1758

Current name: Hooded Merganser, *Lophodytes cucullatus* (Linnaeus)
Wilson's name: Hooded Merganser, *Mergus cucullatus*
Original name: *Mergus cucullatus* Linnaeus 1758

Current name: Red-breasted Merganser, *Mergus serrator* Linnaeus
Wilson's name: Red-breasted Merganser, *Mergus serrator*
Original name: *Mergus serrator* Linnaeus 1758

Current name: American Wigeon, *Anas americana* Gmelin
Wilson's name: American Wigeon, *Anas americana*
Original name: *Anas americana* Gmelin 1789

Current name: Greater Scaup, *Aythya marila* (Linnaeus)
Wilson's name: Scaup Duck, *Anas Marila*
Original name: *Anas Marila* Linnaeus 1761

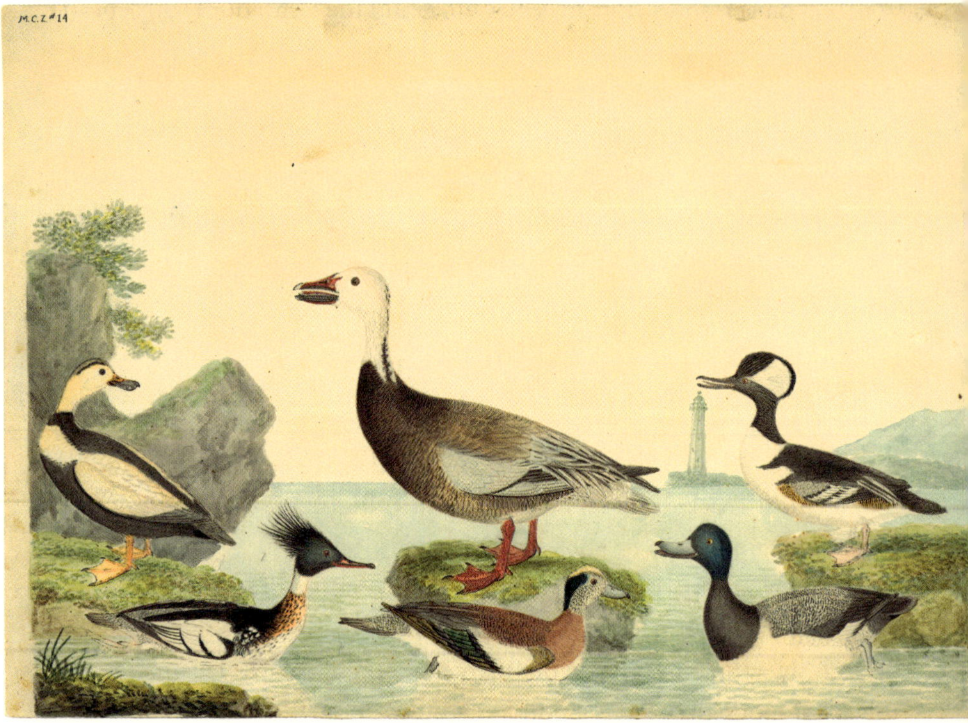

Figure 3.80. Labrador Duck (*upper left*); Snow Goose (dark adult; *upper middle*); Hooded Merganser (*upper right*); Red-breasted Merganser (*lower left*); American Wigeon (*lower middle*); Greater Scaup (*lower right*). (From the archives of the Ernst Mayr Library, Museum of Comparative Zoology, Harvard University.)

Commentary

Here Wilson continues the pattern of providing finished paintings with full backgrounds, except for the sky in this case. The rendering is mostly in pencil with an overlay of watercolors. Ink is used sparingly. Wilson is reasonably successful in conveying the iridescence of the head of the Greater Scaup through the use of a transparent green that allows some of the black to show through. He is similarly successful with his portrayal of the head of the Red-breasted Merganser. The dynamism of this illustration confirms Wilson's maturation as an artist. Three of the six birds have open bills; the wigeon, merganser, and scaup are swimming; and the Labrador Duck's head is turned and seems to be reaching out to the right, giving a hint of movement. (Compare the apparent motion in the duck's posture to the static head reversal in the American Oystercatcher portrayed in Figure 3.72.) Despite the sense of action, all are flat side views, very much like those in modern field guides, reminding us once again of Wilson's goal to make it possible for us to identify our native species.

Pied Duck (Labrador Duck)

This is rather a scarce species on our coasts, and is never met with on fresh-water lakes or rivers.[149] It is called by some gunners the Sand Shoal Duck, from its habit of frequenting sand bars. Its principal food appears to be shell fish, which it procures by diving. The flesh is dry, and partakes considerably of the nature of its food. It is only seen here during winter; most commonly early in the month of March, a few are observed in our market. Of their principal manners, place, or mode of breeding, nothing more is known.[150]

Snow Goose

These birds pass along our coasts, and settle in our rivers, every autumn; among thirty or forty, there are seldom more than six or eight pure white, or old birds. The rest vary so much, that no two are exactly alike; yet all bear the most evident marks in the particular structure of their bills, &c. of being the same identical species. A gradual change so great, as from a bird of this color to one of pure white, must necessarily produce a number of varieties, or differences in the appearance of the plumage, but the form of the bill and legs remains the same, and any peculiarity in either is the surest means we have to detect a species under all its various appearances. It is therefore to be regretted, that the authors above referred to in the synonyms, have paid so little attention to the singular conformation of the bill; for even in their description of the Snow Goose, neither that nor the internal peculiarities, are at all mentioned.[151]

Hooded Merganser

This species on the sea coast is usually called the *Hairy Head.* They are more common however along our lakes and fresh water rivers than near the sea; tracing up creeks, and visiting mill ponds, diving perpetually for their food. In the creeks and rivers of the southern states they are very frequently seen during the winter. Like the *Red-breasted,* they are migratory, the manners, food, and places of resort of both being very much alike.[152]

Red-breasted Merganser

The windpipe of the male of this species is very curious, and differs something from that of the Goosander. About two inches from the mouth, it swells out to four times its common diameter, continuing of that size for about an inch and a half. This swelling is capable of being shortened or extended; it then continues of its first diameter for two inches or more, when it becomes flattish, and almost transparent for [another] two inches; it then swells into a bony labyrinth of more than two inches in length by one and a half in width, over the hollow sides of which is spread a yellowish skin like parchment. The left side of this, fronting the back of the bird, is a hard bone. The divarications come out very regularly from this at the lower end, and enter the lungs.

The intention of Nature in this extraordinary structure is probably to enable the bird to take down a supply of air to support respiration while diving; yet why should the female, who takes the same submarine excursions as the male, be entirely destitute of this apparatus?[153]

American Wigeon

This is a handsomely-marked and sprightly species, very common in winter along our whole coast, from Florida to Rhode Island; but most abundant in Carolina, where it frequents the rice plantations. In Martinico great flocks take short flights from one rice field to another during the rainy season, and are much complained of by the planters. The Widgeon is the constant attendant of the celebrated *Canvass Back Duck,* so abundant in various parts of the Chesapeake Bay, by the aid of whose labor he has ingenuity enough to contrive to make a good subsistence. The Widgeon is extremely fond of the tender roots of that particular species of aquatic plant on which the Canvass back feeds, and for which that duck is in the constant habit of diving. The Widgeon, who never dives, watches the moment of the Canvass back's rising, and before he has his eyes well opened, snatches the delicious morsel from his mouth and makes off. On this account the Canvas backs and Wigeons . . . live in a state of perpetual contention.[154]

Scaup Duck (Greater Scaup)

This Duck is better known among us by the name *Blue Bill.* It is an excellent diver, and, according to Willughby feeds on a certain small kind of shell fish called *Scaup,* whence it has derived its name. It is common both to our fresh water rivers and sea-shores in winter. Those that frequent the latter are generally much the fattest, on account of the greater abundance of food along the coast. It is sometimes abundant in the Delaware, particularly in those places where small snails, its favorite shell fish, abound, feeding also, like most of its tribe, by moonlight.[155]

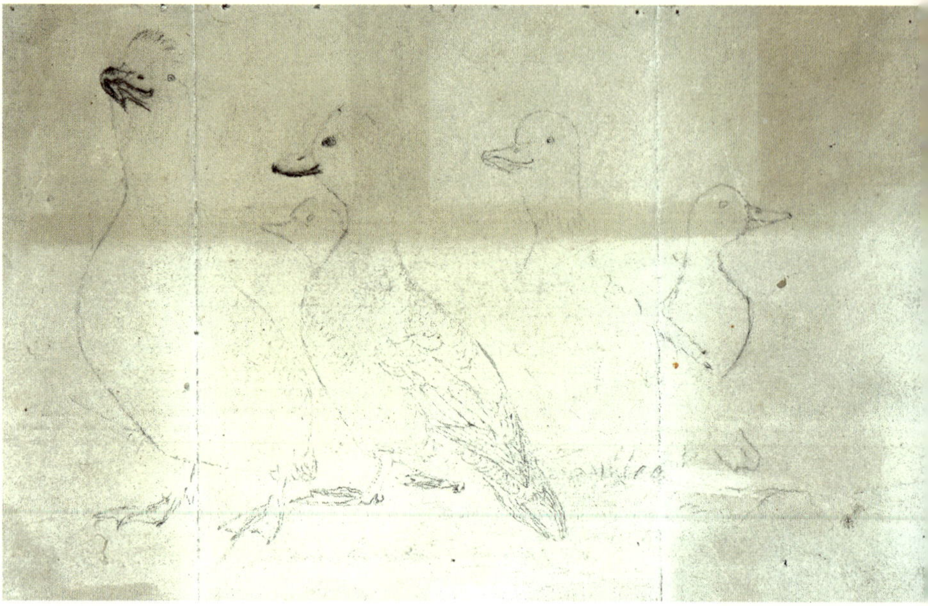

Figure 3.81. Probably a preliminary series of sketches by Alexander Wilson for the now-extinct Labrador Duck. (From the archives of the Paisley Museum, Renfrewshire Council, Scotland.)

Commentary

The bill in the drawings resembles that of the Labrador Duck, and suggests that this may be a preliminary series of sketches for this now-extinct species that was uncommon even in Wilson's time. In plate 69 of *American Ornithology,* the Labrador Duck is in a very different pose, with its head turned to look over its back.

Current name: Long-tailed Duck, *Clangula hyemalis* (Linnaeus)
Wilson's name: Long tailed Duck, *Anas hyemalis*
Original name: *Anas hyemalis* Linnaeus 1758

Current name: Wood Duck, *Aix sponsa* (Linnaeus)
Wilson's name: Summer Duck or Wood Duck, *Anas Sponsa*
Original name: *Anas Sponsa* Linnaeus 1758

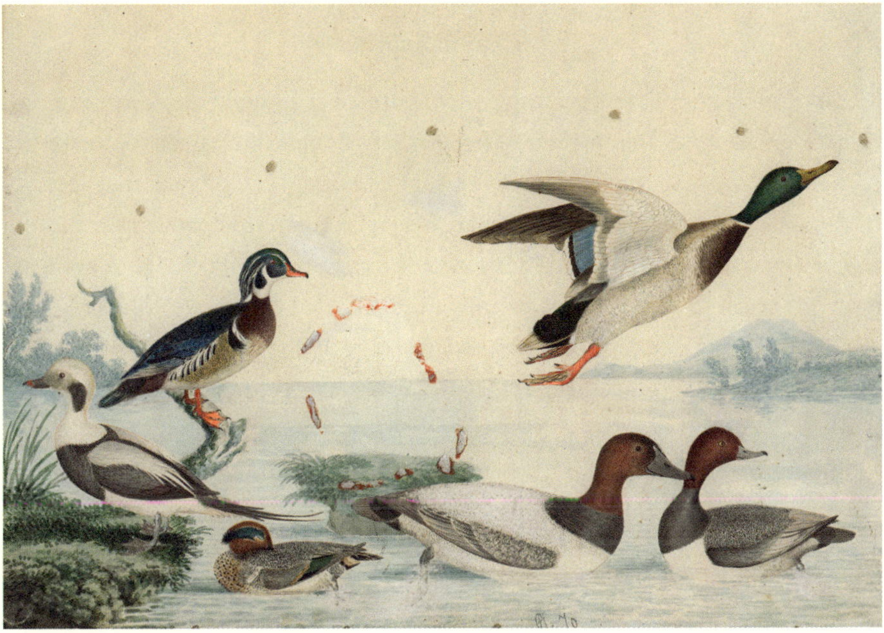

Figure 3.82. Long-tailed Duck (male, *lower far left*); Wood Duck (*upper left*); Mallard (*upper right*); Green-winged Teal (*lower left*); Canvasback (*lower right*); Redhead (*lower far right*). (From the archives of the Ernst Mayr Library, Museum of Comparative Zoology, Harvard University.)

Current name: Mallard, *Anas platyrhynchos* Linnaeus
Wilson's name: Mallard, *Anas platyrhynchos*
Original name: *Anas platyrhynchos* Linnaeus 1758

Current name: Green-winged Teal, *Anas crecca* Linnaeus
Wilson's name: Green-winged Teal, *Anas Crecca*
Original name: *Anas Crecca* Linnaeus 1758

Current name: Canvasback, *Aythya valisineria* (Wilson)
Wilson's name: Canvass-Back Duck, *Anas valisineria*
Original name: *Anas valisineria* Wilson 1814

Current name: Redhead, *Aythya americana* (Eyton)
Wilson's name: Red-headed Duck, *Anas ferina*
Original name: *Fuligula americana* Eyton 1838

Commentary

The composition of this painting incorporates a number of interesting and dynamic features. The Green-winged Teal is pictured asleep with its bill tucked into the feathers of its back. The female Long-tailed Duck that appears in the *American Ornithology* illustration is missing in this painting, although the dried paste and torn paper visible to the right of the Wood Duck indicate that her image had been pasted into the center of the group. No sketch of her exists today, but in *American Ornithology* she is pictured preening her back feathers. Sleeping and preening are characteristic behavioral patterns of wild ducks and their depiction reflects Wilson's pathbreaking commitment to illustrating the natural history of his subjects. In addition, the heads of the Redhead and Canvasback are pictured side by side and facing in the same direction, so as to facilitate comparison—a hint of a style used by modern field guides.

Long-tailed Duck

> The windpipe was very curiously formed; besides the labyrinth, which is nearly as large as the end of the thumb, it has an expansion immediately above that, of double its usual diameter, which continues for an inch and a half; this is flattened on the side next the breast, with an oblong, window-like vacancy in it, crossed with five narrow bars, and covered with a thin transparent skin, like the panes of a window; another thin skin of the same kind is spread over the external side of the labyrinth, which is partly of a circular form. This singular conformation is, as usual, peculiar to the male . . . Some writers suppose the singular voice, or call, of this species, to be occasioned by the remarkable construction of its windpipe; but the fact, that the females are uniformly the most noisy, and yet are entirely destitute of the singularities of this conformation, overthrows the probability of this supposition.[156]

Summer Duck or Wood Duck (Wood Duck)

This most beautiful of all our Ducks, has probably no superior among its whole tribe for richness and variety of colors. It is called the *Wood Duck,* from the circumstance of its breeding in hollow trees; and the *Summer Duck,* from remaining with us chiefly during the summer . . .

This tree had been occupied, probably by the same pair, for four successive years, in breeding time; the person who gave me the information, and whose house was within twenty or thirty yards of the tree, said that he had seen the female, the spring preceding, carry down thirteen young, one by one, in less than ten minutes. She caught them in her bill by the wing or back of the neck, and landed them safely at the foot of the tree, whence she afterwards led them to the water.[157]

Mallard

This is the original stock of the common domesticated duck, reclaimed, time immemorial, from a state of nature, and now become so serviceable to man. In many individuals, the general garb of the tame Drake seems to have undergone little or no alteration; but the stamp of slavery is strongly imprinted in his dull indifferent eye, and groveling gait; while the lofty look, long tapering neck, and sprightly action of the former, bespeak his native spirit and independence.[158]

Green-winged Teal

The naturalists of Europe have designated this little Duck by the name of the American Teal, as being a species different from their own. On an examination, however, of the figure and description of the European Teal by the ingenious and accurate Bewick, and comparing them with the present, no difference whatever appears in the length, extent, color, or markings of either, but what commonly occurs among individuals of any other tribe; both undoubtedly belong to one and the same species.[159]

Canvass-Back Duck (Canvasback)

They are seldom found at a great distance up any of these rivers, or even in the salt-water bay; but in that particular part of tide water where a certain grass-like plant grows, on the roots of which they feed. This plant, which is said to be a species of *valisineria,* grows on fresh water shoals of from seven to nine feet (but never where these are occasionally dry), in long, narrow, grass-like blades of four or five feet in length; the root is white, and has some resemblance to small celery. This grass is in many places so thick that a boat can with difficulty be rowed through it, it so impedes the oars. The shores are lined with large quantities of it torn up by the ducks, and drifted up by the winds, lying, like hay in windrows. Wherever this plant grows in abundance, the Canvass-backs may be expected, either to pay occasional visits or to make it their regular residence during the winter.[160]

Red-headed Duck (Redhead)

This is a common associate of the Canvass-back, frequenting the same places, and feeding on the *stems* of the same grass, the latter eating only the *roots;* its flesh is very little inferior, and is often sold in our markets for the Canvass-back, to those unacquainted with the characteristic marks of each. Anxious as I am to determine precisely whether this species be the Red-headed Widgeon, Pochard, or Dun Bird of England, I have not been able to ascertain the point to my own satisfaction; though I think it very probably the same, the size, extent, and general description of the Pochard agreeing pretty nearly with this.[161]

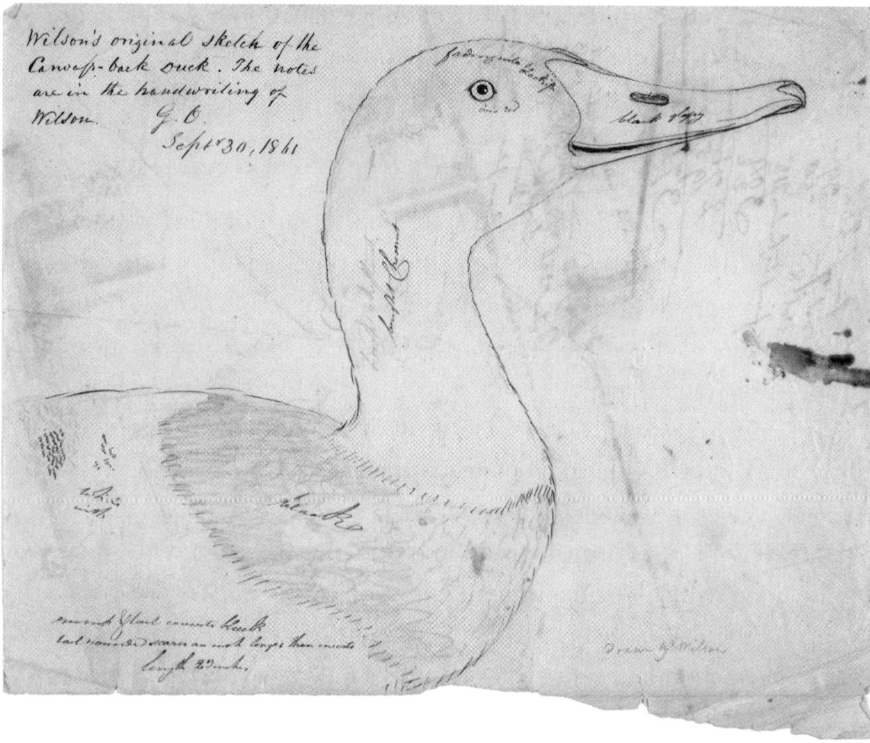

Figure 3.83. Sketch of a Canvasback, with measurements and notes on coloring. (From the archives of the Ernst Mayr Library, Museum of Comparative Zoology, Harvard University.)

Commentary

This sketch of the Canvasback is another example of Wilson's noting on the drawing the colors to be applied in the final painting. Here, too, he includes in the lower left of the drawing comments about the flank coverts and a measurement of their length. The proportions of the duck are excellent, but the lack of feather detail suggests that the sketch was done in the field shortly after the Canvasback was shot and that the notes on color were made before the soft tissues had time to change color. The note in the upper left corner is by George Ord, Wilson's companion on several trips to Cape May and Egg Harbor, New Jersey, from 1811, as indicated here, through 1813. The last trip was made just weeks before Wilson's death.

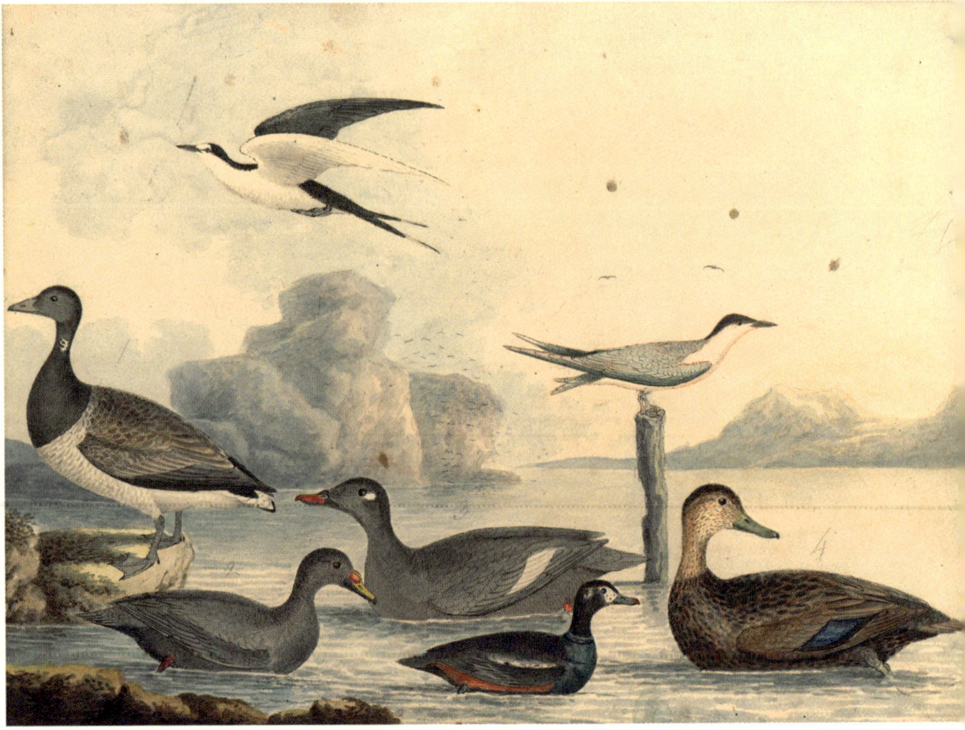

Figure 3.84. Brant (*upper left*); Sooty Tern (*flying*); Black Scoter (*lower left*); White-winged Scoter (*center*); Gull-billed Tern (*upper right*); Harlequin Duck (*lower middle*); American Black Duck (*lower right*). (From the archives of the Ernst Mayr Library, Museum of Comparative Zoology, Harvard University.)

Current name: Brant, *Branta bernicla* (Linnaeus)
Wilson's name: Brant, *Anas bernicla*
Original name: *Anas Bernicla* Linnaeus 1758

Current name: Sooty Tern, *Onychoprion fuscatus* (Linnaeus)
Wilson's name: Sooty Tern, *Sterna fuscata*
Original name: *Sterna fuscata* Linnaeus 1766

Current name: Gull-billed Tern, *Gelochelidon nilotica* (Gmelin)
Wilson's name: Marsh Tern, *Sterna nilotica*
Original name: *Sterna nilotica* Gmelin 1789

Current name: Black Scoter, *Melanitta nigra* (Linnaeus)
Wilson's name: Scoter Duck, *Anas nigra*
Original name: *Anas nigra* Linnaeus 1758

Current name: White-winged Scoter, *Melanitta fusca* (Linnaeus)
Wilson's name: Velvet Duck, *Anas fusca*
Original name: *Anas fusca* Linnaeus 1758

Current name: Harlequin Duck, *Histrionicus histrionicus* (Linnaeus)
Wilson's name: Harlequin Duck, *Anas histrionicus*
Original name: *Anas histrionicus* Linnaeus 1758

Current name: American Black Duck, *Anas rubripes* Brewster
Wilson's name: Dusky Duck, *Anas obscura*
Original name: *Anas obscura* Gmelin 1789

Commentary

By the time this painting was made, backgrounds had become an important feature of Wilson's art. This painting includes an island in the background, with a large number of seabirds in flight above and just off its shore. Presumably these painted birds are meant to suggest terns swarming above shoaling fish near a breeding colony such as those Wilson had observed on his voyages along the coasts of Cuba, the Bahamas, and the southern United States. The four ducks are swimming and one of the terns is flying, giving the composition a strong sense of action.

Brant

> Flocks continue to arrive from the north, and many remain in the bay till December, or until the weather becomes very severe, when these also move off southwardly. During their stay, they feed on the bars at low water, seldom or never in the marshes; their principal food

being a remarkably long and broad-leaved marine plant, of a bright green color, which adheres to stones, and is called by the country people sea cabbage; the leaves of this are sometimes eight or ten inches broad by two or three feet in length; they also eat small shell fish. They never dive, but wade about feeding at low water. During the time of high water they float in the bay in long lines, particularly in calm weather. Their voice is hoarse and honking, and when some hundreds are screaming together, reminds one of a pack of hounds in full cry. They often quarrel among themselves, and with the ducks, driving the latter off their feeding ground.[162]

Sooty Tern

In passing along the northern shores of Cuba and the coast of Florida and Georgia, in the month of July, I observed this species very numerous and noisy, dashing down headlong after small fish. I shot and dissected several, and found their stomachs uniformly filled with fish. I could perceive little or no difference between the colors of the male and female.[163]

Marsh Tern (Gull-billed Tern)

Their voice is sharper and stronger than that of the Common Tern; the bill is differently formed, being shorter, more rounded above, and thicker; the tail is also much shorter, and less forked. They do not associate with others; but keep in small parties by themselves.[164]

Scoter Duck (Black Scoter)

This Duck is but little known along our sea-coast, being more usually met with in the northern than southern districts; and only during the winter. Its food is shell fish, for which it is almost perpetually diving. That small bivalve so often mentioned, small muscles, spout fish, called on the coast, *razor handles,* young clams &c. furnish it with abundant fare; and wherever these are plenty, the Scoter is an occasional visitor. They swim, seemingly at ease, amidst the very roughest of the surf, but fly heavily along the surface, and to no great distance. They rarely penetrate far up our rivers, but seem to prefer the neighborhood of the ocean; differing in this respect from the Cormorant, which often makes extensive visits to the interior.[165]

Velvet Duck (White-winged Scoter)

This and the preceding [Black Scoter] are frequently confounded together as one and the same species by our gunners on the sea coast. The former, however, differs in being of greater size; in having a broad band of white across the wing; a spot of the same under the eye, and in the structure of its bill. The habits of both are very much alike; they visit us only during the winter; feed entirely on shell fish, which they procure by diving; and return to the northern regions early in spring to breed.[166]

Harlequin Duck

This species is very rare on the coasts of the middle and southern states, tho not unfrequently found off those of New England, where it is known by the dignified title of the *Lord,* probably from the elegant crescents and circles of white which ornament its neck and breast. Tho an inhabitant of both continents, little else is known of its particular manners than that it swims and dives well; flies swift, and to a great height; and has a whistling note. Is said to frequent the small rivulets inland from Hudson's Bay, where it breeds. The female lays ten white eggs on the grass; the young are prettily speckled.[167]

Dusky Duck (American Black Duck)

This species is generally known along the sea coast of New Jersey and the neighboring country by the name of *Black Duck,* being the most common and most numerous of all those of its tribe that frequent the salt marshes. It is only partially migratory. Numbers of them remain during the summer, and breed in sequestered places in the marsh, or on the sea islands of the beach. The eggs are eight or ten in number, very nearly resembling those of the domestic duck. Vast numbers, however, regularly migrate farther north on the approach of spring. During their residence here in winter, they frequent the marshes, and the various creeks and inlets with which those extensive flats are intersected. Their principal food consists of those minute snail shells so abundant in the marshes.[168]

Commentary

This sketch of a Brant was done life-sized, as were many of Wilson's preliminary sketches (*a*). Wilson achieves a three-dimensional look on the body and wing by shading the belly, vent, right leg, and the back behind the wing. This sketch shows that although Wilson appreciated how to make his birds look three-dimensional (see also his pencil sketch of the Labrador Duck, Figure 3.81, which includes shading, especially on the bill), he chose not to include this feature in his plates for *American Ornithology.* A similar decision was made by Peterson and others 150 years later when presenting birds in their field guides. As was often the case, Wilson did several drawings of a species, in this case a small version of the Brant (*b*). This alternate pencil sketch of the Brant is very similar in outline to the final plate, although the feather pattern is incomplete. The dark smudge around the sketch indicates that iron oxide

---

Figure 3.85. Life-sized (*a*) and small (*b*) preliminary sketches of a Brant. (From the archives of the Ernst Mayr Library, Museum of Comparative Zoology, Harvard University.)

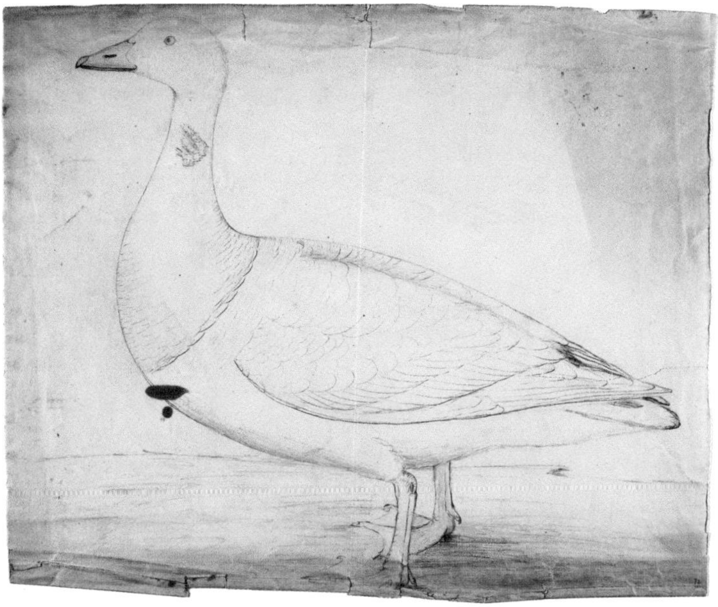

(*a*)

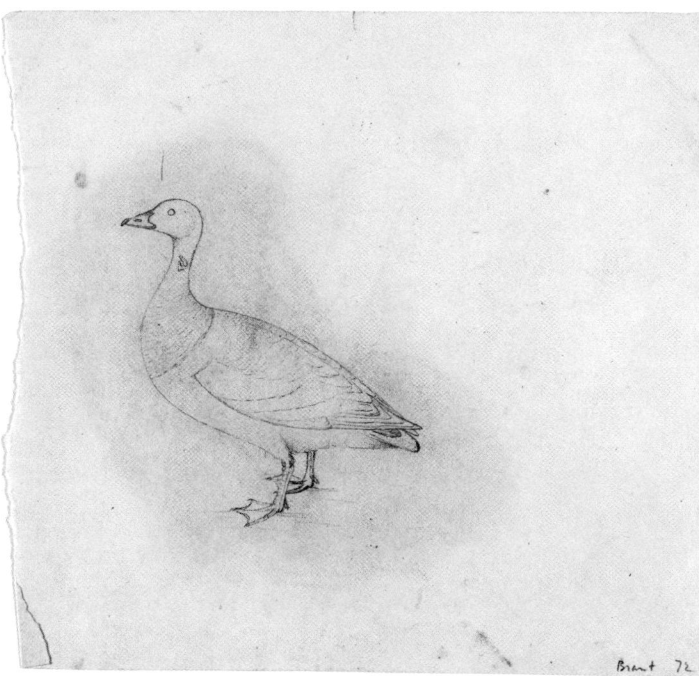

(*b*)

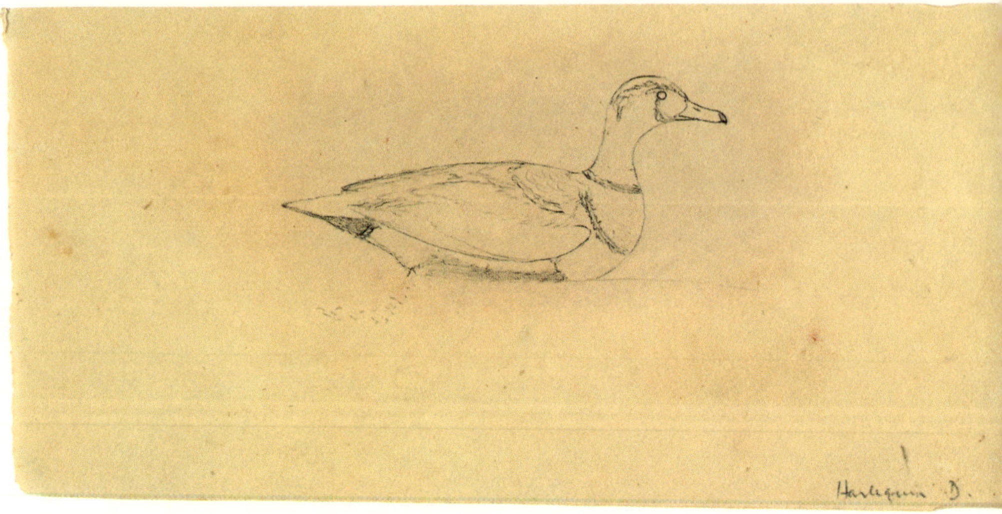

Figure 3.86. Drawing of a Harlequin Duck. (From the archives of the Ernst Mayr Library, Museum of Comparative Zoology, Harvard University.)

was sprinkled on the back side of the paper directly behind the image, which suggests that the image was transferred by tracing it onto the final drawing, rather than by cutting and pasting as in several previous compositions.

Commentary

This pencil drawing of a Harlequin Duck (Figure 3.86) is very similar to the Harlequin in the final composition for *American Ornithology* (Figure 3.84); however, the neck is slightly outstretched and the head slightly more horizontal in the final scene. This gives that Harlequin the appearance of swimming intently toward something it sees to the right of the drawing. The dramatic markings of the Harlequin are more clearly marked in the pencil drawing than in the final composition.

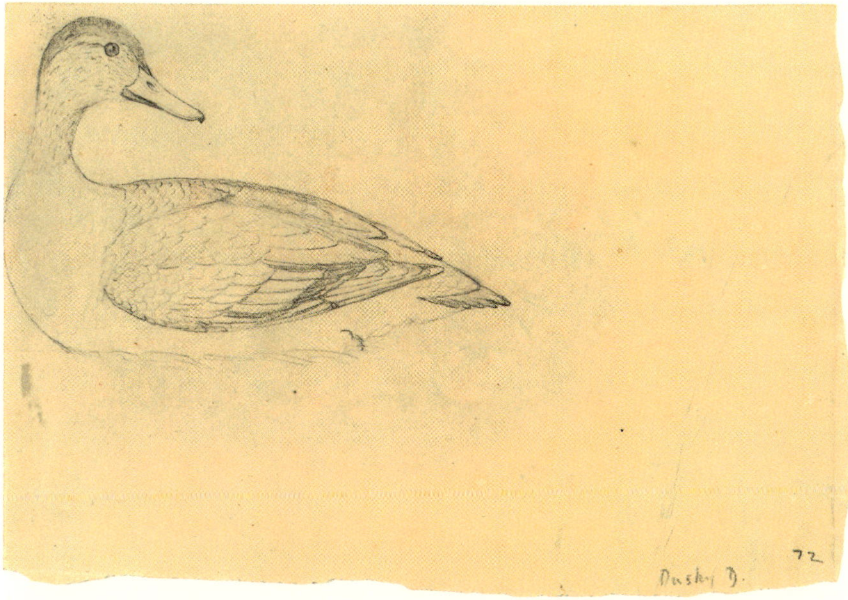

(*a*)

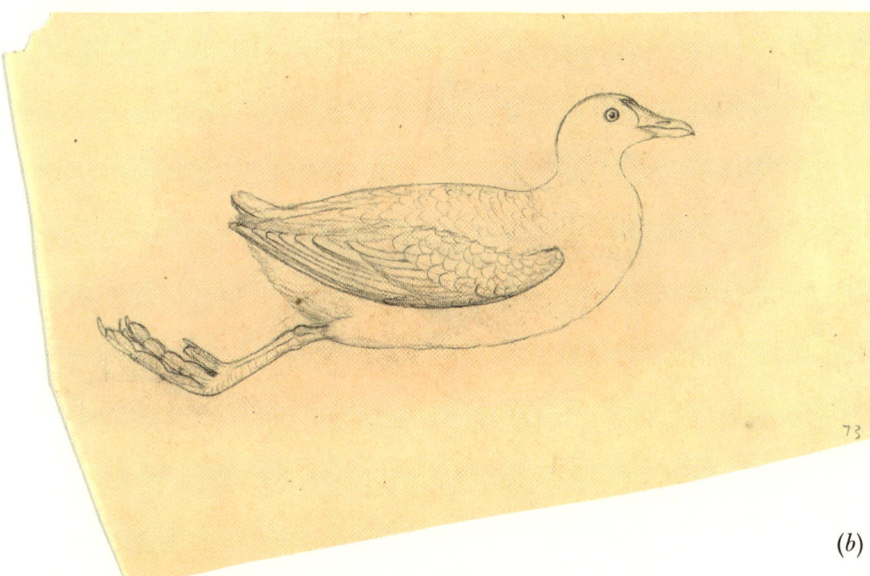

(*b*)

Figure 3.87. A preliminary pencil drawing of an American Black Duck (*a*) for the final painting shown in Figure 3.84 and a preliminary drawing of an American Coot (*b*) for the painting in Figure 3.88. Both paintings are the final drafts for their respective plates. (From the archives of the Ernst Mayr Library, Museum of Comparative Zoology, Harvard University.)

Commentary

Figure 3.87a is probably a preliminary drawing for the American Black Duck in the group shown in Figure 3.84. The head in the final portrait is slightly more raised and the neck slightly straighter, giving the impression that the duck is looking intently at something behind it. The intense gaze is shared by the nearby Harlequin Duck. Together the two ducks give a slight air of mystery to the scene.

The pencil drawing of the American Coot (Figure 3.87b) seems to be the preliminary drawing for the coot in the assembly in Figure 3.88 and plate 73 of *American Ornithology*. The coot's foot is shown in a swimming position, but out of the water, not a naturalistic pose, whereas the foot of the Black Duck is in the water in both the pencil drawing and the group composition (Figure 3.84). Wilson may have decided he did not need to show a duck's webbed foot again whereas he needed to show the coot's unique lobed toes despite the unnatural pose.

Current name: Wilson's Phalarope, *Phalaropus tricolor* Vieillot
Wilson's name: Gray Phalarope, *Tringa lobata*
Original name: *Phalaropus tricolor* Vieillot 1819

Current name: Purple Gallinule, *Porphyrula martinica* (Linnaeus)
Wilson's name: Purple Gallinule, *Porphyrio martinica*
Original name: *Fulica martinica* Linnaeus 1758

Current name: American Coot, *Fulica americana* Gmelin
Wilson's name: Common Coot, *Fulica americana*
Original name: *Fulica americana* Gmelin 1789

Current name: Wilson's Plover, *Charadrius wilsonia* Ord
Wilson's name: Wilson's Plover, *Charadrius wilsonius*
Original name: Wilson's Plover, *Charadrius wilsonius* Ord 1814

Current name: Red Phalarope, *Phalaropus fulicarius* (Linnaeus)
Wilson's name: Red Phalarope, *Tringa Fulicaria*
Original name: *Tringa Fulicaria* Linnaeus 1758

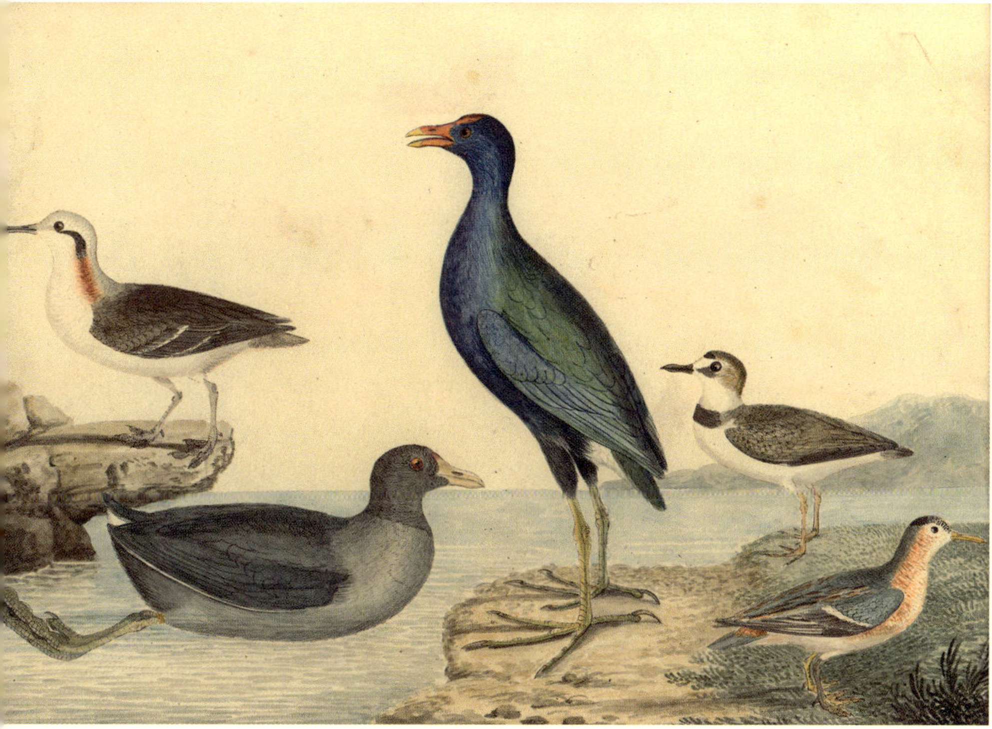

Figure 3.88. Wilson's Phalarope (*upper left*); Purple Gallinule (*center*); American Coot (*lower left*); Wilson's Plover (*upper right*); Red Phalarope (*lower right*). (From the archives of the Ernst Mayr Library, Museum of Comparative Zoology, Harvard University.)

Commentary

Wilson never saw a live phalarope, and the only Wilson's Phalarope he ever saw was a single specimen in Charles Trowbridge's museum in Albany, New York. The only Red Phalarope he saw, the one he used to draw this figure, was sent to him, but so putrefied that the sex could not be determined. Had Wilson known more about the natural history of this species he probably would have depicted it swimming, because they are pelagic birds except when nesting. Long before he made this drawing, Wilson had drawn birds in their correct habitats and shown them behaving in characteristic ways.

Gray Phalarope (Wilson's Phalarope)
(written by Ord)

In the grand and wonderful chain of animated nature, the Phalaropes constitute one of the links between the Waders and the Web-footed tribes, having the form of the Sandpiper with the habits of some of the Ducks; the scalloped membranes on their toes enabling them to swim with facility. They do not appear to be fond of the neighborhood of the ocean, and are generally found in the interior, about ponds and streams of fresh water, where they delight to linger, swimming near the margin in search of seeds and insects. They go in pairs, and we cannot learn that they are anywhere numerous. These circumstances are sufficient to authorize their removal from a tribe to which they have little resemblance, except in their general appearance. Edwards was the first naturalist who introduced them to the world; and although he seems to have been convinced that they ought to constitute a genus of themselves, yet he contented himself with arranging them with the *Tringæ,* a classification certainly neither scientific nor natural.[169]

Purple Gallinule
(written by Ord)

Both the Greeks and Romans, notwithstanding their voracious luxury, abstained from eating the Porphyrion. They brought it from Lybia, from Comagene, and from the Balearic Islands, to be fed and to be placed in their palaces and temples, where it was left at liberty as a guest, whose noble aspect, whose gentle disposition, and whose elegant plumage, merited such honors.[170]

Common Coot (American Coot)
(from a letter by William Bartram)

"The Coot," says William Bartram, "is a native of North America, from Pennsylvania to Florida. They inhabit large rivers, fresh-water inlets or bays, lagoons, &c., where they swim and feed amongst the reeds and grass of the shores; particularly in the river St. Juan, in East Florida; where they are found in immense flocks. They are loquacious and noisy, talking to one another night and day; are constantly on the water, the broad lobated membranes on their toes enabling them to swim and dive like Ducks."[171]

Wilson's Plover
(written by Ord)

Of this neat and prettily-marked species I can find no account, and have concluded that it has hitherto escaped the eye of the naturalist. The bird from which this description was taken, was shot the 13th of May, 1813, on the shore of Cape Island, New Jersey, by my ever-regretted friend; and I have honored it with his name. It was a male, and was accompanied by another of the same sex, and a female, all of which were fortunately obtained.[172]

(written by the naturalist Bonaparte in his *Nomenclature*)

A very rare species established by the Editor [Mr. Ord] and dedicated to Wilson. It is the first homage of the kind paid to the memory of this great and lamented self-taught naturalist. The descriptions of several species in the works of former authors come more or less near to it, but after a careful investigation we are satisfied that it is new.

Red Phalarope
(written by Ord)

> The Red Phalarope is a very rare bird in Pennsylvania; and, as far as we can learn, is but seldom met with in any part of the Union . . . Our figure and description were from this specimen. The person who shot this bird had never seen one of the species before, and was particularly struck with its singular manners. He described it as sitting on the water, dipping in its bill very often, as if feeding, and turning frequently round.[173]

Commentary

The sketches of the Purple Gallinule (Figure 3.89a) and Wilson's Phalarope (Figure 3.89b) are early drafts used in the development of plate 73 in volume 8. The reddish haze surrounding the pencil drawings is the iron oxide spread on the back of each drawing. The images were transferred to another page by tracing the lines of the sketch while holding it in the position. The pressure of the pencil caused the iron oxide to transfer as a mark on the underlying paper or copper plate. We know that no glue was used on these sketches, because there are no brown spots on the drawings. In this case there appears to have been an intermediate draft; the birds in these drawings are smaller than those in the following composite drawing and in the plate that appears in *American Ornithology*.

These and subsequent drawings are the last that Wilson completed before his death on 23 August 1813. Some, such as the phalaropes in Figure 3.88, may not have been intended as final drafts; others, such as the Herring Gull (Figure 3.96) and Tundra Swan (Figure 3.97) are only rough pencil sketches that were never published, for species accounts that were never written. Following Wilson's death, George Ord took up Wilson's pen and wrote or edited the last of Wilson's species accounts. In the passages that accompany these sketches, we have indicated the author. In some cases authorship is uncertain and we have so indicated.

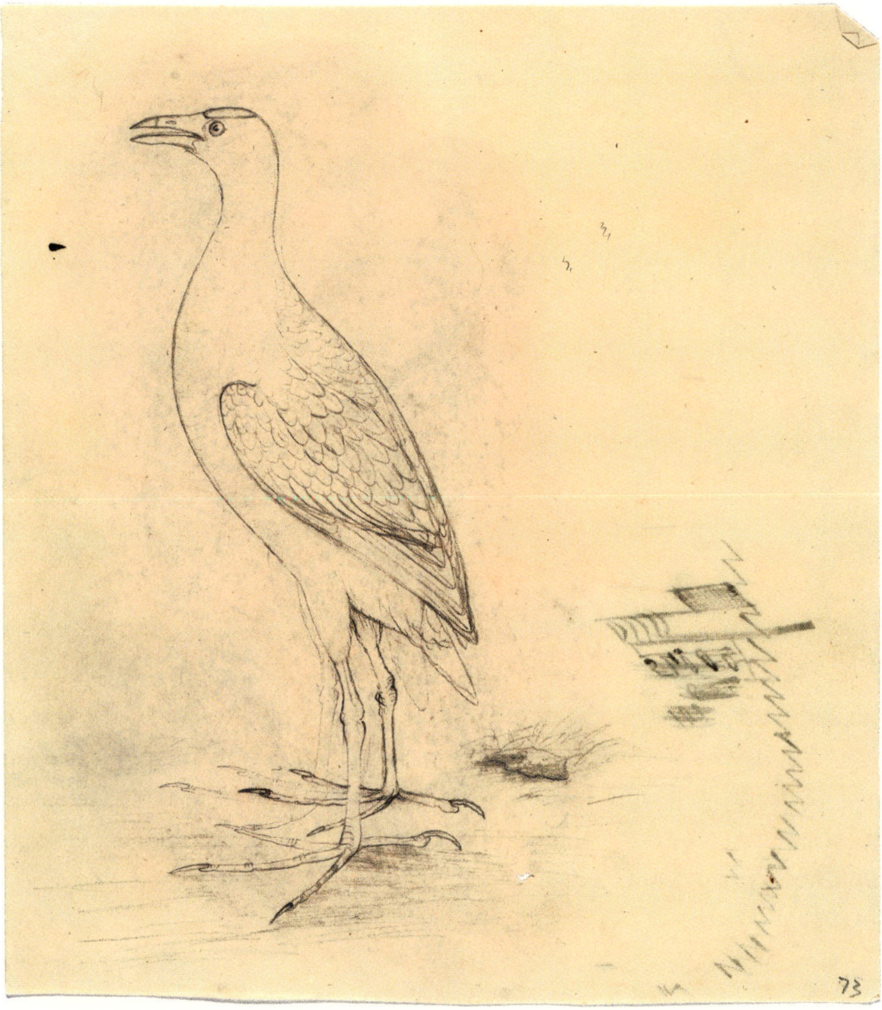

(*a*)

Figure 3.89. Wilson's preliminary sketches of a Purple Gallinule (*a*) and a Wilson's Phalarope (*b*). In each case the shadow of the iron oxide can be seen behind the bird. (From the archives of the Ernst Mayr Library, Museum of Comparative Zoology, Harvard University.)

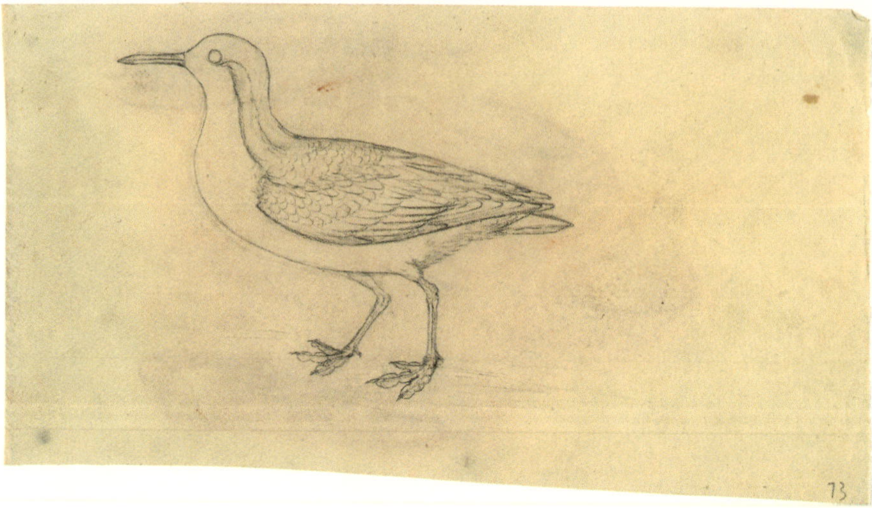

(*b*)

When Ord agreed to edit volumes 8 and 9 of *American Ornithology,* he found fewer and less complete notes than he had expected. In the case of the phalarope the lack of detailed notes and the incomplete state of the sketch led to misidentification. The species is not the Gray Phalarope of Turton as indicated by Ord, but a new species that was later described and given its scientific name by Vieillot. Jardine applied the common name Wilson's Phalarope in honor of its discoverer.

Current name: Anhinga, *Anhinga anhinga* (Linnaeus)
Wilson's name: Black-bellied Darter, *Plotus anhinga*
Original name: *Plotus anhinga* Linnaeus 1766

Current name: Laughing Gull, *Leucophaeus atricilla* (Linnaeus)
Wilson's name: Black-headed Gull, *Larus atricilla*
Original name: *Larus atricilla* Linnaeus 1758

Current name: Dovekie, *Alle alle* (Linnaeus)
Wilson's name: Little Auk, *Alca Alle*
Original name: *Alca Alle* Linnaeus 1758

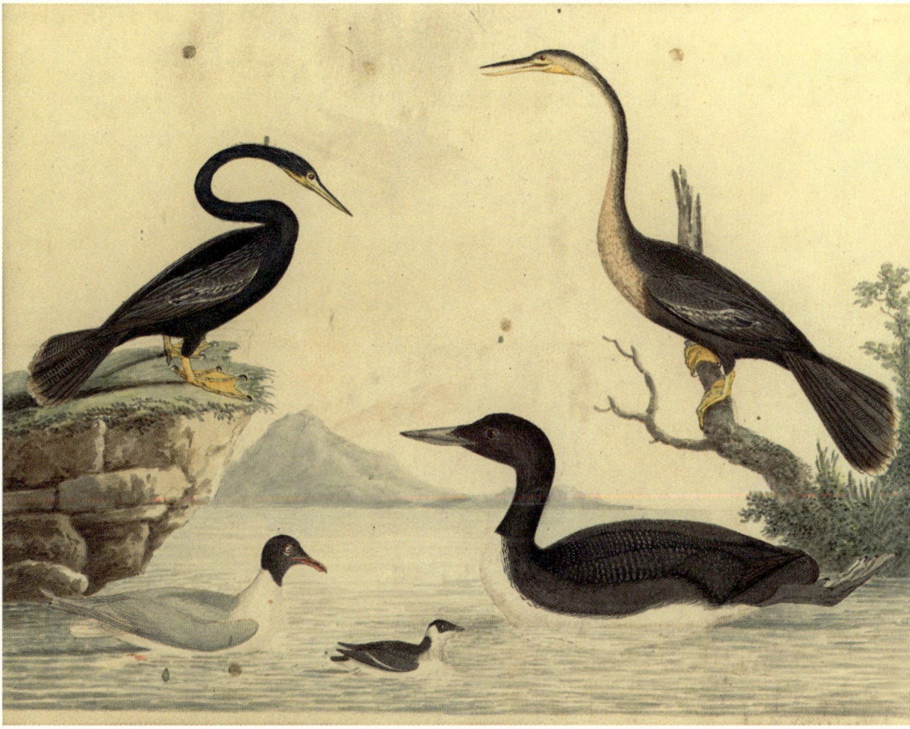

Figure 3.90. Adult male Anhinga (*upper left*); adult female Anhinga (*upper right*); Laughing Gull (*lower left*); Dovekie (*bottom center*); Common Loon (*lower right*). (From the archives of the Ernst Mayr Library, Museum of Comparative Zoology, Harvard University.)

Current name: Common Loon, *Gavia immer* (Brünnich)
Wilson's name: Great Northern Diver, *Colymbus immer*
Original name: *Colymbus immer* Brünnich 1764

Commentary

The Laughing Gull is the only gull species illustrated in Wilson's *American Ornithology,* which suggests that gulls were much less common in colonial times than they are today. Wilson clearly wished to show as much about a bird as he could. The gull, dovekie, and loon are riding higher in the water than is natural, and the loon has one foot out of the water showing its structure and extreme posterior location. The branch on which the female Anhinga is perched might seem to be a throwback to the earlier practice of illustrating birds perched on dead sticks and stumps, but this particular bird's habit of perching on dead branches is specifically mentioned in the species account.

Black-bellied Darter (Anhinga)
(from a letter by John Abbot)

Both the Darters I esteem as but one species. I have now by me a drawing of the male, or black-bellied, only, but have had specimens of both at the same time. I remember that the upper parts of the female were similar to those of the male, except that the color and markings were not so pure and distinct; length, thirty-six inches; extent, forty-six. These birds frequent the ponds, rivers, and creeks, during the summer; build in the trees of the swamps, and those of the islands in the ponds; they construct their nests of sticks; eggs, of a sky blue color.[174]

Black-headed Gull (Laughing Gull)
(written by Ord)

As it respects, in particular, the tribe under review, the zealous inquirer would find himself amply compensated for all his toil, by observing these neat and clean birds coursing along the rivers and coast, enlivening the prospect by their airy movements, now skimming closely over the watery element, watching the motions of the surges, and now rising into the higher regions, sporting with the winds,—while he inhaled the invigorating breezes of the ocean, and listened to the soothing murmurs of its billows.[175]

Little Auk (Dovekie)
(written by Ord)

Of the history of this little stranger, but few particulars are known. With us it is a very rare bird, and, when seen, it is generally in the

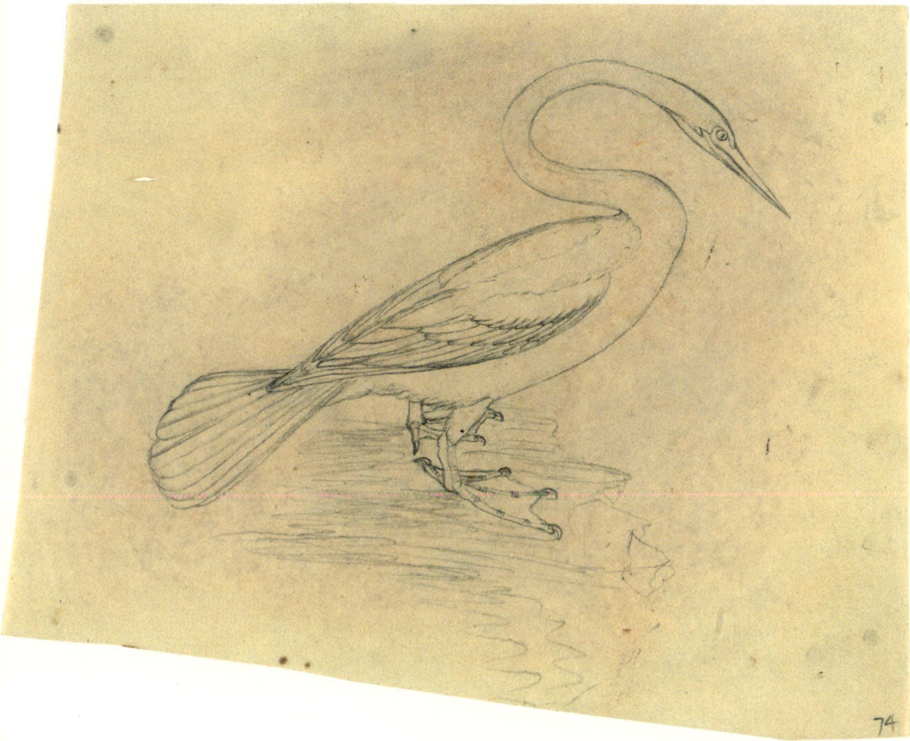

(*a*)

Figure 3.91. Pencil drawings of the Anhinga (*a*) and the Dovekie (*b*) that appear in Figure 3.90. (From the archives of the Ernst Mayr Library, Museum of Comparative Zoology, Harvard University.)

vicinity of the sea. The specimen described was killed at Great Egg Harbor, in the month of December, 1811, and was sent to Wilson as a great curiosity . . . To the anatomist, the internal organization of this species is deserving attention; it is so constructed as to be capable of contracting or dilating itself at pleasure. We know not what Nature intends by this conformation, unless it be to facilitate diving, for which the compressed form is well adapted; and likewise the body, when expanded, will be rendered more buoyant, and fit for the purpose of swimming upon the surface of the water.[176]

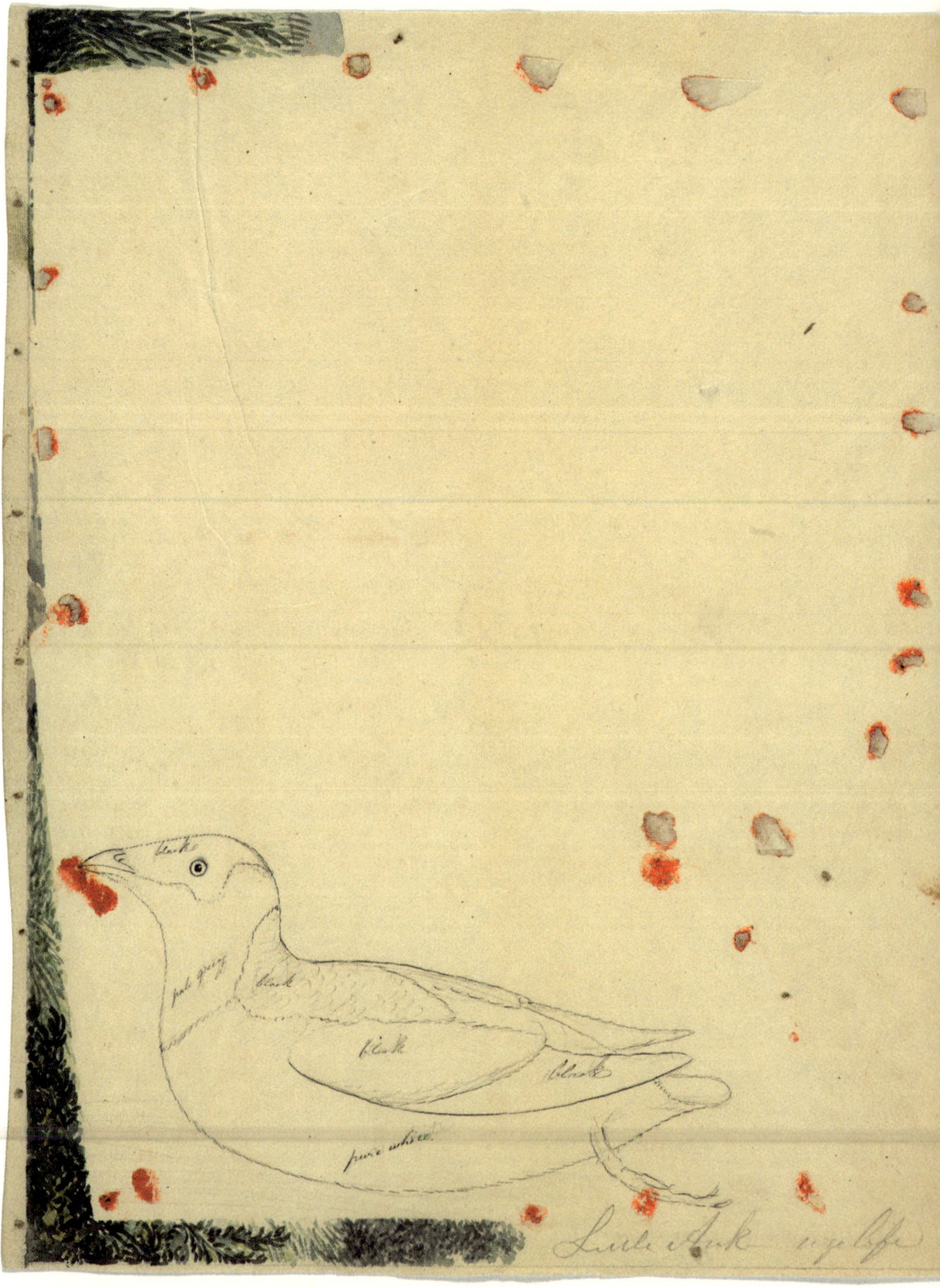

Figure 3.91. (*b*)

Great Northern Diver (Common Loon) (written by Ord)

This bird in Pennsylvania is migratory. In the autumn, it makes its appearance with the various feathered tribes that frequent our waters; and, when the streams are obstructed with ice, it departs for the Southern States. In the months of March and April, it is again seen, and, after lingering a while, it leaves us for the purpose of breeding. The Loons are found along the coast, as well as in the interior; but in the summer, they retire to the fresh-water lakes and ponds.[177]

Commentary

The drawing of the Anhinga (Figure 3.91a) emphasizes the bird's long, snake-like neck and twists the tail just enough to show the feather pattern without destroying the three-dimensional appearance. The pencil sketch of the Dovekie (Figure 3.91b), with colors designated, was pasted over, as indicated by the glue spots along its edges. The edges of the subsequent painting show Wilson's stylized vegetation. The Dovekie is larger than the one on plate 74 in *American Ornithology* and a mirror image, but the posture and proportions are similar.

Commentary

The sketches of a Common Loon (Figure 3.92a) and Laughing Gull (Figure 3.92b) are similar, but not identical, to the final renderings in plate 74 of *American Ornithology*. The loon's head in the drawing is a bit more elevated, which gives it a more elegant line, and the right foot is raised above the water.

The curious life-sized sketch of a Common Loon on three separate pieces of paper (Figure 3.93) has its posterior portion rotated ninety degrees to show the position of the feet relative to the tail from above. The head is in profile. One wonders what Wilson was trying to show and whether, if he had lived a

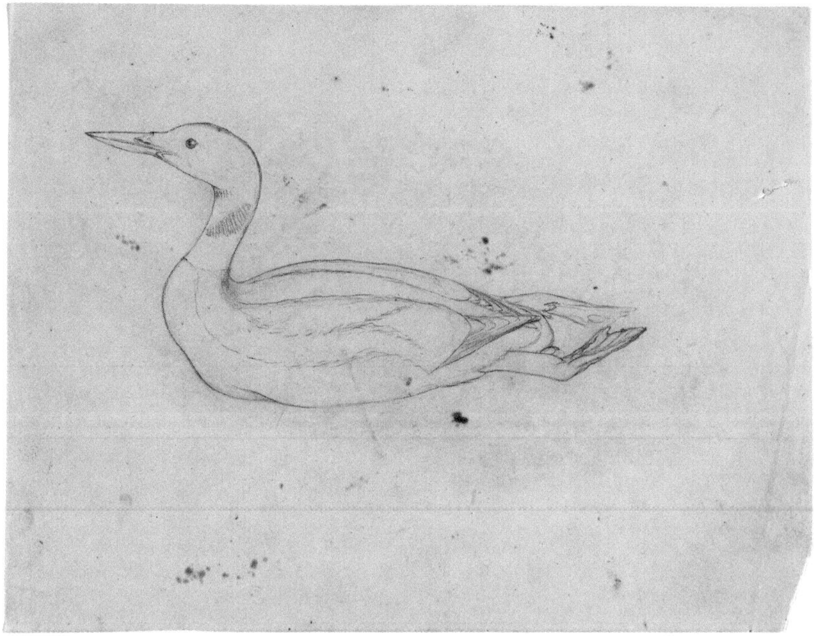

(*a*)

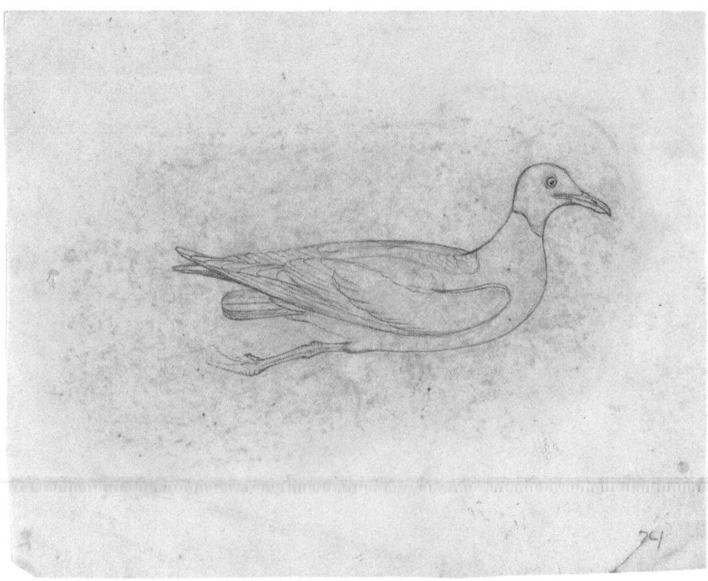

(*b*)

Figure 3.92. Preliminary pencil drawings for the Common Loon (*a*) and the Laughing Gull (*b*) for the final painting in Figure 3.90. (From the archives of the Ernst Mayr Library, Museum of Comparative Zoology, Harvard University.)

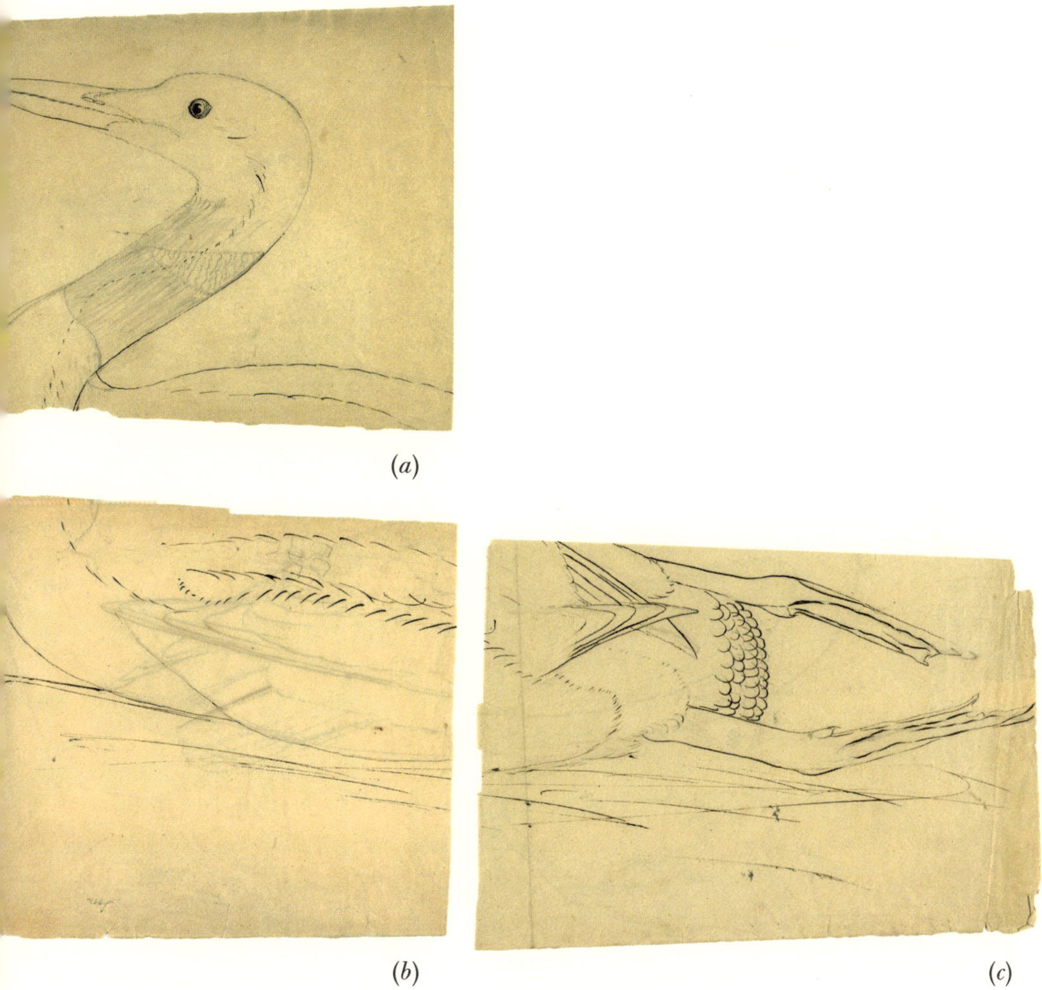

Figure 3.93. A life-sized, composite sketch in pencil and ink of a Common Loon (*a*) head, (*b*) breast, flank, and wing and (*c*) wing tips, feet and tail. (From the archives of the Ernst Mayr Library, Museum of Comparative Zoology, Harvard University.)

few more months, he might have written a description that used the combination of vantage points shown in the sketch.

Current name: Turkey Vulture, *Cathartes aura* (Linnaeus)
Wilson's name: Turkey Vulture or Turkey Buzzard, *Vultur aura*
Original name: *Vultur aura* Linnaeus 1758

Current name: Black Vulture, *Coragyps atrtus* (Bechstein)
Wilson's name: Black Vulture or Carrion Crow, *Vultur atrtus*
Original name: *Vultur atrtus* Bechstein 1793

Current name: Common Raven, *Corvus corax Linnaeus*
Wilson's name: Raven, *Corvus corax*
Original name: *Corvus corax* Linnaeus 1758

Current name: Peregrine Falcon, *Falco peregrinus* Tunstall
Wilson's name: Great-Footed Hawk, *Falco peregrinus*
Original name: *Falco peregrinus* Tunstall 1771

Commentary

This painting is the final product, ready for the engraver, but in the corresponding illustration in *American Ornithology* the Peregrine Falcon has been removed. Instead, an enlarged version of the Peregrine Falcon, shown in virtually the same pose, is presented alone in the dramatic last plate of *American Ornithology*.

Turkey Vulture
(written by Ord)

> These birds, unless when rising from the earth, seldom flap their wings, but sweep along in ogees, and dipping and rising lines, and move with great rapidity. They are often seen in companies, soaring at an immense height, particularly previous to a thunder-storm. Their wings are not spread horizontally, but form a slight angle with the body upwards, the tips having an upward curve. Their sense of

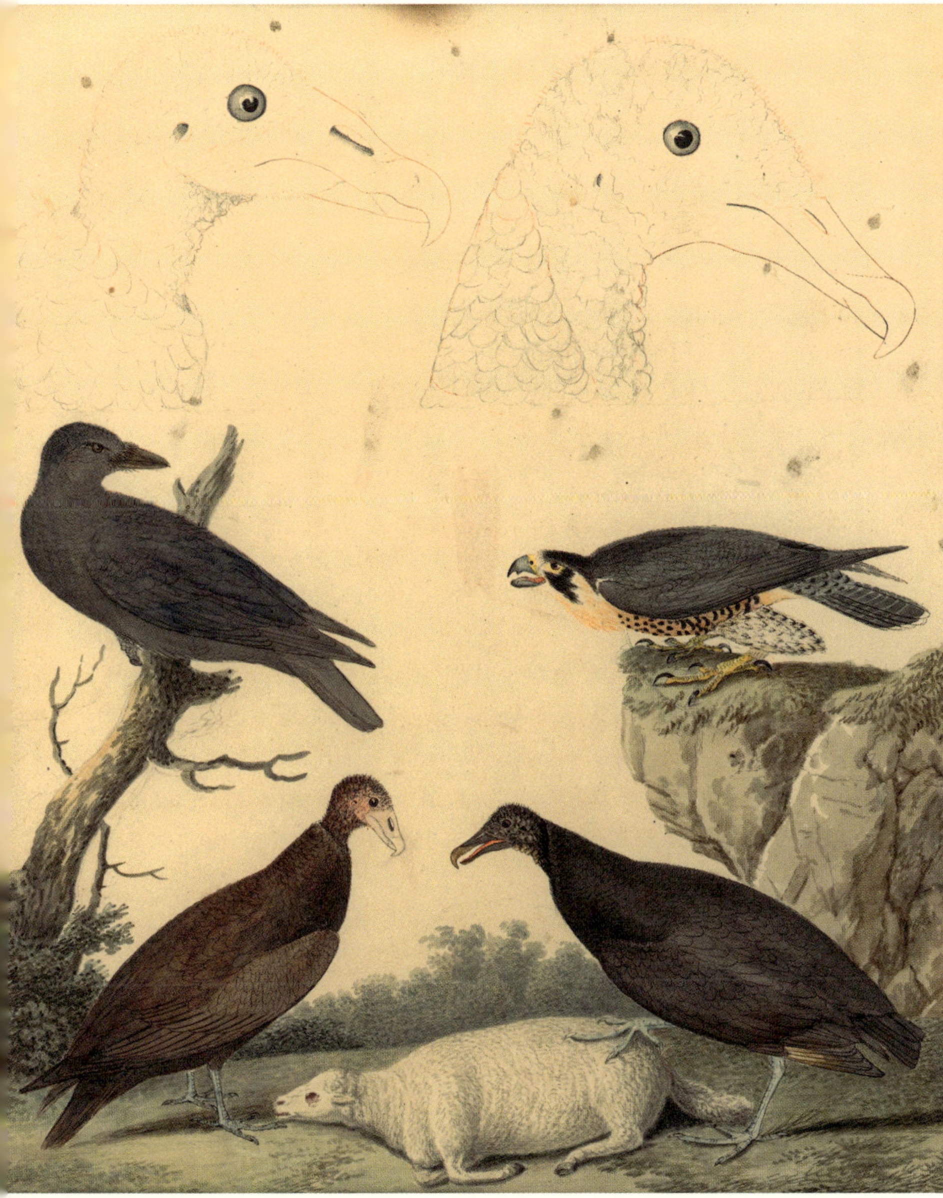

Figure 3.94. Turkey Vulture (*upper left*); Black Vulture (*upper right*); Common Raven (*middle left*); Peregrine Falcon (*middle right*); Turkey Vulture (*lower left*); Black Vulture (*lower right*). (From the archives of the Ernst Mayr Library, Museum of Comparative Zoology, Harvard University.)

smelling is astonishingly exquisite, and they never fail to discover carrion, even when at the distance from it of several miles.[178]

Black Vulture
(written by Ord)

The Black Vultures are indolent, and may be observed in companies, loitering for hours together in one place. They do not associate with the Turkey Buzzards; and are much darker in their plumage than the latter. Their mode of flight also varies from that of the Turkey Buzzard: the black vulture flaps its wings five or six times rapidly, then sails with them extended nearly horizontally; the Turkey Buzzard seldom flaps its wings, and, when sailing, they form an angle with the body upwards. The latter, though found in the vicinity of towns, rarely ventures within them, and then always appearing cautious of the near approach of any one. It is not so impatient of cold as the former, and is likewise less lazy. The Black Vulture, on the ground, hops along very awkwardly; the Turkey Buzzard, though seemingly inactive, moves with an even gait. The latter, unless pressed by hunger, will not eat of a carcass until it becomes putrid; the former is not so fastidious, but devours animal food without distinction.[179]

Raven (Common Raven)
(written by Ord)

The raven is a general inhabitant of the United States, but is more common in the interior. On the lakes, and particularly in the neighborhood of the Falls of the Niagara river, they are numerous; and it is a remarkable fact, that where they so abound, the common Crow [*C. corone*] seldom makes its appearance; being intimidated, it is conjectured, by the superior size and strength of the former, or by an antipathy which the two species manifest towards each other. This I had an opportunity of observing myself, in a journey during the months of August and September, along the lakes Erie and Ontario. The Ravens were seen every day, prowling about in search of the dead fish, which the waves are continually casting ashore, and which afford them an abundance of a favorite food; but I did not see or hear a single Crow within several miles of the lakes, and but very few through the whole of the Genesee country.[180]

Great-Footed Hawk (Peregrine Falcon) (written by Ord)

This noble bird had excited our curiosity for a long time. Every visit which we made to the coast, was rendered doubly interesting by the wonderful stories which we heard of its exploits in fowling, and of its daring enterprise. There was not a gunner along the shore but knew it well; and each could relate something of it which bordered on the marvelous. It was described as darting with the rapidity of an arrow on the Ducks when on the wing, and striking them down with the projecting bone of its breast. Even the Wild Geese were said to be in danger from its attacks, it having been known to sacrifice them to its rapacity . . .

At length in the month of December, 1812, to the unspeakable joy of Mr. Wilson, he received from Egg Harbor a fine specimen of the far-famed Duck Hawk; . . . This species is uncommonly bold and powerful; that it darts on its prey with astonishing velocity; and that it strikes with its formidable feet, permitting the Duck to fall previously to securing it. The circumstance of the hawk's never carrying the Duck off on striking it, has given rise to the belief of that service being performed by means of the breast, which vulgar opinion has armed with a projecting bone, adapted to the purpose. But this cannot be the fact, as the breast bone of this bird does not differ from that of others of the same tribe, which would not admit of so violent a concussion.[181]

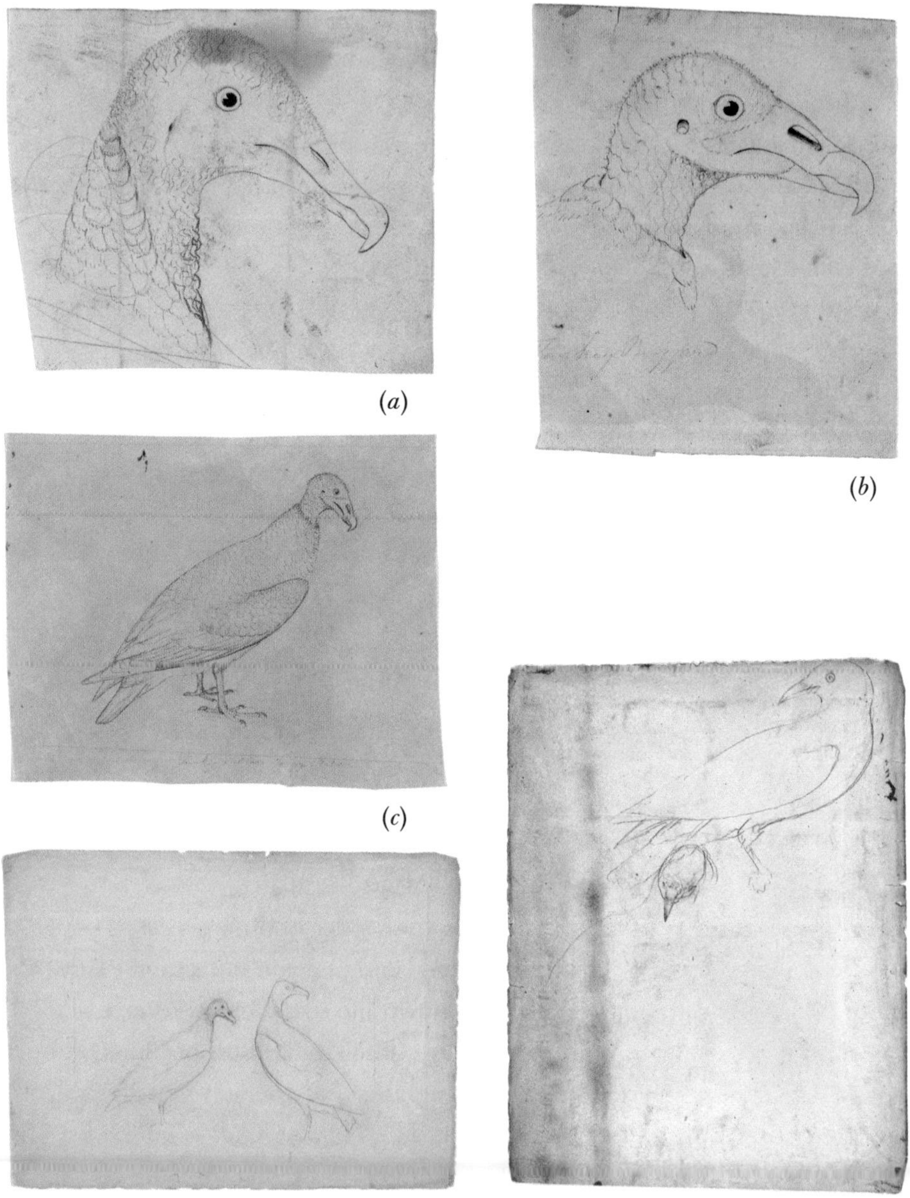

Figure 3.95. These preliminary drawings were the first sketches in what would become a final painting and then a final engraved plate (compare Figure 3.94, which is from volume 9). Note that (*e*) is particularly preliminary and the perspective on the lower bird, looking down on the head, is unusual. (From the archives of the Ernst Mayr Library, Museum of Comparative Zoology, Harvard University.)

Commentary

The heads (*a, b*) and Turkey Vulture (*c*) indicate the development of a final painting ready for the engraver. By contrast, the vulture sketch (*d*) is not fully rendered and suggests a first step in preparation of the final plate. The head of the Black Vulture faces to the rear in one of these images (*d*) but to the front in the final painting. The raven (*e*) is also only an outline, and the final raven in plate 75 in *American Ornithology* is similar but the image is reversed. Wilson also seems to be experimenting with perspective in the head sketch seen beneath the raven's tail. He is looking down on the crown or perhaps the bird is looking down and we see the head from the perspective of the crown. The content and likely timing of the sketch suggest that Wilson was still exploring his artistic abilities even in the last few months of his life.

Current name: Herring Gull, *Larus argentatus* Pontoppidan
Wilson's name: none; never wrote about this species
Original name: *Larus argentatus* Pontoppidan 1763

Commentary

This figure shows drawings that occupy both sides of three pieces of paper. All are rendered in a combination of pencil and pen and ink. On one side of these three pieces are images of a gull drawn life size. On the reverse is the Common Loon shown in our Figure 3.93. The size and proportions of the gull in this drawing are not those of a Laughing Gull, the only gull illustrated and described in *American Ornithology*. The Herring Gull is on the list of birds still to be drawn that Wilson prepared on 13 August 1813, the very day on which he became fatally ill. Had Wilson lived, the last volume, edited and published after his death by George Ord, might have included this gull, probably a Herring Gull.

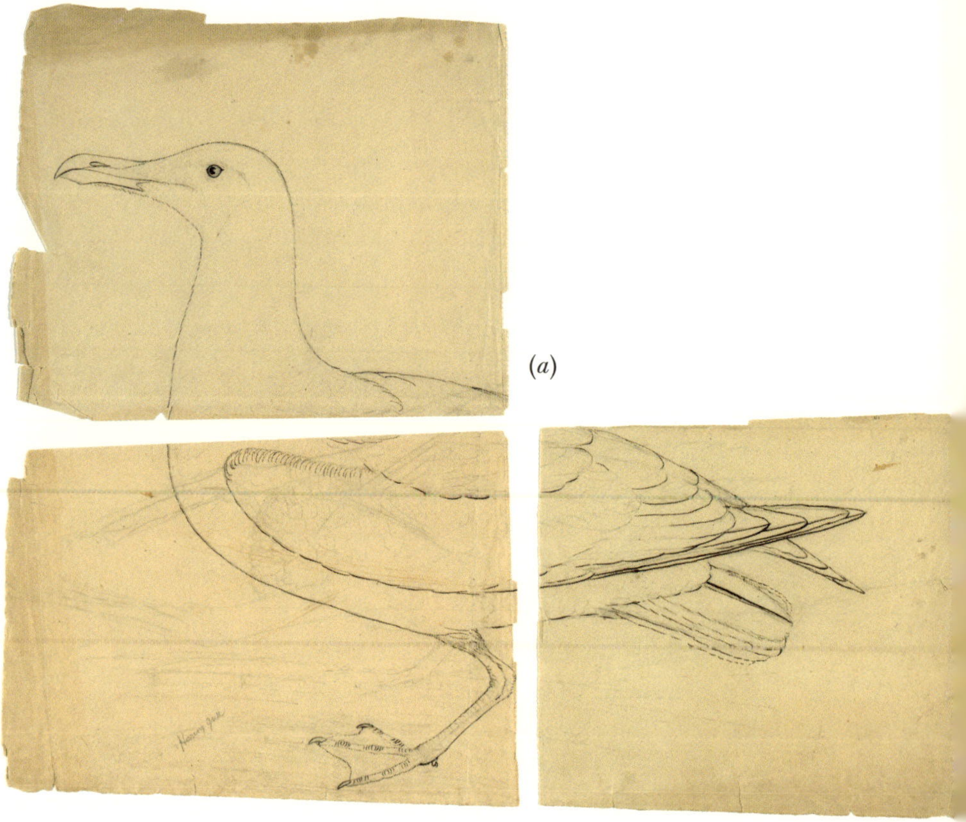

(*a*)

(*b*) (*c*)

Figure 3.96a–c. A life-size drawing of a Herring Gull, (*a*) the head, (*b*) the breast and wing, and (*c*) the wing tips and tail. (From the archives of the Ernst Mayr Library, Museum of Comparative Zoology, Harvard University.)

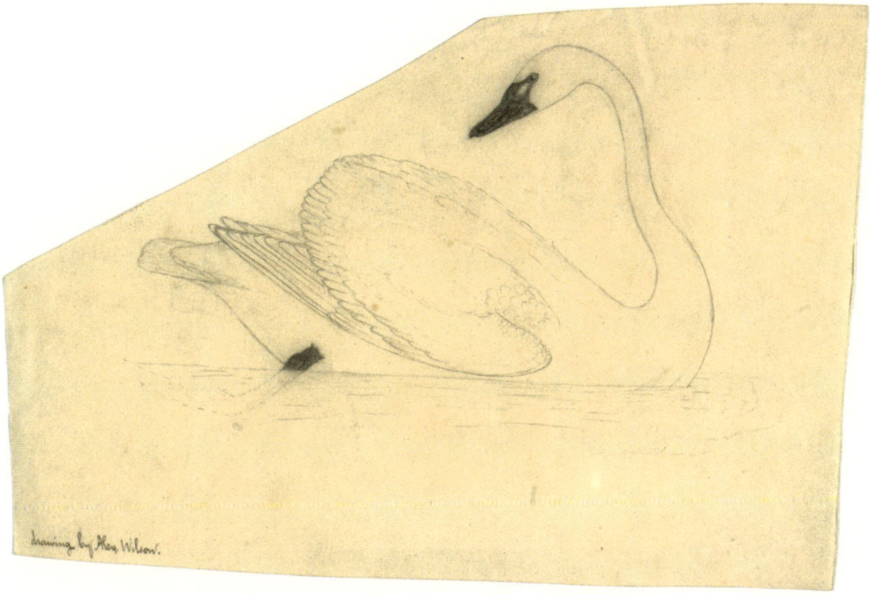

Figure 3.97. Pencil sketch of a Tundra Swan. (From the archives of the Ernst Mayr Library, Museum of Comparative Zoology, Harvard University.)

Current name: Tundra Swan, *Cygnus columbianus* (Ord)
Wilson's name: none; never wrote about this species
Original name: *Cygnus columbianus* Ord 1814

Commentary

This pencil sketch of a Tundra Swan is typical of the strength, simplicity, and elegance of Wilson's sketches. No swan appears in *American Ornithology,* which suggests that Wilson died before he wrote the description of this species or made a final drawing for a plate. It seems likely that it would have appeared in *American Ornithology* had he survived to complete the final volume himself. Indeed, the swan, like the Herring Gull, is on Wilson's list of birds still undrawn made just hours before he became ill. Unfortunately, he did not write down the species of swan, although the sketch leaves little doubt that it is a Tundra Swan.

Innovation in Wilson's Art

Not surprisingly, the early drawings of Alexander Wilson show the influences of his teacher, William Bartram, and of previous and contemporary artists with whose work he was familiar. George Edwards, an English librarian who had done the painting, etching, and hand-coloring of his monumental *A Natural History of Birds* (1743–1751), was one such influence, having illustrated and written descriptions of fifty-seven circumpolar or uniquely North American bird species.[182] Edwards usually worked from poorly preserved specimens or skins and often pictured birds on isolated perches such as tree branches. The use of somewhat dwarfed trees and branches as perches was also favored by John Abbot, with whom Wilson corresponded before and after their meeting in 1809 during Wilson's trip to Georgia.[183] William Bartram, Wilson's principal teacher and mentor, also tended to perch birds on tree branches or on stems and leaves of plants that in life couldn't possibly support the weight of a bird. These early artists often included miniaturized foregrounds in their illustrations. Wilson sent his early work to Bartram and his niece Anne (Nancy) Bartram for criticism, but since only a single example of her bird painting survives, it is not possible to get a measure of her influence on Wilson's early work.

One artist who certainly would have made an impression on Wilson was Mark Catesby, an Englishman who visited America from 1712 to 1719 and again from 1722 to 1726. Catesby published his book *The Natural History of Carolina, Florida, and the Bahama Islands* (1731–1748) in London and included hand-colored etchings of many American birds. He painted his birds as part of decorative patterns of plants and flowers, and his birds were less than perfect renditions. Robert Welker concludes: "Catesby's birds, strictly considered, are seldom painted with complete accuracy. If they lack the glassy stare, the dragonlike claws, and the scaly plumage of medieval bird drawings, nevertheless they also lack persuasiveness in more modern terms. At least seven of his species are superficially unrecognizable. Others are drawn in the distorted poses of birds either patently dead or badly mounted. Seldom does Catesby pay close attention to plumage pattern."[184] Several other early accounts of North American birds, for example, by Latham and Pennant, were simply not illustrated.

Wilson broke new ground with his illustrations of birds that were scientifically accurate and lifelike. A comparison of the details of Wilson's drawings and paintings and those of his contemporaries reveals many differences. The eyes of Wilson's birds are properly positioned in the head, which is not always the case for Bartram and Abbot. Furthermore, the eyes are set properly within the eye socket, not projecting in a pop-eyed stare. There is no prominent semicircular plumage pattern in the area around the ear, a feature particularly common in Bartram's birds. The scapular feathers of Wilson's birds are pointed correctly, unlike those on the birds of Bartram and Abbot. The curves and alignment of the covert feathers are correct in Wilson's bird drawings and paintings, but not in those by Catesby, Edwards, Bartram or Abbot. The scales of the legs and toes are extremely detailed and accurate in Wilson's drawings, but not so in the bird illustrations of his predecessors. Unlike Bartram's birds, Wilson's birds show the hallux relaxed in the raised foot, just as occurs in living birds.

In his early drawings, Wilson, like Bartram and Abbot, drew the tail twisted into an unnatural pose in order to show the color pattern of the tail feathers and any fork or notch in the tail. But as his art matured, Wilson twisted the tails less and drew them with perspective, even if he continued to illustrate tails that were spread or slightly tilted to show the feather pattern and shape. Compare, for example, Wilson's early drawing of the Northern Shrike (Figure 3.20) and later drawings of the Black-billed Magpie (Figure 3.35) or Passenger Pigeon (Figure 3.45), where foreshortening is evident. In his later paintings Wilson often portrayed the bird engaged in characteristic behavior, but his primary objectives remained to emphasize the beauty of birds and facilitate their identification.

Wilson's drawing and painting improved with time. Practice certainly played a role in his improvement, but his later portraits reflect more than better draftsmanship: Wilson also became a more practiced and competent observer. Some of his greatest paintings were of hawks and owls, and are a testament to his artistry—his drawings of the Red-tailed Hawk (Figure 3.50),

Rough-legged Hawk (Figures 3.51 and 3.52), and Northern Saw-whet Owl (Figure 3.35) are technically and artistically superb. The eyes are highlighted and lifelike, the feathering is beautifully rendered, especially in the facial disk of the Northern Saw-whet Owl, and the feet and talons are realistic enough to strike fear into the heart of any rodent. One of the reasons that Wilson portrayed raptors so well may have been his great familiarity with them. In a March 1804 letter to Bartram, Wilson wrote that he was keeping live birds in his rented room, thereby providing ample opportunity for study at close range: "I have had live crows, hawks, and owls . . . so that my room reminds me of Noah's ark."[185] By contrast, some of Wilson's least satisfactory drawings are of shorebirds and seabirds. Wilson lived in the days before binoculars, so he simply had little chance to see shorebirds or seabirds well, and in the case of some—the phalaropes, for example—he never saw the birds alive, a problem he shared with his predecessors Edwards and Catesby. Not surprisingly, the better that Wilson knew a bird and its natural history, the better he was able to portray the bird in a characteristic posture, a natural habitat, and the more convincing and lifelike his portrait.

There has been considerable discussion in the literature about the artistic debt that Wilson owed to his engravers. As Elliott Coues has noted, "It has always been supposed, and apparently vouched for by Wilson's own declarations, that the excellence of his plates was largely due to the skill and care of his engraver. This is not so."[186] We have carefully compared the fully rendered and painted work of Wilson with the engraved and printed versions. In some cases we agree with Coues's conclusion that Wilson was the true artistic mind behind the plates, since the engravers usually followed Wilson's drawings very carefully. The comparison is made problematic, however, by the differences between Wilson's life-sized sketches and his reduced sketches for the plates, which show just how much was lost in the reduction and the translation between the paintings and the plates. In most cases where we have both the large- and small-scale versions, Wilson's life-sized drawings are more artistic and technically accurate—qualities that he apparently found difficult to include in the reductions. This is particularly evident in a comparison of the large- and small-scale drawings of the Brant (Figure 3.85a, b). The slight lump added to the top of the bill in the small version and the slightly thinner neck make the smaller version less accurate. In this case the engraver, Lawson, compounded the problem by making the neck still thinner and the bill misshapen. We think that the eye placement and its setting are superior in Wilson's large drawing of the Clapper Rail (Figure 3.69) as well. And although Wilson's life-sized drawing of the Short-billed Dowitcher (Figure 3.59) is superior to his smaller drawing (Figure 3.60), the engraver faithfully followed the smaller drawing, incorporating Wilson's poor placement of the eye. Overall, Wilson's birds tend to look more lifelike in his larger drawings and paintings than in the final compositions for the plates or the prints in *American Ornithology*.

Although many of Wilson's birds are awkward and posed incorrectly, his best work is magnificent. In fact, Wilson's paintings and drawings are unquestionably superior in technical accuracy and artistic merit to those of his predecessors or contemporaries, and he laid the groundwork for future artist-naturalists, including John James Audubon. Wilson's *American Ornithology* was the first comprehensive work of American natural history, and it was so influential that Elliott Coues would later call the period 1800–1824 the "Wilsonian Epoch" in the *Historical Preface* to his own *Key to North American Birds.*[187] Moreover, the concept and image of the naturalist-illustrator, which Alexander Wilson personified, became strongly associated with the emergence of American ornithology as an independent, scientific discipline.[188]

CHAPTER FOUR

Pioneer Ornithologist

Alexander Wilson was the fortunate victim of two accidents of time and geography. In 1794 his ship, the *Swift,* landed him near Philadelphia, the intellectual and scientific center of the fledgling United States. And in 1802, he accepted the teaching position at the Union School, Kingsessing, which was founded by John Bartram (1699–1777) and located only a short walk from Bartram's home and garden—which was then occupied by William (1739–1823), John's son and the foremost naturalist in the United States, indeed, in the Western Hemisphere. Would Wilson have written the *American Ornithology* were it not for these two events? Perhaps. We know from a letter to his parents written soon after his arrival in the United States that he was smitten with the color and variety of birds seen on his walk to Philadelphia from Newcastle, Delaware, where his ship had landed.[1] But Wilson was still the poet, despite having forsworn poetry as the source of his troubles. In a letter from Milestown written on 11 June 1800, he included a lengthy poem of invitation to his friend Charles Orr that includes only a generic reference to bird song.[2] Yet less than a month later, on 10 July 1800, Wilson writes, again to Orr, a long poem of invitation that includes descriptions of six species of birds by name, and two additional, collective references.[3] Subsequent letters show that over the next two years, Wilson's resolve to forswear poetry gradually weakened while his interest in birds and nature strengthened. The poems he wrote during this time tended to include frequent references to birds, often by name, and to nature. When he lived in Scotland, by contrast, his poetry had focused on people and their interactions, with the notable exception of "The Disconsolate Wren," a sympathetic word picture of the behavior of a wren whose nest has been robbed.

Sometime in 1802, shortly after he accepted the teaching position at the

Union School, Wilson met William Bartram. Also that year Wilson's friend and fellow Scot Alexander Lawson suggested that he take up drawing to fill the lonely hours after school. Wilson began drawing, but William Bartram—naturalist, author, and artist—was the man he turned to for advice and encouragement. Beginning in early 1803 his letters are filled with a rapidly growing passion for drawing and learning about birds. In June 1803 he confided to a friend in Paisley that he was about to begin a collection of the area's finest birds.[4] Bartram became the quiet mentor that the young Alexander Wilson had lacked—a thoughtful, stabilizing influence on his life and dreams. William Bartram was also the knowledgeable teacher to whom Wilson turned with his questions about drawing, painting, and birds; the owner of a library that provided Wilson with access to the accumulated ornithological knowledge of the early nineteenth century; and an internationally respected naturalist who introduced Wilson to the scientists and explorers of the young nation, many of whom were his personal acquaintances and fellow members of the American Philosophical Society. This was a world that Alexander Wilson, with his newly acquired passion for birds, was eager to explore.

Knowledge and Passion

Alexander Wilson has been portrayed as an unlearned man with few resources and little background in ornithology who considered new species to be any birds that William Bartram, his friend and mentor, did not already know.[5] There is an element of truth in this characterization. Wilson's formal education ended when he was ten and he had no training in science, known then as natural philosophy, let alone ornithology. Yet his five years of public schooling, the private tutoring he received, and his three-year apprenticeship as a weaver had provided him with a background similar to that of many of his contemporaries. Furthermore, Wilson read widely, wrote and published poetry, had delivered a widely reprinted political address on the occasion of Jefferson's first inauguration, and had taught himself mathematics sufficiently well to earn money as a surveyor in addition to his wages as a schoolmaster.[6] Wilson was an individual with the intellectual tools for learning and the motivation to teach himself new skills. The unanswerable question in 1803 was did he have the focus and passion to sustain years of exploration in an unknown wilderness full of undescribed birds?

> [to a friend in Paisley]
>
> 1 June 1803
>
> . . . I have had many pursuits since I left Scotland, Mathematics, the German Language, Music, Drawing, &c., and I am now about to make a collection of all our finest birds.[7]

His statement is hardly passionate, but it is the first indication that Wilson had focused on birds.

> *To William Bartram*
>
> [1803]
>
> My Dear Friend,
>
> I send with more diffidence than on any former occasion some further attempts. If from the rough draughts here given you can discover what Birds they are, please to give me their names. Any advice for their amendment from you will be truly welcome.
>
> With sincere esteem and affection,
> I am, Dear Sir,
> Your's,
> Alex Wilson[8]

This letter, unlike the previous one, reveals both Wilson's respect for Bartram's skill and knowledge of birds and his appreciation of how much he, Wilson, had yet to learn. It also conveys a sense of commitment, even passion: "Any advice for their amendment from you will be truly welcome." The tone is respectful, but suggests pleading. "Any" advice, however detailed or picky, will enable him to improve his drawing skills, "any" information, however little, will add to his knowledge about birds. Wilson is clearly the student eagerly seeking the master's help, too concerned with improving his skill and knowledge to guard his ego from criticism. Passion will protect his ego, and this undercurrent of passion is explicit in the following letters:

To William Bartram

20 November 1803

. . . The duty of my profession will not admit me to apply to the study with the assiduity and perseverance I could wish. Chief part of what I do is sketched by candle-light; and for this I am obliged to sacrifice the pleasures of social life, and the agreeable moments which I might enjoy in company with you, and your amiable friend [Bartram's niece Anne, called Nancy by her family and Anna by Wilson].[9]

and to Alexander Lawson, a fellow Scot and good friend, who later became his engraver:

Gray's Ferry, 12 March 1804

. . . I am most earnestly bent on pursuing my plan of making a collection of all the birds in this part of North America. Now I don't want you to throw cold water, as Shakespeare says, on this notion Quixotic as it may appear.

Here is the passionate devotion to learning the necessary skills required to make his collection, even if that commitment means he must give up his former pleasures and spend all his spare time searching for, studying, drawing, and painting birds. And that is exactly what he did. His single-minded devotion to the "collection" would characterize all his remaining days. He would apologize for it again and again in letters to friends and family, but his total commitment never wavered. It ended only with his death.

Alexander Wilson had discovered the passion that would drive him to acquire the skills and knowledge needed to complete his *American Ornithology*. As his commitment and passion increased, he realized that the collection he envisioned would require intellectual and monetary resources. As luck would have it, Wilson arrived in Philadelphia, which was at that time the young nation's center of learning and science. And eight years later, when he became acquainted with William Bartram, he gained access to Bartram's library of natural history, which was one of the best, if not the best, in the country. Moreover, anything Bartram did not have was probably available in the Library Company of Philadelphia—founded almost eighty years earlier by Benjamin Franklin as the first free lending library in the world—or the library at the American Philosophical Society, which he could access through the sponsorship of Bartram, who was a member. Brief descriptions of the books available to and cited by Wilson, and their authors, are provided in Appendix A.

Philadelphia also attracted visiting scientists. John Bachman left a sketchy but important account of a visit by Alexander von Humboldt in May and June of 1804.[10] Bachman was then a boy of fourteen, a theology student with a deep interest in natural history. He mentions that he occasionally accompanied Alexander Wilson on his field excursions and spent his Saturdays and vacations at Bartram's gardens. In June 1804 when Charles Wilson Peale hosted a dinner for Humboldt, Bachman was invited and found among the few guests William Bartram, Alexander Wilson, Alexander Lawson, and George Ord. Bachman's account is important as evidence that within a year of his decision to study birds, Wilson was spending his spare time in the field observing them. Furthermore, he was already included among the natural historians of Philadelphia, already familiar to Charles Wilson Peale with whom he would collaborate, and was known to George Ord, who would edit the final volume of *American Ornithology* and become a controversial figure in promoting Wilson's future reputation. Bachman would become a good friend of Audubon, who was to name Bachman's Warbler in his honor. To the extent that the young Bachman was encouraged by Wilson, he represents another connection between Wilson and Audubon.

Alexander Wilson's Science

Wilson's entire scientific and ornithological career is contained within the pages of *American Ornithology*. Yet its publication not only established American ornithology on the world stage; it also gave the science a taxonomic framework in North America and a distinctive emphasis on natural history, on observing live birds going about their daily activities in the wild. That emphasis was new to science and distinctly American.

Regional Definition and Taxonomic Scope

In 1808, the year in which volume 1 was published, ornithology was largely devoted to the physical description and classification of birds from throughout the world. In many cases the authors described specimens from the collection of a wealthy patron: for example, George Edwards devoted four volumes to describing and illustrating birds in the collection of Sir Hans Sloane, and Comte de Buffon created a multivolume dissertation on the contents of the *Cabinet du Roi.*[11] Frequently these "specimens" were not even preserved birds, but rather pen and ink drawings or paintings of the bird made by an artist who had accompanied an expedition into the field, for example the sketches by Georg Forster, who along with his father accompanied James Cook on his second voyage around the world. Illustrations of plants and animals commissioned by wealthy patrons were an important source of income for both William Bartram, Wilson's mentor, and John Abbot, a fellow naturalist, correspondent, and frequently cited contributor to *American Ornithology.* The text of most ornithological works consisted of brief descriptions of the physical appearance of the species, some of which were so brief as to make identification in the absence of an illustration almost impossible. This is a considerable problem with accounts in Linnaeus's *Systema Naturae,* both the tenth (1758) and twelfth (1766) editions, as well as the 1806 translation of the *Systema Naturae* by William Turton, none of which were illustrated.[12]

Wilson's approach was radically different. First, he focused on a clearly defined geographical area, the United States of 1807–1813. During those six years he visited fifteen of the eighteen states, the four territories, and received three specimens from the Lewis and Clark expedition. He spent a total of nineteen months traveling roughly twelve thousand miles by rowboat and ship, on horseback, in coaches, and on foot.

Among those who preceded Wilson, only Catesby adopted a regional focus, but his region—Virginia, the Carolinas, Florida, and the Bahamas—was discontinuous, his species list was relatively short, and his taxonomy was pre-Linnaean. Among Wilson's contemporaries several adopted a regional approach. Chief among them was Thomas Bewick, author of the two-volume

History of British Birds. An engraver by training, Bewick provided charming black and white scenes of country life with additional black and white illustrations of most, but not all, species to accompany his descriptions of the physical appearance of British birds. His books were readable and popular, but not ornithologically innovative. William Bartram, like Catesby, focused on the Carolinas, but included Georgia and Florida, which made for a more coherent region than that of Catesby. His list of birds was more representative, but he ignored Linnaean taxonomy and his information was not organized into species accounts nor did he provide illustrations. Thomas Jefferson published a list of birds of Virginia, a well-defined region, but his list is essentially that of Catesby with the addition of thirty-three species, some of which are identifiable and some of which are not (Table 4.1). By contrast, Wilson's regional focus was essentially the eastern deciduous forest. Its clear regional definition and ecological coherence were unprecedented.

So too was his coverage of species. Of the estimated 348 species found in the area covered by Wilson, he described and classified 255 species in volumes 1–8 of his *American Ornithology.* Accounts of thirteen additional species were written by George Ord from notes and sketches left by Wilson at the time of his death and were published as volume 9. Thus Wilson described 268 species or 77 percent of the species that occurred in the region.

Wilson also wrote twenty-one additional accounts. Twelve were of species he had described previously, but in these accounts he described plumage characteristics of individuals that differed according to sex (for example, the Common Yellowthroat, *Geothlypis trichas*), age (such as the Red-tailed Hawk, *Buteo jamaicensis*), or season (for instance, the Blackpoll Warbler, *Setophaga striata*) from those described previously. In two further accounts, Wilson concluded that some birds described by earlier authors as distinct species were in fact age or seasonal variants of other species. Thus he correctly identified the Sea Eagle (*Falco ossifragus*) as an immature Bald Eagle (*Haliaeetus leucocephalus*), and documented the Purre (*Tringa cinclus*) as a Sanderling Plover (*Charadrius calidris*) in summer plumage (today both the Purre and the Sanderling Plover are known to be seasonal variants of the Sanderling, *Calidris alba*). He also mistakenly confirmed earlier authors'

TABLE 4.1

JEFFERSON'S ADDITIONS TO CATESBY'S LIST OF BIRDS FROM VIRGINIA, IN THE ORDER LISTED BY JEFFERSON

Jefferson's names	Current names
The Royston crow *(Corvus cornix)*	Hooded Crow (of Europe based on scientific name)
Crane *(Ardea Canadensis)*	Sandhill Crane (based on scientific name)
House swallow *(Hirundo rustica)*	Barn Swallow
Ground swallow *(Hirundo riparia)*	Bank Swallow
Greatest grey eagle	Bald Eagle (?)
Small turkey buzzard with a feathered head	
Greatest owl, or night hawk	Great Horned Owl (?)
Wethawk which feeds flying	Osprey (?)
Raven	Common Raven
Water Pelican of the Missisipi, whose pouch holds a peck	Brown Pelican
Swan	
Loon	
The Cormorant	
Duck and Mallard	Mallard
Widgeon	American Wigeon
Sheldrach, or Canvasback	Canvasback
Black head	Greater or Lesser scaup
Bald Coot	American Coot
Ballcoot	Common Moorhen
Sprigtail	Northern Pintail
Didapper, or Dopchick	Pied-billed Grebe
Spoon-billed duck	Northern Shoveler
Water-witch	species of storm-petrel
Mow-bird	Black-headed Gull
Blue Peter	American Coot
Water wagtail	Spotted Sandpiper (?)
Yellow-legged snipe	Greater or Lesser yellowlegs
Squatting snipe	Common Snipe
Small plover	Killdeer or Semipalmated Plover
Whistling plover	Upland Sandpiper or Black-bellied Plover or American Golden-Plover
Woodcock	American Woodcock
Red bird, with black head, wings and tail	

assumption that the Mottled (*Strix alpina*) and Red (*S. asio*) owls were separate species; currently they are considered plumage variants of the Eastern Screech-Owl (*Megascops asio*). He misidentified as new species four females and one male.[13] The misidentifications and the most probable correct identifications are:

Autumnal Warbler	Bay-breasted Warbler, female (or *fall plumage*)
Blue-green Warbler	Cerulean Warbler, female
Pine-swamp Warbler	Black-throated Blue Warbler, female
Hemlock Warbler	Unknown Warbler, female
Slate-colored Hawk	Sharp-shinned Hawk, male

If we add the additional accounts to those he wrote for each currently recognized species, Alexander Wilson wrote a total of 304 accounts of North American birds between 1807 and 1813. During those same years he spent nineteen months criss-crossing the United States to collect specimens and sell subscriptions, painted at least one individual of every species he described, and supervised all aspects of publication. His achievement is truly monumental.

Concerns of a Taxonomic Pioneer

Wilson was the first American ornithologist to adopt the Linnaean system of names and apply it to every species he described. Admittedly Catesby could not since his work preceded that of Linnaeus, but Jefferson and Bartram published their work after publication of the twelfth edition of Linnaeus's *Systema Naturae* and its translation into English, and after widespread adoption of the system in Europe. Nonetheless, their use of the Linnaean system is haphazard.

Sources: For the order in which Jefferson listed the birds, see T. Jefferson, *Notes on the State of Virginia,* ed. W. Peden (Chapel Hill: University of North Carolina Press, 1982), p. 70. Our best guesses of the species in Jefferson's list are based on D. W. Johnston, *A History of Ornithology in Virginia* (Charlottesville: University of Virginia Press, 2003); American Ornithologists' Union, *Checklist of North American Birds,* 7th ed. (Washington, D.C.: American Ornithologists' Union, 1998) and supplements through 2013; and W. L. McAtee, "Georgian Records in John Latham's 'General History of Birds,'" *Oriole* 11 (1946):1–11.

Wilson adopted John Latham's classification system as described in his *General Synopsis of Birds* (1781–1783) and *Index Ornithologicus* (1790). Latham divided land birds into six orders with sixty-three genera and water birds into three orders with thirty-eight genera (Table 4.2). Wilson's decision to use the Linnaean system as presented by Latham may have been based on its use by Bewick, whose two-volume *History of British Birds* was the only ornithological work he owned, or its use by Thomas Pennant, whose works Wilson cites more often than any other author, including Latham (see Appendix A). Alternatively, it may have been based on language. Linnaeus and Gmelin wrote in Latin, whereas Latham wrote in English. It is doubtful that Wilson read Latin with sufficient fluency to base his taxonomic decisions on a Latin text. Furthermore, Thomas Say, a rising star among the scientific community of Philadelphia, loaned Wilson his copy of Turton's translation of Linnaeus (the thirteenth edition edited by Gmelin), and this book is cited more frequently than the work of William Bartram and almost as frequently as the work of Linnaeus himself (also in Appendix A). Gmelin's thirteenth edition of the *Systema Naturae* and Turton's translation were published after Latham's *Synopsis* and Wilson may have used them to update Latham's taxonomy whenever necessary. In addition Turton's translation would have enabled Wilson to understand Linneaus's tenth and twelfth editions, both of which he cites. However Wilson came to his decision and however he used the various resources he assembled, his work represents the beginning of carefully documented, formal classification in American ornithology.

In 1808, the year in which Wilson published his first volume, the Linnaean system was only fifty years old, and it had been only twenty years since Gmelin had published the thirteenth edition of the *Systema Naturae* and only two years since William Turton's translation of the thirteenth edition. Darwin's presentation of evolution by natural selection was exactly fifty years in the future. Indeed, Charles was little more than a gleam in his parents' eyes in September 1808 when volume 1 of *American Ornithology* rolled off the presses at Robert Carr's printing company. Linnaeus considered species fixed and immutable. He named, described, and grouped species based on the appearance of specimens sent to him, and he often drew his conclusions

TABLE 4.2

THE ORDERS AND GENERA OF BIRDS, BASED ON LATHAM AND ADOPTED BY WILSON

Index Ornithologicus (1790) *Avium Genera*	*Synopsis of Birds* (1781–1783) Genera of Birds
Div. I. Aves Terrestres	**Div. 1. Land Birds**
Ordo I. Accipitres	**Order 1: Rapacious**
1. *Vultur*	Vulture
2. *Falco*	Falcon
3. *Strix*	Owl
Ordo II. Picæ	**Order 2: Pies**
4. *Lanius*	Shrike
5. *Psittacus*	Parrot
6. *Ramphastos*	Toucan
7. *Momotus*	Motmot
8. *Scythrops*	(Channel-billed Cuckoo)
9. *Buceros*	Hornbill
10. *Buphaga*	Beef-eater (Oxpecker)
11. *Crotophaga*	Ani
12. *Callæas*	Wattle Bird
13. *Corvus*	Crow
14. *Coracias*	Roller
15. *Oriolus*	Oriole
16. *Gracula*	Grakle
17. *Paradisæa*	Paradise Bird
18. *Trogon*	Trogon
19. *Bucco*	Barbet
20. *Cuculus*	Cuckoo
21. *Yunx*	Wryneck
22. *Picus*	Woodpecker
23. *Galbula*	Jacamar
24. *Alcedo*	Kingsfisher
25. *Sitta*	Nuthatch
26. *Todus*	Tody
27. *Merops*	Bee-eater
28. *Upupa*	Hoopoe
29. *Certhia*	Creeper
30. *Trochilus*	Humming Bird

TABLE 4.2 (CONTINUED)

THE ORDERS AND GENERA OF BIRDS, BASED ON LATHAM AND ADOPTED BY WILSON

Index Ornithologicus (1790) *Avium Genera*	*Synopsis of Birds* (1781–1783) Genera of Birds
Ordo III. Passeres	**Order 3: Passerine**
31. *Sturnus*	Starling
32. *Turdus*	Thrush
33. *Ampelis*	Chatterer (Waxwings)
34. *Colius*	Coly (Mousebirds)
35. *Loxia*	Grosbeak
36. *Emberiza*	Bunting
37. *Tanagra*	Tanager
38. *Fringilla*	Finch
39. *Phytotoma*	(Plantcutter)
40. *Muscicapa*	Flycatcher
41. *Alauda*	Lark
42. *Motacilla*	Wagtail
43. *Sylvia*	Warbler
44. *Pipra*	Manakin
45. *Parus*	Titmouse
46. *Hirundo*	Swallow
47. *Caprimulgus*	Goatsucker
Ordo IV. Columbe	**Order 4: Columbine**
48. *Columba*	Pigeon
Ordo V. Gallinæ	**Order 5: Gallinaceous**
49. *Pavo*	Peacock
50. *Meleagris*	Turkey
51. *Penelope*	(Guan)
52. *Numida*	Pintado (Guineafowl)
53. *Crax*	Curasso
54. *Phasianus*	Pheasant
55. *Tinamus*	Tinamou
56. *Tetrao*	Grous
57. *Perdix*	Partridge
58. *Psophia*	Trumpeter
59. *Otis*	Bustard

TABLE 4.2 (CONTINUED)
THE ORDERS AND GENERA OF BIRDS, BASED ON LATHAM AND ADOPTED BY WILSON

Index Ornithologicus (1790) *Avium Genera*	*Synopsis of Birds* (1781–1783) **Genera of Birds**
Ordo VI. Struthiones	**Order 6: Struthious**
60. *Didus*	Dodo
61. *Struthio*	African Ostrich
62. *Casuarius*	Cassowary
63. *Rhea*	American Ostrich
Div. II. Aves Aquaticæ	**Div. 2. Water Birds**
Ordo VII. Grallæ	**Order 7: Waders**
64. *Platalea*	Spoonbill
65. *Palamedea*	Screamer
66. *Mycteria*	Jabiru
67. *Cancroma*	Boatbill
68. *Scopus*	Umbre (Hammerkop)
69. *Ardea*	Heron
70. *Tantalus*	Ibis
71. *Numenius*	Curlew
72. *Scolopax*	Snipe
73. *Tringa*	Sandpiper
74. *Charadrius*	Plover
75. *Cursorius*	(Courser)
76. *Hæmatopus*	Oyster-catcher
77. *Glareola*	Pratincole
78. *Rallus*	Rail
79. *Parra*	Jacana
80. *Gallinula*	Gallinule
81. *Vaginalis*	Sheath-bill
Ordo VIII. Pinnatipedes	**Order 8: With Pinnated Feet**
82. *Phalaropus*	Phalarope
83. *Fulica*	Coot
84. *Podicepo*	Grebe
Ordo IX. Palmipedes	**Order 9: Web-footed with Long Legs**
85. *Recurvirostra*	Avoset
86. *Corrira*	Courier
87. *Phœnicopterus*	Flamingo with short legs
88. *Diomedea*	Albatross

TABLE 4.2 (CONTINUED)

THE ORDERS AND GENERA OF BIRDS, BASED ON LATHAM AND ADOPTED BY WILSON

Index Ornithologicus (1790) *Avium Genera*	*Synopsis of Birds* (1781–1783) Genera of Birds
89. *Alca*	Auk
90. *Uria*	Guillemot
91. *Colymbus*	Diver (Loon)
92. *Rhynchops*	Skimmer
93. *Sterna*	Tern
94. *Larus*	Gull
95. *Procellaria*	Petrel
96. *Mergus*	Merganser
97. *Anas*	Duck
98. *Aptenodytes*	Penguin
99. *Pelicanus*	Pelican
100. *Phaeton*	Tropic Bird
101. *Plotus*	Darter (Anhinga)

Note: Wilson used Latham's system for his *American Ornithology* (1808–1814). Where Latham's and Wilson's English names are different from modern names or lacking, the current name is given in parentheses.

from a single representative of the species, still designated the "type specimen." That first specimen was often collected by an explorer in some distant land such as America at considerable risk of life and limb. A tag indicating who collected the bird, as well as where and when, was attached to its leg, and if it was a type specimen, the tag was (and still is) red. Wilson, like Linnaeus, accepted the fixity of species. He wrote: "We have in America many different species of birds that approach so near in resemblance to one another, as not to be distinguished but by the eye of a naturalist, and on a close comparison; these live in the same climate, feed on the same food, and are, I doubt not, the same now as they were five thousand years ago; and, ten thousand years hence, if the species then exist, will be found marked with the same nice discriminations as at present."[14]

Linnaeus considered variations from the type imperfect copies of God's original. The problem Wilson faced was how imperfect did a specimen have to be for it to constitute not an imperfect copy of a known species, but a different original, a new species. Buffon had faced the same problem between twenty and thirty years earlier and had concluded that most North American birds were imperfect forms of similar European species. Wilson, and Jefferson before him, were outraged at Buffon's repeated assertion that the North American avifauna was a motley collection of European has-beens that had settled in the wilds of North America, where the harsh climate and inadequate food had caused a general deterioration of form and behavior. One of the persistent themes in Wilson's species accounts, then, is his comparison of North American forms with similar European species with stress laid on the differences and, often, superiority of the American species. For example his comparison of the song of the Northern Mockingbird (*Mimus polyglottos*) with that of the European Nightingale (*Luscinia megarhynchos*) opens with a passage quoted from Daines Barrington, fellow of the Royal Society of London, in which Barrington compares the Northern Mockingbird favorably with the Nightingale.[15] Wilson gleefully continues: "If, as is here conceded, the Mocking Bird be fully equal to the song of the Nightingale, and, as I can with confidence add, not only to that, but to the song of almost every other bird, besides being capable of exactly imitating various other sounds and voices of animals,—his vocal powers are unquestionably superior to those of the Nightingale, which possesses its own native notes alone."[16]

Wilson's advocacy of North American birds did not blind him to the need for careful decisions based on data. Sometimes Wilson concluded that the differences between the bird he was describing and one described by Linnaeus, Buffon, Latham, Pennant, or one of the other, less frequently cited Europeans (see Appendix B) were not sufficient to warrant designation as a new species. In such cases he was forthright in his comparison and careful to indicate the limits of his data. Take for example, his description of the Winter Wren (*Troglodytes hiemalis*): "In size, color, song, and manners, he approaches nearer to the European Wren (*M. troglodytes*) than any other species we have . . . In short, he possesses almost all the habits of the European species . . . It were much to be wished that the summer residence, nest, and eggs of this bird, were precisely ascertained, which would enable us to determine whether it be, what I strongly suspect it is, the same species as the common domestic Wren of Britain."[17]

In other cases he concluded, after weighing the data, that the North American and European species were distinct, as in his decision on the White-breasted, Black-capped Nuthatch (White-breasted Nuthatch, *Sitta carolinensis*):

> Mr. Pennant considers this bird as a mere variety of the European Nuthatch; but if difference in size, color, and habits, be sufficient characteristics of a distinct species, this bird is certainly entitled to be considered as such. The head and back of the European species is of a uniform bluish gray; the upper parts of the head, neck, and shoulders of ours, are a deep black, glossed with green; the breast and belly of the former is a dull orange, with streaks of chestnut; those parts in the latter are pure white. The European has a line of black passing through the eye, half way down the neck; the present species has nothing of the kind, but appears with the inner webs of the three shortest secondaries and the primaries of a jet black; the latter tipped with white, and the vent and lower parts of the thighs of

> a rust color; the European, therefore, and the present, are evidently two distinct and different species.[18]

Often the decision was complicated by several descriptions of birds that seemed similar, but had been designated as different species by previous authors. Now the question was twofold: first, whether the previous authors were describing different species or age, sex, or plumage variations of the same species, and second, whether the bird Wilson was considering was one of the previously described species; one of its age, sex, or plumage variants; or a new species altogether. The problem is well illustrated in the following passage in which Wilson attempted to deal with the taxonomy of the North American nuthatches, this time in relation to the Red-breasted Nuthatch (*Sitta canadensis*):

> Buffon's *Torchepot de la Canada* (Canada Nuthatch of other European writers) is either a young bird of the present species, in its imperfect plumage, or a different sort, that rarely visits the United States. If the figure *(Pl. enl. 623)* be correctly colored, it must be the latter, as the tail and head appear of the same bluish gray or lead color as the back.[19] The young birds of this species, it may be observed, have also the crown of a lead color during the first season; but the tail-feathers are marked nearly as those of the old ones. Want of precision in the figures and descriptions of these authors makes it difficult to determine; but I think it very probable, that *Sitta Jamaicensis minor,* Briss., the Least Loggerhead of Brown, *Sitta Jamaicensis var. t. st.* Linn., and *Sitta Canadensis* of Linneaus, Gmelin, and Brisson, are names that have been originally applied to different individuals of the species we are now describing.[20]

Wilson faced still another problem. For a number of species he correctly matched his specimen with the description of a previous author, but disagreed with the previous author's choice of genus. Linnaeus, for instance, considered the Golden-winged Woodpecker (Northern Flicker, *Colaptes auratus*) to be a cuckoo and placed it in the genus *Cuculus.* Wilson disagreed and placed it in the genus *Picus,* in which Linnaeus had grouped all the woodpeckers.

> Some European naturalists (and among the rest, Linnaeus himself, in his tenth edition of *Systema Naturae*) have classed this bird with the genus *Cuculus,* or Cuckoo, informing their readers, that it possesses many of the habits of the Cuckoo; that it is almost always on the ground; is never seen to climb trees like the other Woodpeckers, and that its bill is altogether unlike theirs; every one of which assertions, I must say, is incorrect, and could have only proceeded from an entire unacquaintance with the manners of the bird. Except in the article of the bill—and that, as has been before observed, is still a little wedge-formed at the point—it differs in no one characteristic from the rest of its genus.[21]

A few sentences later he describes in detail the bone structure of the tongue and compares it to that of other woodpeckers he has dissected, concluding with:

> In other Woodpeckers we behold the same apparatus, differing a little in different species. In some, these cartilaginous substances reach only to the top of the cranium; in others, they reach to the nostril; and in one species they are wound round the bone of the right eye, which projects considerably more than the left for its accommodation . . . I have also seen them a hundred times alight on the trunk of the tree, though they more frequently alight on the branches; but that they climb, construct like nests, lay the same number and the like-colored eggs, and have the manners and habits of the Woodpeckers, . . . while neither in the form of their body, nor any other part, except in the bill being somewhat bent, and the toes placed two before and two behind, have they the smallest resemblance whatever to the Cuckoo.[22]

The data summarized by Wilson in the two passages just quoted are typical of the care with which he made taxonomic decisions. At no point in *American Ornithology* did he define genus. Nonetheless, he had a sense that birds within a genus should share characteristics and he prioritized these into those that were important, such as those related to foraging, and those that were less so, for instance, the slight bend of the bill and the placement of the toes in the flicker. Equally important, the example of the flicker shows how Wilson used data of different types to make his point. In the case of the flicker, he used both his observation of it climbing trees and digging in wood and his anatomical analysis of the tongue and supporting bone structure to support his conclusion that the flicker was a woodpecker, not a cuckoo. Such care in the collection and use of different types of data to support his conclusions are hallmarks of *American Ornithology*. Wilson's method of argument is hardly new, but he collected more different types of data in support of his argument than had those who preceded him.

What if Wilson concluded that his specimen represented a new species? Naturally there was the excitement of discovery, but he also faced two familiar, but slightly different, problems from those already described. First, he had to decide to what group—that is, to what genus—he should assign his new species. His problem now was to match the general characteristics of his new species with those of a group that shared its general characteristics, but also differed sufficiently to establish the new specimen as a separate species. Well, if it was hard to match specimens to species descriptions, you can imagine that it was that much more difficult to fit a specimen to the often ill-defined characteristics that unified a genus. How much variation could be contained within a genus?

Wilson's predecessors offered little guidance on this question. Linnaeus stated that species were unchanging, and that variation among individuals within a species was the result of imperfections in the individual's development. Buffon attributed these imperfections to climate and diet. But neither Linnaeus nor Buffon defined a genus as anything more than a collection of similar-looking organisms. There was no explanation of the limits of variation within a genus. Wilson clearly gives this problem considerable thought as indicated in the earlier passage on the flicker and in the following quotation, where he considers the probable genus of the Yellow-breasted Chat *(Icteria virens):*

> That the judicious Mr. Pennant, Gmelin, and even Dr. Latham, however, should have arranged it with the Flycatchers, is certainly very extraordinary; as neither in the particular structure of its bill, tongue, feet, nor in its food or manners, has it any affinity whatever to that genus. Some other ornithologists have removed it to the Tanagers; but the bill of the Chat, when compared with that of the Summer Red-Bird, . . . bespeaks it at once to be of a different tribe. Besides, the Tanagers seldom lay more than two or three eggs; the chat usually four: the former build on trees; the latter in low thickets. In short, though this bird will not exactly correspond with any known genus, yet the form of its bill, its food, and many of its habits, would almost justify us in classing it with the genus *Pipra,* (Manakin,) to which family it seems most nearly related.[23]

The Yellow-breasted Chat is currently in the family Parulidae, but is the only species in the genus *Icteria,* a genus that did not exist in 1808 when Wilson made the decision to put it in the genus *Pipra.* That raises an important point. The names and organizational structure of the Linnaean system were relatively new, and Wilson, along with others of his contemporaries, were inclined to try to fit their species into the genera as conceived by Linnaeus rather than create new genera. It was not until Peter Simon Pallas (1741–1811) and Louis Jean Pierre Vieillot (1748–1831), who published while Wilson was publishing, that ornithologists began to substantially expand the number of genera. Linnaeus had divided all birds into six orders and seventy-eight genera, but Wilson based his classification on Latham's system, which assigned birds to nine orders and 101 genera. Wilson made some changes and additions, but very few. This meant that he grouped many species that differed in important ways into a single genus, as in the Yellow-breasted Chat, which he placed with the manakins. As is clear from the account of the chat, Wilson was aware of the poor fit within the genus, but apparently more comfortable with a poor fit than with naming a new genus.

Was Wilson simply unconcerned with names? Absolutely not, as the following letter to William Bartram attests:

> *To William Bartram*
>
> *Philadelphia, 29 April 1807*
>
> The more I read and reflect upon the subject, the more dissatisfied I am with the *specific* names which have been used by almost every writer. A name should, if possible, be expressive of some peculiarity in colour, conformation, or habit; if it will equally apply to two different species, it is certainly an improper one. Is *migratorious* an epithet peculiarly applicable to the robin? Is it not equally so to almost every species of *turdus* we have? *Europea* has been so applied by Pennant to our large *sitta* or nuthatch, which is certainly a different species from the European, the latter being destitute of the black head, neck, and shoulders of ours. Latham calls it *carolinensis,* but it is as much an inhabitant of Pennsylvania and New York as Carolina. The small red-bellied *sitta* is called *canadensis* by Latham, a name equally objectionable with the other. Turdus *minor* seems also improper; in short I consider this part of the business as peculiarly perplexing; and I beg to have your opinion on the matter, particularly with respect to the birds I have mentioned, whether I shall hazard a new nomenclature, or, by copying, sanction what I do not approve of.[24]

Wilson clearly considered his use of names important. Not only is his concern evident in his letter to Bartram, it is evident in the species accounts in his *American Ornithology,* which all begin with his choice of common name and scientific name and are followed by a list of the names given by other authors (and their sources).

Was his reluctance to introduce new genera due to his lack of self-confidence as an ornithologist? Possibly, but consider the pattern of his letters to William Bartram. From 1803 to 1805 they contain numerous questions about the identification of birds in sketches he had sent. Then in September 1805 Wilson wrote to Thomas Jefferson to suggest that a bird Jefferson was unable to identify was, in fact, a Wood Thrush (*Hylocichla mustelina*) or Wood Robin as it was then called. Just two years later, in his letter from April 1807, quoted earlier, Wilson discussed ornithological names and sought Bartram's advice as an equal. And in 1809 he wrote to Bartram to pass along information on a species Bartram had sent to him. These letters show that although initially Wilson may have lacked confidence in his own judgment as an ornithologist, which may have affected volume 1, published in 1808, by 1809 he was demonstrating both a greater knowledge of birds and a well-earned confidence in his knowledge.

In fact, we believe that Wilson's reluctance to establish new genera and his often awkward attempts to fit American species within the existing Linnaean-Latham taxonomy are the product of his overriding interest in the living bird. Wilson was keenly aware of the differences in the taxonomies of his predecessors and the confusion it engendered. In his introduction to volume 1 he writes:

> In order to obtain a more perfect knowledge of Birds, naturalists have divided them into Classes, Orders, Genera, Species and Varieties; but in doing this, scarcely two have agreed on the same mode of arrangement; and this has indeed proved a source of great perplexity to the student. Some have increased the number of orders to an unnecessary extent, multiplied the genera, and out of mere varieties, produced what they supposed to be entire new species. Others, sensible of the impropriety of this, and wishing to simplify the science, as much as possible, have reduced the orders and genera to a few, and have thus thrown birds, whose food, habits, and other characteristical features are widely different, into one and the same tribe, and thereby confounded our perception of that beautiful gradation of affinity and resemblance, which Nature herself seems to have been studious of preserving throughout the whole.[25]

He also suggests at several points in this same introduction that the resolution of this problem depends on observation of the living bird:

> It is also my design to enter more largely than usual into the manners and disposition of each respective species; to become, as it were, their faithful biographer, and to delineate their various peculiarities, in character, song, building, economy, &c. as far as my own observations have extended, or the kindness of others may furnish me with materials . . .
>
> One principal cause of the great diversity of classifications, appears to be owing to the neglect, or want of opportunity, in these writers, of observing the manners of the living birds, in their unconfined state, and in their native countries . . .
>
> It is only by personal intimacy, that we can truly ascertain the character of either, more particularly that of the feathered race; noting their particular haunts, modes of constructing their nests, manner of flight, seasons of migration, favourite food, and numberless other minutiae, which can only be obtained by frequent excursions in the woods and fields, along lakes, shores, and rivers, and requires a degree of patience and perseverance, which nothing but an enthusiastic fondness for the pursuit can inspire.[26]

While these statements from Wilson's introduction constitute his most forceful and complete analysis, the problem of taxonomic disagreements among late eighteenth and early nineteenth century ornithologists and Wilson's proposed solution—observation of the living bird—are emphasized repeatedly within his species accounts. Wilson accepted the necessity of collecting and studying specimens, but he valued observation of the living bird as in the following passage, in which he writes of the Ruby-crowned Kinglet (*Regulus calendula*): "I have often regretted the painful necessity one is under of taking away the lives of such inoffensive, useful little creatures, merely to obtain a more perfect knowledge of the species; for they appear so busy, so active, and unsuspecting, . . . They are remarkably more so in autumn,

. . . and frequently at this season, I have stood under the tree, motionless, to observe them, while they gleaned among the low branches sometimes within a foot or two of my head."[27]

Wilson's emphasis reflects his own interest as described in the opening paragraph of his introduction to volume 1: "Biassed, almost from infancy, by a fondness for birds, and little less than an enthusiast in my researches after them, I feel happy to communicate my observations to others."[28] But it also reflects the philosophy of William Bartram, who found in Alexander Wilson an eager student with the observational skills of a poet. The transformation is acknowledged by Wilson in the following letter written to William Bartram early in their association:

> *Kingsessing, 31 March 1804*
>
> . . . I confess that I was always an enthusiast in my admiration of the rural scenery of Nature; but, since your example and encouragement have set me to attempt to imitate her productions, I see new beauties in every bird, plant, and flower, I contemplate; . . . I am collecting, without injuring my conscience, or wounding my peace of mind, those beautiful specimens of Nature's works that are for ever pleasing. I have had live crows, hawks, and owls—opossums, squirrels, snakes, lizards, &c.[29]

Alexander Wilson recognized the importance of behavioral and ecological observation in making taxonomic decisions. But he also recognized the importance of natural history information for its own sake. All his species accounts contain information based on watching birds in the wild. As is clear in his letter, he also gathered data by closely observing the birds he raised and kept in his rented room. The level of detail and the months, occasionally years, of observation were unprecedented. Reading his species accounts one cannot help but find his curiosity contagious.

"This new species I first discovered . . ."

Wilson believed that he had discovered fifty-one species of North American birds. Of these, twenty are now credited to Wilson by the American

Ornithologists' Union.[30] An additional five species whose discovery Wilson attributed to others are now attributed to him.[31] The earlier descriptions he cites either preceded Linnaeus's tenth edition—for example, the Pine Creeper of Catesby (Pine Warbler, *Setophaga pinus*); lacked a proper Latin name, as in the case of the Field Sparrow (*Spizella pusilla*) of Bartram; or were described incompletely—for instance, the Marsh Wren (*Cistothorus palustris*) of Latham.[32]

And then there is the House Wren (*Troglodytes aedon*), which Wilson attributes to Bartram and the American Ornithologists' Union attributes to Vieillot.[33] Bartram's description is incomplete, hardly more than the name, therefore he is disqualified as the first person to publish a formal description.[34] That leaves Wilson and Vieillot, who both published descriptions and proper, but different, Linnaean names for the House Wren. Vieillot's description was published as part of his *Histoire naturelle des oiseaux de l'Amerique Septentrionale.*[35] The cover page of Vieillot's first volume is dated 1807, which clearly precedes publication of Wilson's first volume in 1808. The first edition of the *Checklist of North American Birds,* published in 1889, lists the House Wren as *Troglodytes aedon* based on Vieillot's description, which it lists as published in 1807, in agreement with the cover page of the *Septentrionale.* The second (1895) and third (1910) editions of the *Checklist* give the same citation. But in 1899, in a footnote, Charles Richmond explained that the French often published the title page and table of contents with the first livraison, a small piece of the complete work. The remaining livraisons were published in installments over subsequent months and years, then later bound in one or a few volumes, which carried the original publication date. Moreover, "among antedated works may be noticed Vieillot's *Histoire naturelle des oiseaux de l'Amerique Septentrionale,* dated 1807, but published in 22 livraisons, of which the first was issued December 1807."[36]

Doubt had crept into the resolution of who had first described the House Wren. In the fourth edition of the *Checklist of North American Birds,* published in 1931, the citation lists 1807 as the publication date for the first description and then "(1808, possibly 1809)." Uncertainty persisted in the fifth edition (1957), which still lists 1807 (1808?). Apparently 1809 had been ruled out. In 1983, the listing was changed to 1808 with 1807 in parentheses. Furthermore, the notes on *Troglodytes aedon* acknowledge the debate: "In view of the uncertainty of publication dates during the year 1808, replacement of the well established name *T. aedon* with *T. domesticus,* based on *Sylvia domestica* Wilson 1808 seems unwarranted."[37]

Burt Monroe, who chaired the taxonomic committee that assembled the checklist and was responsible for deciding which birds were species, what their proper names were, and who discovered them, set out to resolve the uncertainty over publication of the first description of the House Wren.[38] He and M. Ralph Browning confirmed what Richmond had found ninety years earlier: French publishers often printed a cover page that antedated the book in an effort to secure their author's claim of first publication. Furthermore, they found that Vieillot had published the *Septentrionale* as a series of separate parts, each with six plates and accompanying text, published at monthly intervals. When Monroe and Browning checked on the description of the House Wren, they were able to trace its publication to May 1809. Wilson's description, which appeared in volume 1 of the *American Ornithology,* published in September 1808, predates Vieillot's description by nine months. Therefore Wilson was in fact the first person to name and describe the House Wren, which he named *Sylvia domesticus.* If his priority were accepted, the current scientific name would be *Troglodytes domesticus.* But the name will remain as it has been for almost two hundred years, *Troglodytes aedon.* According to the American Ornithologists' Union in 1998, "The International Commission for Zoological Nomenclature has been asked (Bull. Zool. Nomenclature 53: 187–190, 1996) to conserve the widely used specific name *aedon* despite the fact that *T. domesticus* (Wilson, 1808) has priority."[39] So officially Vieillot is credited with first publication, but unofficially we consider Wilson the first ornithologist to have published a species account of the House Wren. Thus Wilson, in volumes 1–8 of his *American Ornithology,* was the first to describe twenty-six species of North American birds.

That leaves thirty-one species Wilson thought he had discovered that are now attributed to other authors. Seven of these were first described by Vieillot and two more by Bechstein, authors who were probably unknown to Wilson. Their claim to priority of publication varies from one month (Yellow-throated Vireo, *Vireo flavifrons*) to four years (Tree Swallow, *Tachycineta bicolor,* and Sharp-shinned Hawk, *Accipiter striatus*). That Wilson was unaware of their work is understandable given the close proximity of publication dates and the difficulty of trans-Atlantic communication during the Napoleonic wars, when trade between the United States and Europe was blocked by the British navy. In addition Vieillot worked without the support of the government or a patron, and Bechstein published in German, a language Wilson read only with difficulty.[40]

Three of the names that Wilson chose for undescribed species were in fact already in use for other species. Perhaps the most unfortunate coincidence of names was *Picus torquatus,* the name Wilson gave to Lewis's Woodpecker. Meriwether Lewis presented the specimen to Wilson, whom he considered the foremost American ornithologist. The two men, so like each other, became close friends. On his trip west in 1810 Wilson visited the site of Lewis's death the previous October along the Natchez Trace in Tennessee and attempted to determine whether Lewis had died of suicide as rumored or been murdered. Wilson's investigation, although published, was inconclusive, but he named the woodpecker Lewis's Woodpecker in honor of his friend's gift. Much later when the duplication of scientific names was discovered, Gray honored Wilson's intention by naming the species *Picus lewis* in 1849.[41] Since that time the genus has been changed and it is now listed as Lewis's Woodpecker (*Melanerpes lewis*).

The other two species that Wilson named inappropriately were given names Wilson knew were already in use, in one case a name that he himself had already used. George Ord corrected the mistake in his edition of Wilson's work published in 1824, but Vieillot, once again, had published first: a year earlier he had renamed the Broad-winged Hawk *Sparvius platypterus,* and although the genus name has since changed, Vieillot has received credit for the species name. Equally inexplicable was Wilson's use of the same scientific name for the Veery as had been applied to the Wood Thrush by both Pennant and Latham, the two most frequently cited authors in Wilson's work. The Veery was distinct from the Wood Thrush and was undescribed as Wilson asserted, but he did not supply a unique Latin name as required by Linnaean taxonomic rules. In 1817 Stephens supplied the unique name and so is recognized as the person who provided the first scientific description.

Six species that Wilson claimed to describe for the first time had been described by Linneaus, five by Gmelin, and one by Latham, all of whom Wilson cited frequently. Four species of the thirty-one that Wilson thought he had described for the first time were females of species for which he had already described the male, and another was a male of a species for which Wilson had previously described the female. The final two disallowed species were the Blue-Mountain Warbler (*Sylvia montana*) and the Small-headed Flycatcher (*Muscicapa minuta*). No one has been able to match the plates or the descriptions to currently recognized species. Audubon painted both and claimed to have seen both, but his paintings do not permit identification any better than Wilson's drawings or written descriptions.

The frequency of duplicate names, misidentified species, and failures to cite Linneaus, Gmelin, and Latham occur more frequently when the interval between volumes is short.[42] The trend is most clear if mistakes are considered as a proportion of new species described. Six species among the twenty-four (25 percent) that Wilson believed to be new in the first three volumes are in error compared with fifteen (56 percent) among the twenty-seven species he believed to be new in the last five volumes. The interval between volumes for the first three volumes was sixteen months, whereas the interval between volumes for the last five volumes was six months.[43] The time available for preparation and the increasing proportion of errors indicate that Wilson was less thorough in his preparation and editing of the later volumes. From 1811 through 1813, the years in which these volumes were prepared and published, Wilson was under increasing pressure from Samuel Bradford, his publisher and underwriter, to publish rapidly so as to provide the cash needed for preparation and publication of the later volumes. The embargo on European trade enacted by Congress and signed into law by Jefferson in 1807, combined with British and French interference with American shipping, had severely stressed the U.S. economy and effectively eliminated the cash reserves with which Bradford was financing *American Ornithology*. Without the reserves, he and Wilson depended on the cash generated by subscribers, who paid twelve dollars for each book delivered during the previous year. Volume 1 was published in 1808, and much of the income that it generated was needed to pay the bills incurred to begin the series. Volume 2 was not published until 1810 and volume 3 was delayed until early in 1811, which meant only half the expected cash from subscribers arrived in 1810. As mentioned earlier, this shortfall forced Wilson to lay off his colorists and personally undertake their duties in addition to those of author, editor, illustrator, and salesman. To get an idea of the magnitude of this job, consider that each volume contained nine plates, except volume 1 which contained ten plates, with an average of five birds per plate. That means he had to paint forty-five birds in each copy and there were 450 copies, one for each subscriber, which means 20,250 birds to be hand-colored for each volume. Added to the workload and the financial worry was Wilson's declining health during the preparation of volumes 6, 7, and 8.

Time, economic pressures, workload, and poor health contributed to many of Wilson's taxonomic errors. Nonetheless, the twenty errors we have detailed affect only 7.5 percent of the 268 species described by Wilson and are less than the twenty-five species (House Wren excepted) first described by him and still attributed to him.

Wilson's achievement stands out even more in the context of early nineteenth-century ornithology. In the first decade of the new century about twenty new species of birds were described each year. That number began to rise starting in 1813 and reached a peak of 130–150 new species described each year from 1830 to 1850, before declining gradually as fewer and fewer species remained to be discovered.[44] During 1808–1813, ninety-six North American species were discovered and described, of which Wilson is credited with twenty-six (House Wren included), or 27 percent, more than any other single author.

Wilson's taxonomic achievement is all the more remarkable when one considers that he worked without the financial and intellectual support available to the European collectors of his time. Wilson appreciated and frequently acknowledged his intellectual debt to Charles Wilson Peale, who like Wilson, advocated Linnaean taxonomy and opposed Buffon's theory of degeneration of American forms; to Thomas Jefferson, also an opponent of Buffon's theory; and to William Bartram, America's foremost naturalist. Nonetheless, Wilson lacked the intellectual stimulation that the scientific debates among Johann Karl Wilhelm Illiger, Coenraad Jacob Temminck, Etienne Geoffroy Saint-Hilaire, Georges Cuvier, and John Latham provided his European contemporaries. Wilson acknowledged his frequent use of the Peale Museum—indeed he contributed 279 specimens to the collection, including twenty-four type specimens (Table 4.3)—but this resource, valuable though it was, could hardly compare with the museums of Berlin, London, Leiden, or Paris.[45] Wilson's isolation was only partly the result of the trans-Atlantic distance and the time required for sailing ships to make the crossing. More important, England and France were in a perpetual state of war and each was engaged in preventing the other from benefiting from American raw materials and the prices that Americans would pay for goods (such as books and scientific journals) manufactured in Europe. Add to that the embargo imposed by President Jefferson and Congress, which continued under President Madison, and the result was little to no scientific communication between Europe and the United States. This also meant that few expeditions were sent from Europe to explore biologically unknown areas. Consequently few new species of birds were being described during the time that Wilson wrote his *American Ornithology*.[46]

Isolated as he was, Wilson had to make do with the poorly codified rules of ornithological taxonomy employed by Gmelin, Latham, and Pennant more than twenty years earlier. The system as used by these authors provided no category for family, for example ducks, thrushes, or warblers.[47] Because there was no clear taxonomic level between order and genus, Wilson expanded the boundaries of existing genera to encompass all North American species. His frustration with inappropriate generic boundaries, however, is evident in his thoughtful account of the Blackpoll Warbler: "This bird may be considered as occupying an intermediate station between the Flycatchers and the Warblers, having the manners of the former, and the bill, partially, of the latter. The nice gradations by which nature passes from one species to another,

TABLE 4.3

TYPE SPECIMENS DEPOSITED IN THE PEALE MUSEUM BY ALEXANDER WILSON, THEIR CURRENT LOCATION, IF KNOWN, AND OUR NOTES AND THOSE OF WALTER FAXON

Species	Peale No.	Notes
(House Wren)	7283	Technically this is not a type specimen since the initial description is attributed to Vieillot, but this is the specimen used by Wilson in his description, which antedates Vieillot's description.
Field Sparrow	6560	MCZ, "Very likely the type."
American Tree Sparrow	6575	
Marsh Wren	7282	
Song Sparrow	6573	MCZ, "This specimen is one of those large, heavily marked Song Sparrows, of a pronounced rufous tint, such as pass through eastern Massachusetts in small numbers in the spring, along with the Fox Sparrows. It agrees well with Wilson's figure, if some allowance is made for adapting the bird to its place on the plate. Mr. Outram Bangs is quite confident that it is the individual figured by Wilson."[a]
Pine Siskin	6577	
Cerulean Warbler	7309	
Pine Warbler	7312	
Bay-breasted Warbler	7311	
Western Tanager	6236	
Clark's Nutcracker	1371	
Lewis's Woodpecker	2020	MCZ, "A single venerable looking specimen, probably either the type, which was No. 2020 of the Peale Museum (Lewis and Clark Expedition), or else one of the two individuals shot by T. R. Peale near the Rocky Mountains, on the Long Expedition, I presume."[b] Because Wilson used a scientific name that was already in use, his description is not taxonomically valid. When Gray renamed the species, however, he based the description on Wilson's illustration in volume 3 of *American Ornithology*, which is based on this specimen.

TABLE 4.3 (CONTINUED)
TYPE SPECIMENS DEPOSITED IN THE PEALE MUSEUM BY ALEXANDER WILSON, THEIR CURRENT LOCATION, IF KNOWN, AND OUR NOTES AND THOSE OF WALTER FAXON

Species	Peale No.	Notes
Magnolia Warbler	7783	
Grasshopper Sparrow	6585	MCZ
Mississippi Kite	403	Acad.[c]
Tennessee Warbler	7787	
Kentucky Warbler	7786	
Wilson's Warbler	7785	MCZ, "Probable type"
Nashville Warbler	7789	
Black-billed Cuckoo	1854	
Fish Crow	1369	
Whip-poor-will	7721, male 7722, female	
Solitary Sandpiper	7763	
Canvasback	2816	
Broad-winged Hawk	407	Acad.[d] Wilson gave the same name to the Broadwinged Hawk and the Slate-colored Hawk (male, Sharp-shinned Hawk). When Vieillot renamed the Broadwinged Hawk, he based his description on this specimen, pictured in Wilson's volume 6.

Note: Faxon's comments are from his "Early Editions of Wilson's *Ornithology*," *Auk* 18 (1901): 216–218. MCZ stands for Museum of Comparative Zoology, Harvard University.

[a] O. Bangs, "Wilson's Song Sparrow," *Proceedings of the New England Zoological Club* 4 (1912): 86.

[b] C. L. Bonaparte, "Observations on the Nomenclature of Wilson's Ornithology," *Journal of the Academy of Natural Sciences of Philadelphia* 3 (1824): 340–371.

[c] Committee on Publication, Untitled Notes on the Dispersal of Wilson's Specimens, section 11 (Philadelphia: Proceedings of the Academy of Natural Sciences, 1899).

[d] Ibid.

even in this department of the great chain of beings, will forever baffle all the artificial rules and systems of man. And this truth every fresh discovery must impress more forcibly on the mind of the observing naturalist."[48] Here Wilson recognizes the problem of continuity of variation among species that ultimately led Darwin to formulate the theory of evolution.

One factor that added to Wilson's consternation was his pioneering attempt to incorporate the bird's behavior into his choice of genus. Often this resulted in seemingly contradictory evidence, as in his account of the American Redstart (*Setophaga ruticilla*):

> Though this bird has been classed by several of our most respectable ornithologists among the Warblers, yet in no species are the characteristics of the genus *Muscicapa* more decisively marked; and, in fact, it is one of the most expert fly-catchers of its tribe. It is almost perpetually in motion, and will pursue a retreating party of flies from the tops of the tallest trees, in an almost perpendicular, but zig-zag direction, to the ground, while the clicking of its bill is distinctly heard; and I doubt not but it often secures ten or twelve of these in a descent of three or four seconds . . . [and its] bill, of the true *Muscicapa* form, [is] triangular at the base, beset with long bristles, and notched near the point.[49]

Wilson's splendid observation of the redstart's flycatching behavior and his detailed description of the bill and rictal bristles led him quite reasonably to disagree with previous authors (such as Linnaeus, Latham, and Gmelin) who had offered little detail. But rather than create a new genus within the warblers, most of which were in the genus *Sylvia,* Wilson simply followed the practice of most taxonomists of his day and forced the species into an existing genus. Gmelin too had introduced many new species in his thirteenth edition of Linnaeus's *Systema Naturae,* but not a single new genus.[50] Yet there were limits to this system. Frustration among European taxonomists with the artificiality of large and unwieldy genera led in the early part of the nineteenth century to the introduction of many new genera and to the revision of the generic relationships of many species. Indeed, Vieillot led the way beginning in 1808 when he established a number of new genera for North American birds in his *Septentrionale,* and Bonaparte helped accelerate the trend in his revision of Wilson's *American Ornithology* (1825–1833). Introduction of a family category that grouped similar genera and was included in an order of similar families also contributed to the resolution of many problematic relationships, but these changes occurred after Wilson's death. At the time Wilson wrote, taxonomists arranged species within Linnaeus's generic divisions, even when they recognized important structural, and in Wilson's case behavioral and ecological, differences among the species in the genus.

Coming as it did during a period of minimal taxonomic discovery, Wilson's work helped to stimulate the subsequent surge in ornithological exploration, particularly in North America where his influence was acknowledged by Charles Lucien Bonaparte, John James Audubon, and Spencer Baird. The latter directed the ornithological exploration of the United States west of the Mississippi while serving as secretary (the highest administrator) of the Smithsonian Institution. In fact, Wilson is second only to Linnaeus in the number of North American species first described as a proportion of the number of species described worldwide during the period in which he worked.[51] Wilson's discoveries of North American birds comprised almost 25 percent of all North American species described during the years in which he worked.

More Than Just New Species

Wilson's contributions to ornithological systematics go far beyond describing and naming new birds. He resolved a number of confusing situations in which previous taxonomists had mixed two species together or subdivided a single species into two or more species. Buffon, Latham, Pennant, and Catesby, for instance, had mixed together male and female Baltimore (*Icterus galbula*) and Orchard (*I. spurius*) orioles of different ages to produce no fewer than six different descriptions of the orioles (Table 4.4). Wilson was able to resolve the confusion through a detailed study of the bird's physical features; song; nest construction and materials; color of the eggs, which are illustrated

TABLE 4.4

WILSON'S CLASSIFICATIONS FOR VARIOUS ORIOLES

Species name in work(s) consulted by Wilson	Author(s)	Wilson's decision and currently accepted identification
Baltimore Oriole	Comte de Buffon[a], Latham[b]	Male—male Baltimore Oriole Female—male Orchard Oriole
Spurious Oriole	Comte de Buffon, Latham	Male—female Baltimore Oriole Female—immature male Orchard Oriole
Baltimore Oriole	Pennant[c]	Male—male Baltimore Oriole Female—immature male Baltimore Oriole
Spurious Oriole	Pennant	Male—male Orchard Oriole Female—immature male Orchard Oriole
Baltimore Oriole	Catesby[d]	Male—male Baltimore Oriole Female—none described
Spurious Oriole	Catesby	Male—male Orchard Oriole Female—immature male Orchard Oriole

Note: When Wilson sat down to classify the orioles of North America, he faced the following classifications by his predecessors. He sorted it out by raising and keeping several orioles for up to two years.

a. G.-L. Leclerc Comte de Buffon, *Histoire Naturelle, Générale et Particulière: Histoire Naturelle des Oiseayx* (Paris: Royale imprimerie, 1769).

b. J. Latham, Index Ornithologicus, Sive Systema Ornithologiae: Complectens Avium Divisionem in Classes, Ordines, Genera, Species, Ipsarumque Varietates: Adjectis Synonymis, Locis, Dexdriptionibus, &c, vol. 2 (London: Leigh and Sotheby, 1790).

c. T. Pennant, *Arctic Zoology*, 2d ed. (London: Robert Faulder, 1792).

d. M. Catesby, *The Natural History of Carolina, Florida and the Bahama Islands* (London, 1731), available in a facsimile edition edited by A. Feduccia and published by University of North Carolina Press, 1985.

in his plate (see Figure 3.3); and by keeping birds through several molts to observe the progression of plumage changes.

Division of Bobolinks (*Dolichonyx oryzivorus*) into two species by Catesby had been the source of confusion for almost a hundred years. The mixup hinged on the mysteries of where the male Bobolinks went during fall migration and where the breeding grounds of the Rice Bunting were located. Rice Buntings were abundant consumers of the rice crop in the Carolinas during their fall migration south, but otherwise unknown. By keeping a live male Bobolink in his room, Wilson found that in late June it sang less and less and soon began to molt from its striking white and gold back and black front to the tan and brown striped pattern of the female Bobolink and what was thought to be the Rice Bunting. He also dissected hundreds of Rice Buntings in September and October, when hunters were shooting them to protect the rice fields and sending cartloads to market in Philadelphia and elsewhere. Wilson found to his surprise that the testes of the males had shrunk to:

> no larger than the smallest pin's head, and in young birds of the first year can scarcely be discovered; though in the spring [based on Wilson's dissections] their magnitude in each is a least one hundred times greater. To an unacquaintance with this extraordinary circumstance, I am persuaded, has been owing the mistake of Mr. Catesby, that the females only return in the fall; for the same opinion I long entertained myself [as did Bartram], till a more particular examination showed me the source of my mistake. Since [then], I have opened and examined many hundreds of these birds, . . . and . . . found about as many males as females among them.[52]

By careful dissection Wilson not only resolved the mystery surrounding the Bobolink and Rice Bunting, but also described the seasonal change in the size of avian testes for the first time.

Despite his reluctance to create new genera, Wilson was not reluctant to reclassify species if he considered the classification of earlier authors mistaken. The example of the Northern Flicker initially classified as a cuckoo and the redstart mistakenly reclassified as a flycatcher based on its behavior have already been discussed; another example is the Blue Yellow-backed Warbler (Northern Parula, *Setophaga americana*), which Linnaeus placed with the tits in the genus *Parus*. Wilson explains his reasons for moving the species to the warblers, genus *Sylvia* in his taxonomy: "Notwithstanding the respectability of the above authorities, I must continue to consider this bird as a species of Warbler. Its habits, indeed, partake something of the Titmouse; but the form of its bill is decidedly that of the *Sylvia* genus."[53]

Wilson's designations of genus and species show that he succeeded in one of his most important goals for the *American Ornithology* project. He began *American Ornithology* by advocating taxonomic study based on everything about the bird: its appearance, its anatomy, and all aspects of its life history. His 304 species accounts offered just this kind of information: detailed physical descriptions of specimens together with information on their anatomy, behavior, clutch size, egg color and shape, ecology, migration, nest site selection, nest structure, song, and much more. Wilson, like Linnaeus, believed in the fixity of species, and so in the fixity of the characteristics that defined a species, but unlike Linnaeus whose definition was based on appearance, Wilson's definition encompassed the living bird, including its appearance to be sure, but also its anatomy and physiology (for example, activity level) and everything about its natural history. This was an important expansion of the species concept as practiced in the late eighteenth and early nineteenth centuries and represented an important shift in the paradigm of ornithology. Wilson established field observation of birds as a distinctive characteristic of American ornithology.

Expanding Scientific Boundaries

Much as he had expanded the definition of a species to include more than just its appearance, so too did Wilson expand ornithology to include more than just the physical description and classification of species. Many of his 304 species accounts are substantial and follow a standard, though flexible, format that foreshadows the species accounts of Arthur Cleveland Bent, published in twenty-three volumes between 1919 and 1968, and the more recent

"Life Histories of Birds of North America" species accounts now available on line. Wilson's accounts open with the English and scientific names of the species, followed by citations of previous taxonomic works, which often include the names given by earlier writers. Typically a few introductory comments begin the text portion of the species account. These comments often include a brief mention of the habitat or of local names for the species, and always feature the dimensions of the bird—bill tip to tail tip, and tip to tip of the fully extended wings—followed by a physical description that concludes with colors of the soft tissues. Wilson was well aware that these colors faded quickly after death, and he made notes on their color in the field soon after shooting the bird (see later and Figure 3.4). The detail of Wilson's physical descriptions is unmatched by his contemporaries, and among his predecessors similar attention to detail is found only in Francis Willughby's descriptions of European birds published in 1676.

Most species accounts also include a description of the bird's song and calls; nest, eggs, incubation, and nestling periods; habitat preference; migratory behavior with arrival and departure dates; geographic distribution; and some account of its behavior and ecology. Many of the accounts refer to correspondents for some of the information and many also cite references such as Pennant's *Arctic Zoology,* which was an important source of information on distribution of the more northern species. Common birds such as the Blue Jay are covered in greater depth than new species such as the Tennessee Warbler (*Oreothylpis peregrina*), although even new species receive some coverage of habitat, ecology, and behavior in addition to a detailed physical description.

Another innovation was Wilson's use of the text to explain and elaborate on particular aspects of his illustrations, such as the plumage. In his description of the plumages of the Baltimore and Orchard Orioles, for example, he suggests that the reader compare the two figures: "To convince the foreigner that the present is a distinct species from the Baltimore, it might be sufficient to refer to the representation of the latter, in Fig. 3, and to Fig. 14, of this work."[54] He also uses his plates to illustrate behavior such as in the following from the accounts of the Cow-Pen Bunting (Brown-headed Cowbird, *Molothrus ater*) and the Maryland Yellow-throat (Common Yellowthroat, *Geoth-*

lypis trichas): "Those that pass in May and June are frequently observed loitering singly about solitary thickets, reconnoitering, no doubt, for proper nurses, to whose care they may commit the hatching of their eggs, and the rearing of their helpless orphans. Among the birds selected for this duty are the following, all of which are figured and described in this volume: . . . the Maryland Yellow-Throat, on the ground, at the roots of brier bushes."[55] And later: "The individual from which the figure in the plate was taken, was the actual nurse of the young Cow-Pen Bunting, which it is represented in the act of feeding."[56]

In the eighteenth century, most collections were based either on sketches made by naturalists such as Georg Forster, John Abbot, or William Bartram, or on specimens preserved in alcohol, often brandy, which was not infrequently drunk by sailors transporting the specimens from the site of collection to the residence of a wealthy European patron like Sir Hans Sloane. Alcohol leached the color from the feathers, although the physical proportions and anatomy remained intact. Another method of preservation was drying, but this not only distorted the specimen; it also did not prevent its eventual consumption by insects and microorganisms. Patrons receiving such specimens would thus hire someone to illustrate and describe them before they were consumed. Sir Hans Sloane hired George Edwards for this purpose and the result was the four-volume *Natural History of Uncommon Birds,* which simply referred to whatever Sir Hans Sloane had in his cabinets, and the three-volume *Gleanings of Natural History,* which described what Sloane acquired after Edwards wrote the previous four volumes.

Unlike the wealthy patrons of Europe, Peale wanted specimens for his museum, which was open to all those who paid a small fee to view the specimens of animals, plants, and artifacts. One of Peale's challenges, then, was to preserve the animal specimens in such a way that he could both exhibit them and preserve them (so that he was not under constant pressure to find new material).

Wilson's need for specimens derived from a very different concern: a need to confirm the flood of observations he received from the public, some of which referred to birds by their regional names, with actual physical data.

With publication of his first volume, Wilson became the authority on North American birds and began to receive letters from throughout the country. Some were from people who had questions, but many were from individuals who wished to contribute their own observations: "the author has been honoured with communications of facts, from various quarters of the United States, . . . For all these he returns his most grateful acknowledgments; and gladly indulges the idea that they will become more and more frequent."[57]

Since at least September 1808 Wilson had worked to establish a network of correspondents so that, as he wrote from New England to a friend in Philadelphia, "scarcely a wren or tit shall be able to pass along, from New York to Canada, but I shall get intelligence of it."[58] He understood that to research and write about so many species dispersed over such a large area he needed more than just his personal observations, that "unless assisted by the experience and observations of others, a thousand interesting facts and minutiæ of character, would unavoidably escape him, which might otherwise have formed the most valuable part of his publication."[59]

With this statement in the introduction to the second volume of *American Ornithology,* Wilson became the first to recognize the importance of citizen science, the first to publicly encourage correspondents to send information, and the first to set standards so that he could verify the information he received. As the following explanation suggests, verification was a little more complicated in 1810 than it is today. From the preface of volume 2 Wilson writes to future correspondents that "as the provincial names of many of our birds are so multiplied, and frequently so local, as to be altogether unknown in other districts; and, as the communications of those unacquainted with the scientific names and arrangement, render it sometimes very difficult to determine what particular species is really meant; if, in addition to well authenticated facts, preserved skins of such birds as are supposed rare or new, could be conveniently transmitted to the author, the obligation would be greatly increased and properly acknowledged."[60]

To facilitate sending "preserved skins," and perhaps to avoid opening a package of putrified flesh, Wilson printed the following recipe for the preparation of study skins as a footnote to the preface of volume 2:

> As soon as the bird is shot, let *memoranda* be taken of the length, the breadth (measuring from tip to tip of the expanded wings), color of the eyes, bill, legs and feet, and such particulars of its manners &c as may be known [see Figure 3.67 for an example of Wilson's field notes as just described by him]. Make a longitudinal incision under the wing, sufficiently large to admit the body to be taken out; disjoint the wing close to the body under the skin, and endeavour with a pair of scissors or penknife to reach the neck, which cut off; pass the skin carefully over the other wing, which also disjoint and separate from the body, then over the whole body and thighs, which last cut off close to the knees; lastly, separate the whole skin from the body at the roots of the tail feathers, which must not be injured. Return to the neck and carefully pass the skin to, and beyond, the eyes, which scoop out; cut off the neck close to the scull, penetrate this way with your knife into the brain, which scrape completely out; dissect all the fleshy parts from the head, wings and skin; rub the whole in side with a solution of arsenic, sprinkle some of the same into the cavity of the brain, throat, &c; stuff the vacuity of the brain and eyes with cotton, to their full dimensions; return the skin carefully back, arranging the eye lids and plumage; stuff the whole with cotton to its proper size and form, sew up the longitudinal incision, and, having carefully arranged the whole plumage, sprinkle it outwardly with a little powdered arsenic; place it in a close box into which some camphor has been put, and cover it with cotton or ground tobacco. In the whole operation the greatest care must be taken not to soil the plumage with blood.
>
> If arsenic cannot conveniently be had, common salt may be substituted.[61]

This is the first published methodology for the preparation of a study skin and probably represents a collaboration between Peale, who most likely provided the chemical knowledge, and Wilson, who almost certainly contributed the anatomical knowledge. We know that at least one correspondent complied with Wilson's request, and was gratefully acknowledged. When the family of William Dunbar, one of Wilson's correspondents, sent information on a Roseate Spoonbill (*Platalea ajaja*) collected near Natchez, Mississippi, they sent a preserved specimen from which Wilson drew the bird shown in Figure 3.70. After Wilson prepared the final plate, the specimen was deposited in the Peale Museum (No. 3553). The specimen now resides in the Museum of Comparative Zoology at Harvard University and is clearly identifiable, by the unusual coloring of its face, as the Roseate Spoonbill from which Wilson's figure was painted (Figure 3.71).[62]

Through his association with Charles Wilson Peale and his museum, Wilson was able to provide not only pictures of each species, but also study skins of 255 of the 283 species he described. Of these, twenty-three species were represented by two or three specimens and twenty-four were type specimens. Six of the type specimens are in the Museum of Comparative Zoology at Harvard and two reside in the collection at the Academy of Natural Sciences in Philadelphia (see Table 4.3).

The Scientific Process

Wilson, who was not a trained scientist, was unexpectedly good at generating hypotheses based on his observations and then testing those hypotheses. A particularly interesting example occurs in his account of the Mississippi Kite (*Ictinia mississippiensis*), which is interesting because his idea seems so unlikely, even to him:

> I frequently remarked this Hawk sailing about in easy circles, and at a considerable height in the air, generally in company with the Turkey Buzzards, whose manner of flight it so exactly imitates as to seem the same species, only in miniature, or seen at a more immense height . . . They may perhaps be engaged, at such times, . . . in pursuit of their respective food;—one, that he may reconnoiter a vast extent of surface below, and trace the tainted atmosphere to his favorite carrion; the other in search of those large beetles, or coleopterous insects, that are known often to wing the higher regions of the

> air; . . . For several miles, as I passed near Bayou Manchak, the trees were swarming with a kind of *Cicada,* or locust, that made a deafening noise; and here I observed numbers of the Hawk now before us sweeping about among the trees like Swallows, evidently in pursuit of these locusts; so that insects, it would appear, are the principal food of this species.[63]

Although he has proposed the hypothesis that insects are the principal food of this hawk, Wilson found it hard to believe given the powerful, hooked bill and equally powerful talons. Wilson himself experienced this hawk's ferocity:

> This Hawk, though wounded and precipitated from a vast height, exhibited, in his distress, symptoms of great strength and an almost unconquerable spirit. I no sooner approached to pick him up than he instantly gave battle, striking rapidly with his claws, wheeling round and round as he lay partly on his rump, and defending himself with great vigilance and dexterity; while his dark, red eye sparkled with rage. Notwithstanding all my caution in seizing him to carry him home, he struck his hind claw into my hand with such force as to penetrate into the bone. Anxious to preserve his life, I endeavored gently to disengage it; but this made him only contract it the more powerfully, causing such pain that I had no other alternative but that of cutting the sinew of his heel with my penknife.[64]

Wilson probably cut the Achilles tendon, thereby causing the hawk's grip to relax enough so that he could extract the talon from his hand.

Wilson's "home" at the time was the estate of William Dunbar, a fellow Scottish immigrant and naturalist living in Natchez, Mississippi. Although confined to bed and mortally ill, Dunbar had heard that Wilson was in Natchez and insisted that he stay at "The Forest," Dunbar's plantation, while writing about and sketching birds of the area. Wilson gratefully accepted the invitation and after the incident with the Mississippi Kite received care for both the injured bird and his punctured hand, which must have healed without incident because this is the only mention of the injury in either *American Ornithology* or his letters from Natchez and New Orleans.

Wilson's observations and experience with this newly discovered species led him to hypothesize that despite the bird's power and weaponry, it fed largely on insects, which it caught in the air. He proceeded to test his hypothesis in the most direct way possible. He shot and dissected several kites, finding that "those large beetles, or coleopterous insects, that are known often to wing the higher regions of the air; and which, in the three individuals of this species of hawk, which I examined by dissection, were the only substances found in their stomachs."[65]

The process of observing birds, interpreting their characteristics and behavior, formulating a hypothesis, and testing that hypothesis may seem simplistic to modern readers. Certainly the process as described by Wilson is not difficult to grasp, and the steps he took to complete it were not difficult. But at the time, no other naturalists were applying this level of rigor to their observations. Many, like Gmelin and Latham, were not even making observations and the few that were tended to observe, but not test their interpretations (for example, Forster, Pallas, Pennant, and Vieillot). Furthermore, Wilson addressed ornithological misinformation throughout *American Ornithology* by testing each myth, whether it was the disappearance of male Bobolinks or the hibernation of swallows in the mud at the bottom of ponds. He questioned the assertions of previous authors and tested their statements, rather than merely repeating them. In doing so he brought a new level of rigor to field biology generally and ornithology specifically.

Watching and Counting

A major innovation in Wilson's approach to field observation and description of species was his use of quantification. When he began teaching school, he spent evenings teaching himself mathematics. Initially this was a matter of improving his ability to teach his students, but he later taught himself surveying. Ultimately he became proficient enough to survey property boundaries for members of the Milestown community where he lived and taught.

Wilson's most straightforward use of quantification was in his descriptions of behavior such as nest construction by the Orchard Oriole (*Icterus spurius*):

> They usually suspend their nest from the twigs of the apple-tree; and often from the extremities of the outward branches. It is formed exteriorly of a particular species of long, tough, and flexible grass, knit, or sewed through and through in a thousand directions, as if actually done with a needle. This nest is hemispherical, three inches deep by four in breadth; the concavity scarcely two inches deep by two in diameter. I had the curiosity to detach one of the fibres, or stalks of dried grass, from the nest, and found it to measure thirteen inches in length, and in that distance was thirty-four times hooked through and returned, winding round and round the nest![66]

Not all species accounts contain such detailed, quantitative observation, but many do—it is characteristic of Wilson's species accounts, not an isolated example. No other ornithological or natural history work published prior to or contemporaneous with *American Ornithology* contains this level of quantitative detail.

Wilson also quantifies migration dates, distances flown, and flock sizes. Certainly his most famous estimate is that of a flock of the now extinct Passenger Pigeon (*Ectopistes migratorius*), that flew over him continuously for four hours as he rode on horseback toward Frankfort, Kentucky:

> If we suppose this column to have been one mile in breadth, (and I believe it to have been much more,) and that it moved at the rate of one mile in a minute, four hours, the time it continued passing, would make its whole length two hundred and forty miles. Again, supposing that each square yard of this moving body comprehended three Pigeons, the square yards in the whole space, multiplied by three, would give two thousand two hundred and thirty millions, two hundred and seventy-two thousand Pigeons!—an almost inconceivable multitude, and yet probably far below the actual amount.[67]

Those who have looked at the historical records of Passenger Pigeons agree with Wilson's concluding thought that 2,230,272,000 is "probably far below the actual amount." Such a number of Passenger Pigeons is hard to imagine, but Wilson's presentation helps us to understand how reasonable the number actually is. More importantly his calculation and his presentation of the steps used to arrive at his staggering estimate are unprecedented in the ornithological works of either his predecessors or his contemporaries. The quoted passage seems so utterly simple from our twenty-first-century perspective, but in 1812 it represented a startling innovation.

Finally, Wilson undertook the first breeding bird census ever done. He introduced the concept with the following estimate:

> "From the twenty-first day of March to the first of May, it might with truth be asserted, that at least one hundred million of birds enter Pennsylvania from the south; part on their way farther north, and part to reside during the season. This is no extravagant computation, since it is allowing only about four hundred individuals to each square mile; tho even those resident for the summer would probably average many more."[68]

We might be willing to accept this estimate, but Wilson did not rely on the credulity of his readers nor on their good will. He spent the spring and summer of 1811 as a guest of William Bartram, living in the Bartram home, researching and writing volumes 4 and 5, and conducting a census of the nesting birds. Here in Wilson's own words are the data:

> In Mr. Bartram's botanic garden, and the adjoining buildings, comprehending an extent of little more than eight acres, the author has ascertained, during his present summer residence there, that not less than fifty-one pair of birds took up their abode and built their nests within that space.
>
> 5 pair of House Wrens,
> 1 pair of Common Pee-wees (Eastern Phoebe),

2 pair of Baltimore Orioles,
2 pair of Orchard Orioles,
3 pair of Summer Yellow Warblers,
5 pair of Catbirds,
1 pair of White-eyed Flycatchers,
2 pair of Warbling Flycatchers (Vireos),
1 pair of Robins,
1 pair of Swamp Sparrows,
1 pair of Wood Pee-wees,
1 pair of Indigo Buntings,
1 pair of Yellow-breasted Chats,
4 pair of Purple Grakles,
5 pair of Song Sparrows,
3 pair of Chipping Sparrows,
2 pair of Chimney Swallows (Swifts),
1 pair of Purple Martins,
10 pair of Barn Swallows[69]

Wilson extrapolated from this modest census to the population of migrants in an estimated 3,200,000 acres of suitable land in the state of Pennsylvania to get an estimate of 153,000,000 birds, which makes his previous estimate of 100,000,000 conservative. What is important is that this is the first breeding bird census ever done. The concept was new. Furthermore, his extrapolation from his census sample to the population of migratory and resident birds inhabiting Pennsylvania was the first estimate ever made of a wildlife population.

Economic Ornithology

Ornithologists, including Catesby and before him, Lawson, had included comments about the value of some species as food, but Wilson was the first ornithologist to write about the economic impact of birds beyond their value as human food.[70] For example, when writing about White- and Red-breasted Nuthatches he provides a glowing account of their contribution to the health of our trees: "Both these little birds, from the great quantity of destructive insects and larvae they destroy, both under the bark and among the tender buds of our fruit and forest trees, are entitled to, and truly deserving of, our esteem and protection."[71] Writing about the Orchard Oriole, which nests in orchards, he comments: "The Orchard Oriole, though partly a dependant on the industry of the farmer, is no sneaking pilferer, but an open and truly beneficent friend. To all those countless multitudes of destructive bugs and caterpillars that infest the fruit-trees in spring and summer, preying on the leaves, blossoms, and embryo of the fruit, he is a deadly enemy."[72] Though some may raise questions about the oriole, Wilson's report suggests that they, like the nuthatches, offer unmitigated benefit. Wilson emphasizes the point by showing the male Orchard Oriole about to grab a caterpillar from an apple leaf in the illustration in volume 1 of *American Ornithology* (see Figure 3.3).

The Red-headed Woodpecker is a more difficult case, but here Wilson truly enters into the field of economic ornithology by pleading for a balanced approach.

> But let us not condemn the species unheard: . . . If their merits and usefulness be found, on examination, to preponderate against their vices, let us avail ourselves of the former, while we guard as well as we can against the latter. Though this bird occasionally regales himself on fruit, yet his natural and most useful food is insects, particularly those numerous and destructive species that penetrate the bark and body of the tree to deposit their eggs and larvæ, the latter of which are well known to make immense havock.[73]

Although not quantitative, Wilson urged a balanced, ecological approach to the question of crop damage. Just prior to this passage, he described the "vicious" attitude of many people toward Red-headed Woodpeckers and the existence of bounties in many communities and some states. In that context he urged taking a balanced approach to the question of fruit destruction by the woodpeckers. Both the reference to their varied diet and the effort to look at the issue broadly represented a new attitude toward the natural world. But Wilson did not always advocate for preserving the bird. For example, he described the destruction of crops by the Bobolink and offered no alternative to shooting them to protect the harvest.

Wilson's discussion of economic ornithology, even for species with seemingly little economic impact, for example the Common Yellowthroat, was part of his attempt to change Americans' attitude toward birds. He set out to convince readers of the value of our native birds, that they were part of our cultural heritage as Americans. Wilson also set out to persuade his readers that birds in particular, and nature in general, were not adversaries to be conquered and destroyed, but potential allies with whom to live and work. This view was radically different from the perspective of those who wrote about nature in the previous century. Wilson understood the power of an economic incentive and his insight was to show that birds could contribute to our economic well-being. This vision sparked a debate that continues today in which participants often refer to the economic benefits to be derived from saving habitats and birds.

Wilson is properly recognized as the Father of American Ornithology on several counts. He was the first American to describe and classify the birds of North America within the Linnaean system. He also managed to include in *American Ornithology,* his most influential work, descriptions of almost 80 percent of the bird species present in the United States in 1812, thereby establishing the foundation for all future description and taxonomy of North American birds.

Before and during Wilson's life, American ornithology was undertaken by Europeans who either depended on preserved specimens or made relatively brief trips to parts of North America. Because Wilson lived in the United States and visited most of the country, he was able to write about living birds. His emphasis on observation and description of live birds introduced a new perspective into ornithology, one shaped by his years as a poet and observer of human nature. Wilson's ability to generate hypotheses and test them by careful observation of wild birds was little short of revolutionary.

Also unusual was his extraordinary ability to keep birds in captivity and observe their plumage changes. In this way he resolved many errors made by earlier taxonomists who had to base their decisions on specimens brought back from North America.

Wilson was also the first ornithologist to introduce quantification into the study of live birds, and he brought the concept of economic ornithology to the discipline. Prior to his work authors had written about the food value of those species large enough to have such value, but Wilson expanded this idea to include the ways that birds both benefit humans through their control of insects and indirectly harm them by eating crops. Furthermore, he recognized that some birds might require a cost/benefit analysis since they ate pests or weed seeds, but also consumed some human crops. He even provided analyses for some of these species to show how carefully we should draw our conclusions. In his approach to economic ornithology and elsewhere in discussing differences in the natural history of species, he introduced several of the concepts on which ecology is based.

The influence of Alexander Wilson's accomplishments is profound. His taxonomy, his quantification, his introduction of economic considerations, and his scientific approach to behavioral observation are clear in his writings. These advances would chart the course for ornithology and generations of ornithologists yet to be born.

CHAPTER FIVE

Wilson's Legacy

On 23 August 1813, at age forty-seven, Alexander Wilson died in the home of William Jones, where he had lived for several years. The next day, after a simple service, he was buried in the graveyard of Philadelphia's Gloria Dei Church, better known as Old Swedes Church, in the Jones family plot. The eighth volume of *American Ornithology* was written, edited, illustrated, and in production. The ninth volume was in preparation, but consisted only of a few drawings for the plates, some text, and a few notes. George Ord decided to edit the ninth and final volume, and although he was greatly surprised at how few notes Wilson had left, he pulled together what existed and added the first biography of the author. In 1814 the ninth and last volume was published, literally closing the book on the life and work of Alexander Wilson.

Born a Scot, Wilson was, at the time of his death, emotionally and legally an American citizen. He was, unquestionably, the first American ornithologist and is regarded by ornithologists as the Father of American Ornithology. Nonetheless, Wilson's reputation among the general public and even among birders and ornithologists is dwarfed by the reputation of John James Audubon. Stop someone on the street or ask the person next to you in the coffee shop who the father of American ornithology is and the answer will almost certainly be Audubon. Unquestionably Audubon's romantic portraits of American wildlife and the growth of national and state Audubon societies in the twentieth and twenty-first centuries have done much to promote our appreciation, knowledge, and concern for birds. What then is Wilson's legacy and why do ornithologists consider him the Father of American Ornithology?

Before Wilson, several Europeans visited some part of North America and drew its birds. John White came to North America in 1587 as the governor of the ill-fated colony on Roanoke Island. He was the first European to draw American birds, even though his drawings of thirty-two species remained packed in a family attic until the eighteenth century, when they were discovered by a descendant and eventually, in 1753, passed into the collection of the British Museum.[1] White's work is of historical interest, but had little to no effect on the genesis of modern ornithology.

John Lawson first journeyed to the British colonies before 1709, the date of publication of his book *A New Voyage to Carolina.* The book contains a list and descriptions of 129 species.[2] Mark Catesby visited southeastern North America and the Bahamas twice, in 1712–1719 and 1722–1726. During his first visit he explored the coastal plain as had White and Lawson, but also traveled inland to the Appalachians. After his second trip he settled in England to write and produce the two-volume *Natural History of Carolina, Florida, and the Bahama Islands,* in which he illustrated 109 species of birds. The volumes were published in 1732 and 1747. Catesby's list of birds and their descriptions are based on Lawson's book, but Catesby illustrated the species, which gave his work far greater impact than Lawson's.[3] Of Catesby's 220 plates, seven were copied directly from White. All were cited by Linnaeus, and the catfish and gar were cited as type specimens.

Two persons born in North America also contributed to American ornithology before Wilson. William Bartram (Upland Sandpiper [*Bartramia longicauda*]) published a list of species in his *Travels,* but the list is essentially the same as Catesby's—either because Bartram traveled through the same area or because he used Catesby's list as a reference. Thomas Jefferson published the first state bird list in his *Notes on the State of Virginia,* but his list, too, is very similar to Bartram's—and whether this is because Jefferson collected in the same area or used Bartram's list as his reference cannot be known; Jefferson and Bartram were contemporaries and correspondents.

All of these lists suffer from a common problem: none use Linnaean taxonomy. White, Lawson, and Catesby published before the tenth edition of the Carl Linnaeus's *Systema Naturae* (1758), which marks the starting point of modern systematic naming of species. Bartram and Jefferson published later, but did not accept Linnaean classification or names. Why Bartram failed to use the new system is unclear, but in Jefferson's case his failure to adopt Linnaean names may have been symptomatic of his rejection of European influences.

Wilson, unlike the preceding individuals, is considered the Father of American Ornithology for many reasons, but chief among them are that he collected and described birds from most states and territories in the United States of 1804–1813; he classified species according to the Linnaean system and placed new species within the system; he illustrated all the species he described so that readers could identify the birds that they saw; and he introduced a truly scientific approach to ornithology—dissection to explore dietary and morphological detail, and used behavioral, ecological, and quantitative observations. Whereas his predecessors had been sponsored by a wealthy patron, Wilson proved that there was enough public interest to successfully underwrite a large publishing project in natural history through an economic model based on subscriptions by many essentially middle-class individuals.

Certainly Wilson helped create an appetite among Americans for knowledge about their birds. Less well known today, even among professional ornithologists, is that *American Ornithology* was the first major scientific work published in the United States. As such it caused European scientists to sit up and take notice. Baron Cuvier, the French anatomist and paleontologist, who was probably the most prominent scientist in France, if not all Europe, in the early nineteenth century, wrote of Wilson, "He has treated of American birds better than those of Europe have yet been treated."[4]

A deeper appreciation of Wilson's importance in the development of ornithology in the United States may be gained by considering the men who were inspired by him and followed his example of exploring North America and describing its birds. Included with each ornithologist listed are, in parentheses, birth and death dates and the species that now bear his name.

Audubon, John James (1785–1861, Audubon's Shearwater [*Puffinus lherminieri*], Audubon's Oriole [*Icterus graduacauda*]). Audubon met and interacted with Wilson for several days in March 1810 in Louisville, Kentucky. Wilson described his meeting with Audubon in journal entries written while in Louisville on his way south and west to the Mississippi River, Natchez, and New Orleans:

> **March 17** (Saturday).—Rained and hailed all last night, set off at eight o'clock, after emptying my boat of the deluge of water. Rowed hard all day; at noon recruited myself with some biscuits, cheese and American wine. Reach the falls—night sets in—hear the roaring of the rapids. After excessive hard work arrive at Beargrass creek, and fasten my boat to a Kentucky one. Take my baggage and grope my way to Louisville—put up at the Indian Queen tavern, and gladly sit down to rest myself.
>
> **March 18.**—Rose quite refreshed. Found a number of land speculators here. Titles to lands in Kentucky subject to great disputes.
>
> **March 19.**—Rambling round town with my gun. Examined Mr. A's drawings in crayons—very good. Saw two new birds he had, both *Motacillae.*
>
> **March 20.**—Set out this afternoon with the gun—killed nothing new. People in taverns here devour their meals. Many shopkeepers board in tavern—also boatmen, land speculators, merchants, &c. *No naturalist to keep me company.*
>
> **March 21.**—Went out this afternoon shooting with Mr. A. Saw a number of Sandhill cranes. Pigeons numerous.
>
> **March 23.**—Packed up my things which I left in the care of a merchant here to be sent on to Lexington.[5]

On March 19, Wilson delivered his letters of introduction to the prominent citizens of Louisville and spent the rest of the week waiting for responses that he could translate into subscriptions to *American Ornithology.* He received no responses, so upon the recommendation of Benjamin Bakewell, a prosperous glass manufacturer whom he had met in Pittsburgh, he paid a visit to Audubon. When Wilson showed Bakewell volumes 1 and 2, Bakewell, the uncle of Audubon's wife, mentioned the artistic ability of the young Frenchman and suggested that Wilson visit Audubon when he passed through Louisville.

According to Audubon's account written some sixteen years later, Wilson showed him volumes 1 and 2 of *American Ornithology* and explained how he could subscribe. As Audubon was about to sign for a subscription, his partner called to him in French from the back of the store. "My dear Audubon, why do you want to subscribe?" Audubon paused and Ferdinand Rozier continued, "Your drawings are far better." Audubon admitted that his vanity was aroused. He knew he was the better artist. Rozier spoke again, still in French, "and you must know as much about birds as he." Audubon noted that Wilson looked displeased and wondered if he understood French, because Rozier's last comment was untrue and he knew it. Audubon put the pen down and declined to subscribe. Wilson eased the tension by asking Audubon if he had his own drawings of birds, which suggests that he indeed had understood some of the preceding conversation between Audubon and Rozier. Audubon retrieved his portfolio from the back of the store, put it on the table at which Wilson was seated, and let him look through it. Wilson was impressed and said so. Two flycatchers, possibly new species, caught Wilson's attention and he arranged to go hunting with Audubon later in the week. Wilson then asked if he planned to publish his work. Audubon said no and Wilson seemed surprised at this answer.[6] Thus far Audubon's recollection agrees with Wilson's journal and they did hunt together. Audubon's account then diverges from Wilson's account in his journal and in a letter to Lawson, written when he reached Lexington. According to Wilson's journal, it seems that the men did not hunt throughout the week nor did they obtain birds that Wilson had never seen. We know that when Wilson did find a new species he would shoot it and prepare a specimen, which was ultimately deposited in the Peale Museum in Philadelphia. The museum is no longer present, but a portion of Wilson's collection is now at Harvard and there are no specimens from the Louisville area nor does Peale mention receiving any. In addition, Audubon wrote that he offered to send Wilson notes and sketches of birds on the condition that he receive credit for whatever contributions he made, adding that Wilson declined his offer. There is no record of the conversation in Wilson's papers and Audubon's claim is extremely unlikely. For years Wilson had been at pains to establish correspondents wherever he went for the express purpose of helping him learn about the habits of birds throughout the United States. He already had a flourishing correspondence with John Abbot, for instance, who was an excellent artist and an observant naturalist. He cites Abbot's contributions many times in *American Ornithology.* Furthermore, he published a request for information to readers of his books in the preface to volume 2 along with directions for preparing specimens to document any unusual information. William Dunbar of Natchez, Mississippi, responded to this request by sending Wilson a specimen of a Roseate Spoonbill (Figure 3.71), and Wilson not only acknowledged Dunbar for his contributions, but also visited him when he was in Natchez in 1810. Wilson published a further request for information in the preface to volume 3.

We do know that before he left Louisville, Wilson gave Audubon a synopsis of his *American Ornithology.* The synopsis was essentially a checklist that Wilson had compiled from previous works on American birds and was intended as an advertisement for his books. Audubon found it useful and kept it for decades.

However truthful Audubon's recollection was, the two men appear to have parted on good terms and never met again. Except for the brief mention of Audubon in his journal, which is now lost, Wilson never mentioned the artist in his correspondence nor in *American Ornithology.* But he does mention his time in Louisville in a letter to Alexander Lawson that was later published in *Port Folio.*

> **March 23** . . . I bade adieu to Louisville, to which place I had four letters of recommendation, and was taught to expect much of everything there; but neither received one act of civility from those to whom I was recommended, one subscriber, nor one new bird; though I delivered my letters, ransacked the woods repeatedly, and visited all the characters likely to subscribe. Science or literature has not one friend in this place. Everyone is so intent on making money that they can talk of nothing else; and they absolutely devour their meals that they may return the sooner to their business. Their manners correspond with their features.[7]

Wilson's harsh account reflects his disappointing sales in Louisville; he had clearly expected a better response given his letters of recommendation to individuals who had the means to subscribe. For his part, Audubon was undoubtedly upset by the unflattering reference to Louisville and its inhabitants, and one wonders if he may have altered his account of his meeting with Wilson to show that Wilson had rejected him first. Yet despite Wilson's disappointment and Audubon's unhappiness, there was no argument. Wilson was long dead when Audubon penned his recollection of their meeting.

If their meeting seems to have had little effect on Wilson, it had an enormous influence on Audubon. By his own account Audubon had no plan to publish when he first met Wilson and looked at volumes 1 and 2 of *American Ornithology*. At the time, he was a storekeeper who painted portraits and gave drawing lessons as a way of earning extra cash.[8] He was a gifted artist with no focus for his talent. When he saw Wilson's books, and when Rozier indicated that he could produce illustrations that were at least as good and write a natural history text to complement the art, a career as an artist-naturalist opened before him.[9]

Audubon's career change would take time. He and Rozier closed their store, packed their goods and belongings on a flatboat, and steered west along the Ohio River to Henderson, Kentucky, a community of about 159 persons. There they moved into an abandoned log cabin and opened a store. Audubon hunted and fished, at first only to provide food for his wife, two sons, and partner, but by the end of May he was also collecting and sketching birds. Although he had previously written captions to his sketches in French, he now wrote them in English and began to keep detailed notes, also in English.[10] In this way Audubon began the collection of bird paintings for the publication that was taking shape in his mind. It was not quite three months since his conversation with Wilson.

In April 1824, Audubon arrived in Philadelphia hoping to support himself by giving private lessons in sketching with pastels and painting with watercolors. He also hoped to improve his oil painting technique by taking lessons from one of the Peales. Audubon knew that Wilson had published *American Ornithology* in Philadelphia, and he had his portfolio with him to explore the possibility of publication. He stayed with an old friend, James Mease, a physician, geologist, and encyclopedist. As Mease's guest, Audubon had access to the libraries of the Philadelphia Athenaeum and the American Philosophical Society, which contained, among other ornithologies, all nine volumes of Wilson's *American Ornithology.* During that first week Mease introduced him to Thomas Sully, America's most famous portrait painter, who agreed to help Audubon improve his oil painting technique in return for drawing lessons for his daughter. Later that same week Mease introduced Audubon to Charles Lucien Bonaparte, nephew of the deceased emperor, who was revising and updating *American Ornithology.* Bonaparte took Audubon to an evening meeting of the Academy of Natural Sciences and there, on a Saturday evening in March 1824, the claims and counter claims between Audubon and "Wilson" began. But Wilson had been dead for eleven years. He played no part in the evening or the events to follow.

At the meeting, Audubon opened his portfolio and invited members of the Academy of Natural Sciences to view his birds. Bonaparte and Thomas Say, an explorer, artist, and the Father of American Entomology, were impressed. All was well. Then George Ord, vice president of the academy, editor of the ninth volume of *American Ornithology,* and one of the executors of Wilson's will, joined the group. He took an immediate dislike to the work. He objected to the co-mingling of birds and plants, to the unnatural positions of many of the birds, and to the anatomical inaccuracies forced by their melodramatic positions. At the time Ord was planning a three-volume edition of Wilson's work and probably saw Audubon as a competitor. He blocked Audubon's election to the Academy.

Two days later, on Monday, Bonaparte took Audubon to meet Titian Peale, who was working on the illustrations for Bonaparte's revision and expansion of Wilson's *American Ornithology.* After looking at Peale's work, Audubon pronounced it lifeless, lacking the animation that was the essence of the living bird. Peale was stung by the criticism, but Bonaparte, still impressed by what he had seen of Audubon's work, began to think of collaborating with the talented, if brash, artist-naturalist.

On Wednesday, Bonaparte once again picked up Audubon in his carriage and took him to meet Alexander Lawson, the foremost engraver in Philadelphia, who had engraved Wilson's volumes and was now engraving Rembrandt Peale's drawings for Bonaparte's revision of *American Ornithology*. Lawson thought that Audubon's sketches were extraordinary for an untrained backwoodsman (he had judged Audubon by his clothes and long hair), but "not true to nature, and anatomically incorrect." A second engraver whom they consulted, Gideon Fairman, had no such objections, but did suggest that Audubon go to England to have his work engraved.

Several days later Audubon visited Lawson's shop, where the two men argued about Audubon's request that Lawson engrave his drawing of a Great-tailed Grackle (*Quiscalus mexicanus*). The argument ended bitterly.

In June, Rembrandt Peale visited Audubon, viewed his drawings, expressed his admiration for them, and invited Audubon to his studio to view a portrait of George Washington. Audubon went to Peale's studio, but considered the portrait inferior to the work of Thomas Sully, with whom he was studying oil painting. At this point, then, Audubon had managed to alienate just about everyone who had known Wilson and wished to preserve his legacy.

Bonaparte and the French consul in Philadelphia were pressuring Audubon to have his drawings engraved in France, but Henry McMurtrie, a prominent physician and naturalist who visited Audubon and admired his drawings, advised him to have his work engraved in England. Eventually Audubon would heed Fairman's and McMurtrie's advice and go to England, where *The Birds of America* would be engraved and published and where he would collaborate with William MacGillivray to write his *Ornithological Biography*.

Wilson's visit had spurred Audubon to focus his extraordinary artistic talent on birds and to plan a publication on the birds of America. Unfortunately, George Ord's jealous guardianship of Wilson's legacy blocked Audubon in Philadelphia. Equally unfortunate was Audubon's failure to accept Wilson's legacy, a decision that added to the appearance of discord between the two ornithologists.

Audubon's rejection of Wilson's legacy is most evident in his plagiarism of a number of Wilson's birds in *The Birds of America*. Among the best known are Audubon's drawing of the Yellow-shafted Flicker (*Colaptes auratus*). The females bickering at the top of the stump are unquestionably Audubon's. But the three males, two partially hidden and one perched on an open branch at the lower right, are tracings of Wilson's male flicker from plate 3, volume 1 of *American Ornithology* (see Figure 3.17). The addition is undoubtedly Audubon's, not one made by Richard Havell, Jr., Audubon's engraver, since the original drawing has had five inches of paper added at the bottom to accommodate two of the males. Audubon's drawing of a Mississippi Kite (*Ictinia mississippiensis*) corresponds exactly to the kite at the top of the branch in Havell's engraving for plate 117 in *Birds of America*. The Mississippi Kite lower on the branch, however, is the mirror image of Wilson's Mississippi Kite from his volume 3, plate 25. Was the lower kite Havell's addition or Audubon's? It is not in Audubon's original drawing. Audubon's drawing of the Red-winged Blackbird (*Agelaius phoeniceus*) is another case in point. It contains four individuals. The female at the bottom is almost identical to the female Red-winged Blackbird in Wilson's volume 4, plate 30 (or in this book, Figure 3.34; juvenile at the bottom and male at the top). Audubon's male is in the same position as Wilson's male, but its wings and tail are somewhat more spread out and the feathers are bent as if with the force of its dive. Audubon's male is more dynamic, although the position is unmistakably the same as Wilson's male.

Such plagiarism was not uncommon at the time. Recall Catesby's copies of White's birds and fish, two of which became type specimens in Linnaeus's *Systema Naturae*. But unfortunately, Audubon's desire to overshadow Wilson led to more serious transgressions. In the summer of 1812, Ord brought Wilson an unknown small yellow and brown bird, which Wilson drew and squeezed into a plate already filled with hawks and owls. He described the new species and named it the Small-headed Flycatcher. In 1838 Audubon described a similar bird for his *Ornithological Biography* and claimed that Wilson had drawn the bird from a sketch that he, Audubon, had loaned Wilson when the two were together in Louisville in 1810. Ord responded that it wasn't possible: he had been with Wilson when the bird was shot. Lawson added that he had engraved the bird, using a stuffed specimen as a reference. The amusing part of the spat is that no Small-headed Flycatcher has ever been found. The species does not exist. Wilson's drawing seems to be that of a female Black-throated Green Warbler (*Setophaga virens*) in remarkably dull plumage, possibly because it was a juvenile. Since Audubon had no specimen he had to copy Wilson, but rather than credit Wilson with the discovery, erroneous as it turned out to be, he tried to claim the discovery for himself.

Ord's most important contribution to ornithology was his discovery of the Cape May Warbler (*Setophaga tigrina*), which Wilson drew and added to a plate with a Broad-winged Hawk (*Buteo platypterus*). In his account, Wilson notes that "the new and beautiful little species was discovered in a maple swamp, in Cape May County, not far from the coast, by Mr. George Ord of Philadelphia, who accompanied me on a shooting expedition to that quarter in the month of May last."[11] Since Wilson was writing in the summer of 1812, "May last" was May 1812. The discovery was announced in print in volume 7 of *American Ornithology,* published in 1813. Then, in 1835, Sir William Jardine and Charles Lucien Bonaparte noticed a bird in Audubon's *Ornithological Biography* that appeared identical to the Cape May Warbler of Wilson and Ord. Audubon issued a counterclaim that he had shot his bird in Henderson, Kentucky, in 1811, but could produce only his drawing as evidence. Ord rushed to change the date of his discovery to May 1811, but this only produced a confusion of dates given Wilson's accounts of his travels. In the end the controversy was resolved in favor of Ord's discovery and Wilson was credited with the first description of the Cape May Warbler.

Less confusing and more amusing is Audubon's claim to have been the first to paint a Ruffed Grouse (*Bonasa umbellus*). Neither Wilson nor Audubon claimed to have discovered the species, but Audubon painted a Ruffed Grouse (Figure 5.1), then signed and dated it 1805 so that he could claim the first illustration. If he had painted it in 1805, he would have had priority. Wilson painted a Ruffed Grouse (Figure 5.2) and used the unsigned, undated painting as the basis for plate 49 in volume 6 of the *American Ornithology,* published in 1812. A cursory look at the grouse shows that one is a copy of the other that has been reversed. If you could look at the original Audubon painting, which is in the Harvard Museum of Comparative Zoology library, you would see that Audubon's date is faked. The watermark in the paper is

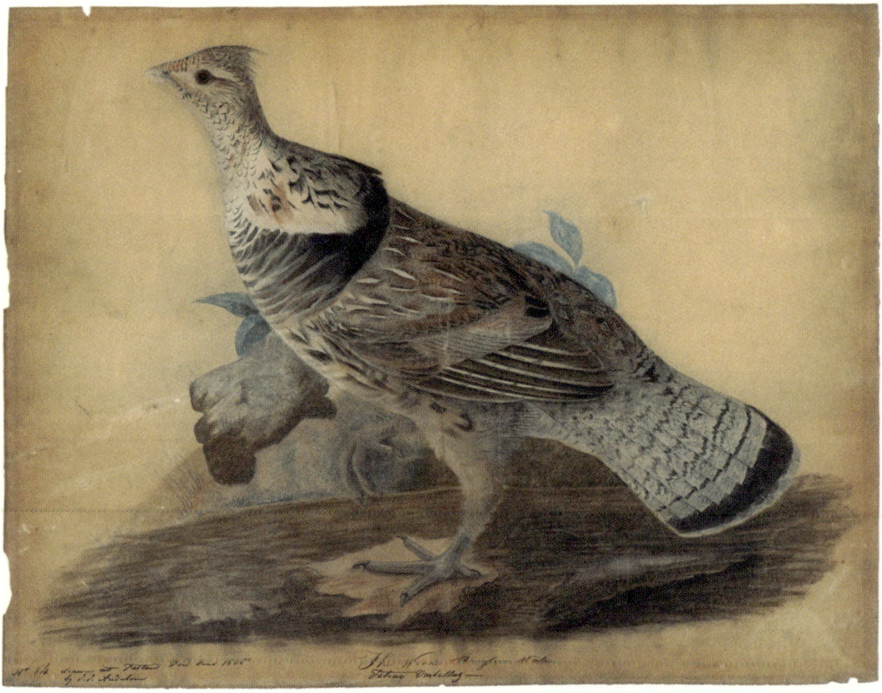

Figure 5.1. Audubon's painting of the Ruffed Grouse (From the archives of the Ernst Mayr Library, Museum of Comparative Zoology, Harvard University.)

1810. Audubon's Ruffed Grouse was painted on a piece of paper that did not exist in 1805. We do not know the date on which it was painted, but John Bell, a taxidermist to whom Audubon gave the painting, met Audubon on a trip up the Missouri River in the 1840s and received the painting shortly thereafter.[12] Whatever the date, it was almost surely later than 1812, probably much later.

Several conclusions can be drawn from these events. First, Wilson and Audubon agree that they met in Louisville in 1810, looked at each other's artwork, hunted together, and parted with expressions of goodwill on both sides. Second, Wilson does not cite Audubon in his text because Audubon did not offer him any notes. Third, Audubon had no plans to publish his work before meeting Wilson in March 1810, and although he began to focus his considerable artistic talent on birds and to record his observations of birds later that year, he continued to pursue business partnerships with

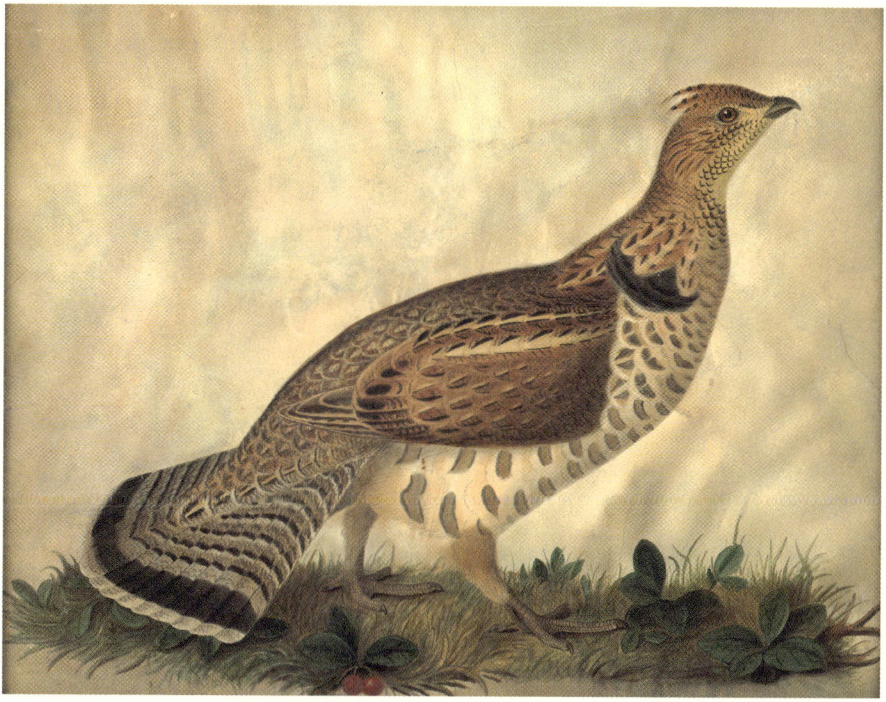

Figure 5.2. This is a draft painting of Wilson's Ruffed Grouse, but is in the same pose as in plate 49, volume 6, of *American Ornithology* (1812). Audubon's Ruffed Grouse is the mirror image of this bird, and may have been obtained by tracing Wilson's Grouse, turning the paper over, and rubbing the back to transfer the pencil lines to a fresh sheet of paper. (From the archives of the Ernst Mayr Library, Museum of Comparative Zoology, Harvard University.)

Rozier and his brother-in-law Thomas Bakewell for another nine years, well beyond 1813, the year of Wilson's death. The timeline suggests that Audubon was inspired to undertake his great work by his interaction with Wilson and that he began work on his sketches and notes as soon as possible and increasingly as he was able to extract himself from his business partnerships. Wilson was an ever-present influence in the background of Audubon's magnificent depiction of American birds. Furthermore, Audubon added species to Wilson's list and followed Wilson's and Bonaparte's lead by adhering to the Linnaean classification system and by inventing new names within that system as needed. Like Wilson before him, Audubon had and acknowledged many correspondents who contributed notes and species to *The Birds of America.* Finally and importantly, Audubon, like Wilson, emphasized the behavior, ecology, migration, and distribution of birds. Audubon and Wilson shared a passion for study of the living bird and knowledge of its life—a passion that was their distinctly American contribution to early ornithology.

It is indeed unfortunate that George Ord, to whom Wilson entrusted his legacy, could not comprehend that Wilson's achievement was beyond eclipse. Ord simply could not understand that Wilson—naturalist, scientist, and artist—had not only established scientific ornithology in the United States, but had also set a new standard for the illustration of nature, had set a new standard for scientific description, and had elevated nature writing to a distinctly American literary tradition.

It is equally unfortunate that Audubon, the quintessential artist-naturalist, could not see that acknowledging Wilson's pioneering work would only enhance his own unique contribution to ornithology and American culture. Perhaps Ord's unreasonable and crushing opposition to Audubon's plans tainted Audubon's view of Wilson. One has to wonder how their reputations and American ornithology might be different today if Audubon had been able to acknowledge Wilson's importance to his own career and success.

George Ord (1781–1863) played an important role in Wilson's life beginning in 1810, but more importantly he became the self-appointed guardian of Wilson's reputation after the ornithologist's death in 1813 (Figure 5.3).[13] We know nothing of Ord's birth, except that he was born in 1781 into a wealthy family. He was well educated somewhere. As a young man he collected obscure words but got into a quarrel with Noah Webster, author of the first American dictionary, and sent his collection of words to the editor of a new edition of Dr. Johnson's dictionary of the English language.[14] The pattern begins to emerge.

In the early 1800s, Ord was a rich young man who filled his leisure time, as was fashionable, by dabbling in philosophy, science, and literature. His interest in birds was that of a natural philosopher who watched birds for relaxation and had a reasonable knowledge of the habits of the birds he hunt-

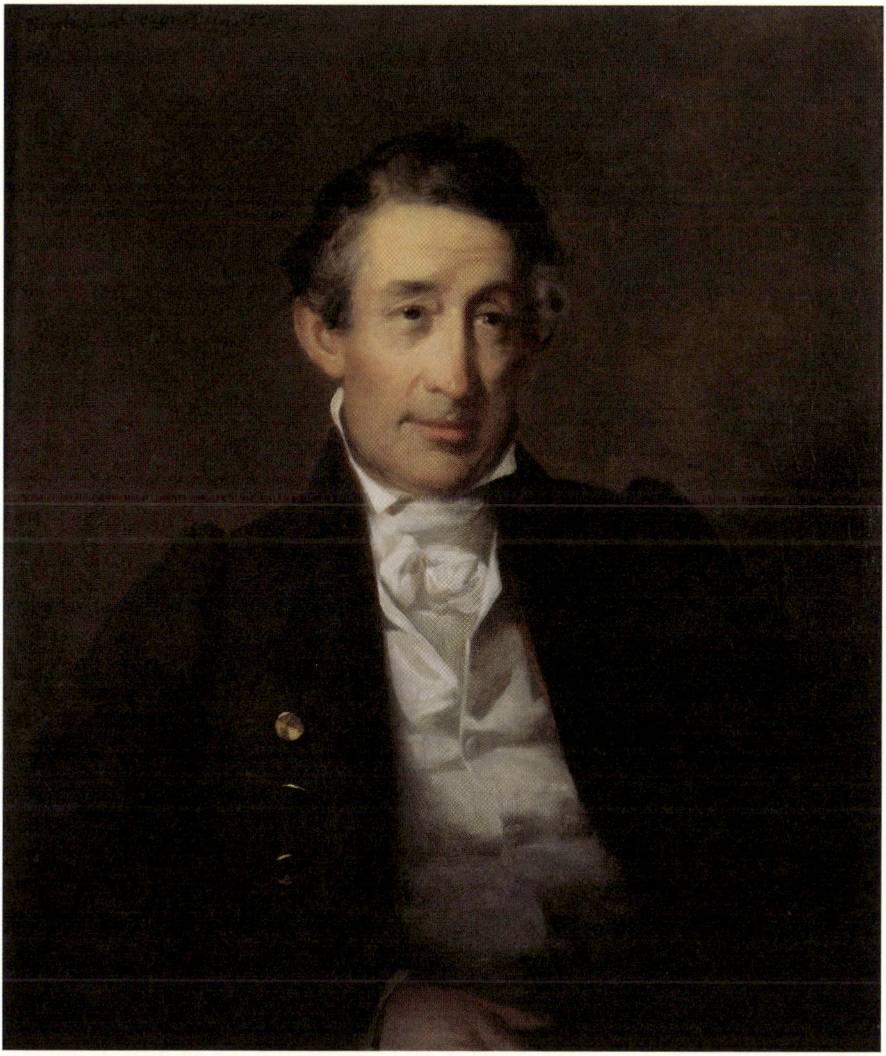

Figure 5.3. Portrait of George Ord, Wilson's controversial friend, painted by John Neagle in 1829 (From the Stewart Library, Academy of Natural Sciences of Philadelphia [Drexel University], Coll 2011-050).

ed. For George Ord, birds were a pleasant distraction that filled the idle hours that defined his life. When he read Wilson's books he questioned Wilson's presumption to think he could describe all the birds of the United States and his authority for writing in such detail about their biology. When he met Wilson, however, he came to realize that Wilson was passionately convinced that the study of wildlife, especially birds, was an integral part of American culture and therefore of practical concern to all. To George Ord this man, who was devoted body and soul to the discovery and description of birds, was incomprehensible. Yet within months Ord himself became completely absorbed in the study of birds. It was an absorption that continued the rest of his life. As a result of his association with Wilson, he is credited with the discovery of the Cape May Warbler and Wilson's Plover (*Charadrius wilsonia*), both of which were described in volume 8 of *American Ornithology*. He was also elected to the American Philosophical Society, serving as its secretary for nine years and its president for seven years. His jealous "protection" of Wilson's reputation and his pointless but ruthless antagonism toward Audubon were unfortunate for both Audubon and Ord. To give balance to our picture of the man, it must be stated that he was soft-spoken and kindly, but seemingly unable to form relationships that were not centered on him.[15] Wilson affected his life greatly, though probably not to the benefit of Wilson's legacy.

Wilson's contributions to Audubon's accomplishment were among his most important, but not his only, contributions to the development of modern ornithology.

Charles Lucien Bonaparte (1803–1857; Bonaparte's Gull [*Larus philadelphia*]) came to the United States in 1823, by which time he had already established himself as a European ornithologist.[16] He grew up in Italy and spoke Italian, French, and English. He read and wrote Latin, had begun his own natural history collection, and had discovered a new species, the Moustached Warbler, near Rome. On his voyage to the United States he observed storm-petrels following the ship and realized that there were two species, so after watching them closely he collected both. He found that one, Leach's Petrel, was more common near Europe and the other was more common near North America. He compared the two species to each other and to two other species in his first scientific paper, which he read before the Academy of Natural Sciences in Philadelphia on 13 January 1824. Bonaparte determined that the Stormy Petrel (*Procellaria pelagica*) described by Wilson was, in fact, a new species, and he renamed it Wilson's Petrel (*Procellaria Wilsonii*) in honor of the American ornithologist he so greatly admired. As it turned out the species had been previously described by Heinrich Kuhl so Kuhl's scientific name *Oceanites oceanicus* took precedence, but the common name remains Wilson's Storm-Petrel.

Bonaparte was deeply impressed with Wilson's accomplishment, but also realized that American ornithology had expanded greatly in the eleven years since Wilson's death. He decided to update and expand Wilson's *American Ornithology* and enlisted the aid of members of the Academy. George Ord used his influence to help Bonaparte gain access to the major American collections. How different from Ord's initial response to Audubon.

Bonaparte's genius lay in his ability to define taxonomic relationships among species, and he extensively revised Wilson's classification of families, genera, and species, thereby greatly updating the usefulness of Wilson's original work. Thomas Say (Say's Phoebe [*Sayornis saya*]), Titian Peale, William Cooper (Cooper's Hawk [*Accipiter cooperii*]), John James Audubon, and George Ord all contributed field observations to Bonaparte's text. Yes, both Audubon and Ord. Indeed, when Audubon arrived in Philadelphia in March 1824, Bonaparte was among the first to meet him. He was deeply impressed with Audubon's drawings and took him to a meeting of the academy (where Audubon was immediately rebuffed by George Ord). Bonaparte also introduced Audubon to Lawson, who was engraving the illustrations for Bonaparte's revision. Lawson also rebuffed Audubon. Bonaparte was nonplussed. He asked Audubon to help illustrate his revision of *American Ornithology,* but could get Lawson to engrave only one of Audubon's sketches for the revised edition, that of a female Great-tailed Grackle.

Bonaparte's revision of Wilson's work is important, because it changed *American Ornithology* from a historically important, but dated starting point for ornithology in America, to a living document that twenty years after its first publication had become a current list of American species, their taxonomy, and species descriptions. It became the predecessor of the *Checklist of North American Birds,* which would be published by the American Ornithologists' Union and is still maintained by the organization in hard copy and online. Bonaparte together with Wilson also created the taxonomic foundation on which Audubon would base his *Birds of America.* Bonaparte was thus an important player in securing Wilson's legacy as the Father of American Ornithology by establishing his *American Ornithology* as the point of origin for the scientific treatment of North American birds.

Wilson's *American Ornithology* was the first guide to the birds of North America and for decades it was the standard reference for anyone working with birds, but the books were large and there were nine volumes in the complete set: they were not a suitable "field guide." Bonaparte's revision expanded the information available, but made no progress toward recasting the material as a "field guide."

Thomas Nuttall (1786–1859, Common Poorwill [*Phalaenoptilus nuttallii*], Nuttall's Woodpecker [*Picoidis nuttallii*], Yellow-billed Magpie [*Pica nuttallii*]), unlike Bonaparte, saw the need for a field guide and answered that need with his *Manual of the Ornithology of the United States and Canada.* He added a few species to Bonaparte's list and made some taxonomic revisions, but the importance of his *Manual* is its size: despite giving identification and life history information for "all" species of North American birds, the book was both small enough to fit in a large pocket and inexpensive. Nuttall's *Manual* went through dozens of editions that extended its life into the early twentieth century. Nuttall's text accompanied by Wilson's pictures (most species were illustrated with woodcuts of Wilson's birds) gave Americans their first book that could be taken into the field to help identify local birds.

Thomas Brewer (1814–1880; Brewer's Sparrow [*Spizella breweri*], Brewer's Blackbird [*Euphagus cyanocephalus*]), contributed observations and specimens to Audubon while just a teenager and was cited for his contributions in Audubon's *Ornithological Biography.*[17] In 1840, as a sometime doctor, active politician, author, and editor, he brought out a new edition of Wilson's *American Ornithology* that included descriptions and taxonomy of all the North American species described by Wilson, Bonaparte, and Audubon. Wilson's illustrations were reproduced as small black and white prints, which brought the cost of the book within the means of thousands. Wilson's dream of describing all the birds of the United States was moving closer to completion, and his plan to publish an inexpensive edition to follow the original *American Ornithology* was now a reality.

Spencer Fullerton Baird (1823–1887; Baird's Sandpiper [*Calidris bairdii*], Baird's Sparrow [*Ammodramus bairdii*], Kauai Creeper [*Oreomystix bairdi*]) was the first man to fall under the spell of Wilson's *American Ornithology* and then build his career from the encounter.[18] While a student at Dickinson College, from which he graduated at the age of sixteen, Baird received a copy of Wilson's *American Ornithology* on loan and promptly cut all his classes to study it. The next year he went to New York to study medicine, but his passion was natural history, particularly birds. In New York he met Audubon, with whom he had already corresponded, and John Bell (*Bell's Vireo*), a taxidermist who prepared skins for Audubon. Bell was so impressed with the young Baird that he invited him to meet the other naturalists who gathered at his shop to discuss natural history.[19] By now Baird had begun his own natural history collection and medicine was not on his radar. He accepted a professorship at Dickinson College. When the Smithsonian was founded and the position of assistant secretary opened, he applied, was appointed, and moved to Washington with his bird collection, which became the nucleus of the national collection. Over the next twenty years he would remake natural history and particularly ornithology in North America. His father-in-law, Colonel Sylvester Churchill, was the inspector general of the army, and with his approval Baird was able to recruit members of the Pacific Railroad survey teams to collect natural history specimens. He also approached Jefferson Davis, then secretary of war, to enlist the help of army officers being posted to the American West. These "missionaries," as he called them, became a remarkably successful group of collectors for the national museum. Baird instructed and supplied them, wrote letters praising their work and their many discoveries, and pointed out as needed how they could improve their sampling, shipping, or notes.[20]

The *Pacific Railroad Reports* were published in twelve volumes between 1855 and 1859. The ninth volume, edited by Baird with assistance from John Cassin (Cassin's Auklet [*Ptychoramphus aleuticus*], Cassin's Finch [*Haemorhous cassinii*], Cassin's Kingbird [*Tyrannus vociferans*], Cassin's

Sparrow [*Pueceaea cassinii*], Cassin's Vireo [*Vireo cassinii*] and George Lawrence (Lawrence's Goldfinch [*Spinus lawrencei*]), was on birds. The men revised the report and, in 1860, reissued it as *Birds of North America.* In their revision they overhauled the nomenclature and constructed a taxonomic framework that was more complete and reliable than any that had preceded it, including those of Wilson, Bonaparte, Audubon, and Brewer. Baird had not only carried Wilson's original effort to classify North America's birds into the mid-nineteenth century, but had also expanded the list of known species to include all the states and territories of the United States from the Atlantic to the Pacific. Wilson had inspired the teenage Baird to dream of ornithology and the fulfillment of Baird's dream would have pleased Wilson immensely. The study of birds was now an integral part of the national scientific enterprise, and birds were an important collection within the national museum.

Elliot Coues (1842–1899, Grace's Warbler [*Setophga graciae*, after Elliot's sister, Grace]) wrote of Wilson's *American Ornithology* that it was remarkable for its vision, its breadth, and its accuracy. In 1874 Coues published his own *Field Ornithology, Comprising a Manual of Instruction for Procuring, Preparing and Preserving Birds, and a Check List of North American Birds.* The *Check List,* which was most of the publication, was devoted to classification and naming. It brought Baird's list up-to-date and straightened out the inevitable confusions that result from the discovery of new species. In 1883 Coues became one of the founders of the American Ornithologists' Union and his checklist became the basis for the Union's checklists, which were published periodically and are now available electronically with annual updates. Alexander Wilson's goal had been fulfilled. The birds of America were being discovered, classified, and described. Ornithology was a science with a future that would continue into the twentieth century and beyond.

Wilson's desire to describe and classify all of America's birds may have reached fruition in the American Ornithologists' Union, but it is in the Wilson Ornithological Society that his emphasis on study of the living bird became the dominant philosophy. The society was founded originally as a few widely scattered young birders who were part of a young ornithologists' association, then as the Wilson Ornithological Chapter of the Agassiz Association. When the Agassiz Association dissolved, the chapter pottered along for a few years to emerge in 1902 as the Wilson Ornithological Club. The group coalesced around Lynds Jones, professor of natural history at Oberlin College, who edited and published the *Wilson Bulletin* for thirty-five years. Today the Wilson Ornithological Society is one of several North American scientific societies devoted to the study of birds, but its emphasis remains the observational study of birds and the encouragement of young ornithologists.

Field Guides.—Wilson conceived *American Ornithology* as a guide to the birds of North America. That is why he not only described each species, but also pictured it, typically from the side and with the tail partially rotated to show the pattern on the dorsal surface. Wilson's use of two-dimensional space is very similar to that of Roger Tory Peterson and, more recently, David Allen Sibley.[21] Compare the head of Wilson's Pileated Woodpecker (*Dryocopus pileatus*) in Figure 3.30 to the one in your Peterson or Sibley field guide and you will see at once a remarkable similarity of style and purpose. The heads are in profile and essentially two-dimensional with an emphasis on color and pattern. Peterson and Sibley have produced outstanding, three-dimensional bird portraits. But the purpose of the two-dimensional portraits in their field guides, like those in Wilson's *American Ornithology,* is to facilitate identification by the reader. The anatomical detail of Wilson's illustrations, whether the alignment of the feathers, pattern of scales on the feet, or the color and pattern of color of the plumage, was unprecedented in 1808. One need only compare Bartram's painting (Fig. 2.6) of the Field Sparrow (*Spizella pusilla*) to Wilson's Fox-colored Sparrow (*Passerella iliaca*, Figure 3.25) or Seaside Sparrow (*Ammodramus maritimus*, Figure 3.35) to appreciate the difference in detail and the greater value of Wilson's portraits to the would-be bird watcher. Like most bird portraits that preceded his, and most bird portraits in today's field guides, Wilson's birds are often perched on branches and stumps. Many of Wilson's later illustrations, however, while remaining two-dimensional to maximize their usefulness for field identification, include helpful behavioral and ecological details: see the Orchard Oriole (*Icterus spurius*) reaching for a caterpillar on the underside of an apple leaf, the eagle clutching its fish on the bank of a river, or the female Common

Yellowthroat (*Geotlypis trichas*) feeding a juvenile Brown-headed Cowbird (*Molothrus ater*). These appropriate and accurate details are very different from, say, those in Bartram's Field Sparrow perched on a gentian that could not possibly hold its weight or Catesby's Baltimore Oriole standing in front of a lily shown with its roots, bulb, stem, leaves, and blossom. Both illustrations are artistic but they are not behaviorally or ecologically meaningful.

The frequent criticism that Wilson's birds are inferior to those of Audubon misses the point. Audubon was an artist who painted birds. Wilson was an ornithologist who used illustrations to accurately portray the bird and, often, its ecology and behavior. His purpose was not art. His purpose was to accurately convey information that would help the reader identify the bird in the field. This concept, so utterly commonplace today with our multitudes of field guides, was new in 1808 when volume 1 of the *American Ornithology* appeared.

Literature.—Wilson's influence on literature is difficult to document, but at the time he wrote there were few American authors, and among them only Bartram had written sympathetically about the wilderness. Wilson followed in his mentor's footsteps. For him the wilderness and its magnificent birds were an American heritage that deserved to be written about, and he wrote about them lyrically. Gone was the dread. Gone was the wolf in sheep's clothing. Wilson treasured and celebrated the beauty of America's birds and their wilderness home. Perhaps Ralph Waldo Emerson, Henry David Thoreau, Emily Dickinson, Walt Whitman, Aldo Leopold, and Loren Eisley were inspired by Wilson's writings and outlook. If not, they certainly shared his vision and built it into a lasting legacy in American literature.

Today the vision of Alexander Wilson lives on among those who feed and enjoy birds around their homes, among those who join local bird clubs and Audubon chapters, among those who spend their weekends and vacations watching birds, and among America's ornithologists, who have a profound appreciation for the importance of observation to the study of birds. Alexander Wilson opened our eyes to our natural heritage and we are all his beneficiaries. Let us strive to see what he saw and value it as passionately.

Appendixes

Notes

Selected Bibliography

Acknowledgments

Index

APPENDIX A

On the Shoulders of Giants: Wilson's Predecessors

Wilson was fortunate to have three important libraries available to him: William Bartram's private library, the library of the American Philosophical Society, and that of the Library Company of Philadelphia, the first lending library in the world, founded in 1731 by Benjamin Franklin. All included books with information on birds (see Table A.1). To give some idea of what was and was not known when Wilson began to write, we offer brief synopses of the authors whose work was cited by Alexander Wilson, and whom Wilson knew only through their published work. Our treatment is strictly alphabetical. The dates of birth and death, if known, and any species named after the person are given in parentheses after his name. (Table A.2 provides an overview of all authors cited by Wilson.)

Albin, Eleazar (ca. 1713–1759). Albin was a professional artist and illustrator and an amateur naturalist who published four books on or relating to birds. His *Natural History of Birds* was published in three volumes, in 1731, 1734, and 1738. His contribution to American ornithology, however, is slight. Albin illustrated and described seven North American species, none for the first time. Several of his descriptions closely resemble those of Catesby, but his anecdotal remarks often include descriptions of the stomach contents and internal structures, which was unusual at the time. Albin was the first to include the eggs of birds in his illustrations. Wilson followed Albin's example for a number of species and illustrated the chicks of several species as well. Modern readers will find Albin's notes on the medicinal uses of birds amusing. In addition to his three books on birds, he published accounts of the lice and fleas that live among the plumage in his *Natural History of Spiders* (1736).

TABLE A.1

A LIST OF BOOKS AVAILABLE TO ALEXANDER WILSON RELATING TO ZOOLOGY, ESPECIALLY ORNITHOLOGY, IN THE LIBRARIES OF THE AMERICAN PHILOSOPHICAL SOCIETY, LIBRARY COMPANY OF PHILADELPHIA, AND WILLIAM BARTRAM

Author	Date of Publication	Title	Location
Anon.	1688	*Memoirs for a Natural History of Animals, Transl. from French by Alexander Pitfield*	Am. Phil. S.
Audebert, Jean Baptiste, and Louis Jean Pierre Vieillot	1802	*Histoire naturelle des oiseaux dorés*	Am. Phil. S.
Bartram, John	1751	*Observations on the Inhabitants, Climate, Soil, Rivers, Productions, Animals, and Other Matters Worthy of Notice, Made by John Bartram in his Travels from Pensilvania to Onondago . . .*	Am. Phil. S.
Bartram, William	1791	*Travels through North and South Carolina, Georgia, East and West Florida*	Am. Phil. S.
Brisson, Mathurin Jacques	1756	*Le règne animal divisé en IX classes*	Am. Phil. S.
Buffon, George Louis Leclerc	1772	*Histoire naturelle des oiseaux* (5 vols.)	Am. Phil. S.
Catesby, Mark	Vol. 1, 1732; Vol. 2, 1747	*Natural History of Carolina, Florida, and the Bahama Islands*	Bartram
Cuvier, Georges	1798	*Tableau élémentaire de l'histoire naturelle des animaux*	Am. Phil. S.
Darwin, Erasmus	1800	*Phytologia*	Bartram

Edwards, George	Vol. 1, 1743; Vol. 2, 1747; Vol. 3, 1750; Vol. 4, 1751	*Natural History of Uncommon Birds*	Bartram
Edwards, George	Vol. 1, 1758; Vol. 2, 1760; Vol. 3, 1764	*Gleanings of Natural History* (3 vols. 1758, 1760, 1764)	
Forster, Johann Reinhold	1771	*A Catalogue of the Animals of North America*	Am. Phil. S.
Hermann, Johann	1783	*Tabula affinitatum animalium*	Am. Phil. S.
Lacépéde, Bernard-Germaine de	1801	*Discours d'ouverture et de clôture du cours de zoologie*	Am. Phil. S.
Latham, John	1781–1783	*General Synopsis of Birds*	Library Co.
Lettsom, John Coakley	1799	*Naturalist's and traveller's companion.* 3rd ed.	Am. Phil. S.
Linnaeus, Carolus	1740	*Caroli Linnaei Naturae Curiosorum Dioscoridis Secundi Systema Naturae . . .*	Bartram
Linnaeus, Carolus	1758	*Systema naturae*	Bartram
Pallas, Peter Simon	1776	*Naturgeschichte merkwürdiger Thiere*	Am. Phil. S.
Peale, Charles Wilson	1796	*A Scientific and Descriptive Catalogue*	Bartram
Pennant, Thomas	1785	*Arctic Zoology* (3 vols.)	Am. Phil. S.
Sloane, Sir Hans	1725	*A Voyage to the Islands Madera, Barbados, Nieves, S. Christophers and Jamaica* (2 vols.)	Bartram
Vieillot, Louis Jean Pierre	1808	*Histoire naturelle des oiseaux l'Amerique Septentrionale* (2 vols.)	Am. Phil. S.
Willughby, Francis	1678	*The Ornithology of Francis Willughby, Translated into English with Additions, by John Ray*	Am. Phil. S.

TABLE A.2

AUTHORS CITED BY ALEXANDER WILSON WITH A SUMMARY OF THEIR CONTRIBUTIONS, IF KNOWN, TO THE DESCRIPTION OF NORTH AMERICAN BIRDS

Author	Citations by Wilson	Nationality	Dates	Time in North Ame
Albin, E.	5	British	1713(?)–1759	none
Aldrovandus, U.	1	Italian	1522 [or 1527]–1605	none
Barrère, P.	1	French	1690–1755	1722–172
Bartram, W.	62	United States	1739–1823	life
Belknap, J.	1	United States	1744–1798	life
Belon, P.	1	French	1517–1564	none
Bewick, T.	32	British	1753–1828	none
Brisson, M. J.	82	French	1723–1806	none
Brünnich, M. T.	3	Danish	1737–1827	none
Buffon, G-L. L.	151	French	1707–1788	none
Catesby, M.	75	British	1682–1749	1712–171 1722–172
Charlevoix, P. F. X.	1	French	1682–1761	1705–170 1720–?
Clavigero, F. S.	2	Italian	1731–1793	1731–176
Du Pratz, L. P.	1	French	1695–1775	1718–173
Edwards, G.	55	British	1694–1773	none
Fernandez[d]	2	unknown	unknown	
Feuillee	2	unknown	unknown	
Forster, J. R.	7	German	1729–1798	none
Frisch, J. L.	1	unknown	1666–1743	
Gmelin, J. F.	1	German	1748–1804	none
Hearne, S.	1	British	1745–1792	1769–178
Hernández, F.	1	Spanish	1514–1578	1570–157

North nerican pecies scribed	North American species named	Ornithological Works
7	0	*A Natural History of Birds* (3 vols. 1731, 1734, 1738
4	0	*Ornithologia* (3 vols. 1599, 1600, 1603)
		Ornithologiae Specimen novum (1745)
215	0	*Travels in the Carolinas, Georgia, East and West Florida* (1791)
116	0	*History of New Hampshire* (3 vols. 1784, 1791, 1792)
~200	0	*Histoire de la Nature des Oyseaux* (1555)
0	0	History of British Birds (2 vols. 1797, 1804)
0	0	*Ornithologie* (1760)
2	1	*Ornithologia Borealis* (1764)
		Histoire Naturelle des Oiseaux (9 vols. 1770–1786)
109	0 (75)[b]	*Natural History of Carolina, Florida, and the Bahama Islands* (2 vols. 1732, 1747)
~20	0	*Journal of a Voyage to North America* (1761)
200+	0	*History of Mexico* Italian 1780–1781, English 1787
?[c]	0	*History of Louisiana* (1758)
90	0	*Natural History of Uncommon Birds* (4 vols. 1743, 1747, 1750, 1751), *Gleanings of Natural History* (3 vols. 1758, 1760, 1764)
	5	*Catalogue of the Animals of North America* (1771), "An account of the birds sent from Hudson's Bay" in the *Phil. Trans. Roy. Soc.* 62. (1772)
?	?	13th edition of *Systema Naturae* (1788–1793)
53	0	*A Journey from Prince of Wale's Fort in Hudson Bay to the Northern Ocean* (1795)
19[e]	0	*Historiae animalium et mineralium novae Hispaniae* (1651)

TABLE A.2 (CONTINUED)

AUTHORS CITED BY ALEXANDER WILSON WITH A SUMMARY OF THEIR CONTRIBUTIONS, IF KNOWN, TO THE DESCRIPTION OF NORTH AMERICAN BIRDS

Author	Citations by Wilson	Nationality	Dates	Time in North Ame
Kalm, P.	1	Swedish	1716–1779	1747–175
Klein, J. T.	2		1685–1759	none
Latham, J.	173	British	1740–1837	none
Lawson, J.	2	British	d. 1711	1700–17
			1709–1711	
Leems	1	unknown	unknown	
Linnaeus, C.	66	Swedish	1707–1778	none
Lorenzi	1	unknown	unknown	none
Marcgraf, G.[f]	1	German	1610–1644	1638–16
Marten, H.	1	British	1729–1790	1750–178
McKenzie, A.	3	Canadian	1764–1820	life
Molina, J. I.[g]	1	Italian	1740–1829	1740–176
Müller, P. L. S.	3	Dutch	1725–1776	0
Pallas, P. S.	1	German	1741–1811	0
Pennant, T.	184	British	1726–1798	0
Ray, J.	10	British	1627–1705	0
Salerne	1			
Sloane, H.	6	British	1660–1753	0
Turton, W	64	British	1762–1835	0
Ulloa	1			
Willughby, F.	10	British	1635–1672	0

[a] Barrère's time was spent in French Guiana.

[b] Catesby's work preceded use of the binomial Latin names to designate species, therefore he is not considered the author of the species name. Yet his descriptions and illustrations of seventy-five species were used by Linnaeus to describe and name those species. In that sense Mark Catesby is responsible for the names of seventy-five species of North American birds.

[c] Many of Du Pratz's descriptions cannot be identified.

orth erican ecies scribed	North American species named	Ornithological Works
some	0	*En Resa til Norra America* (1753–1761)
0	0	*Historiae Avium Prodromus* (1750)
some	0	*General Synopsis of Birds* (1781–1783), *Index Ornithologicus* (1790), *General History of Birds* (1821–1828)
~100	0	*A New Voyage to Carolina* (1709)
		Systema Naturae 10th edition (1758), 12th edition (1766)
133	0	*Historia naturalis Brasiliae* (1648)
27	0	
a few	0	*Voyages from Montreal* (1801)
33	0	*Saggio sulla Storia Naturale del Chili* (1782)
		Des Ritters Carl von Linné vollständiges Natursystem. (1773)
		Spicilegia zoologica (1767), *Reisen durch verschiedene Provinzen des Rusischen Reiches* (3 vols. 1771, 1773, 1776), *Zoographia Rosso-Asiatica* (3 vols. 1811–1813)
94 including species)	0	*The British Zoology* (1761–1766), *Indian Zoology* (1769), *Genera of Birds* (1773), *Arctic Zoology* (vol. 2, 1785)
0	0	*Ornithologiae libri tres* (1678)
0	0	*A Voyage to the Islands Madera, Barbados, Nieves, S. Christophers, and Jamaica* (2 vols. 1707, 1725)
0	0	*A General System of Nature . . . By Sir Charles Linné; Translated from Gmelin, Frabricius, Willdenow, . . .* (1806)
0	0	*Ornithologiae libri tres* (1678)

[d] We can find no record of ornithological contributions by a person with the surname Fernandez and suggest that the reference may be to Francisco Hernández.

[e] Hernández described some 230 species of birds from Mexico, of which nineteen also occur in the United States.

[f] Marcgraf's seven years were spent in Brazil.

[g] Molina was born and lived in Chile until 1768 when Jesuits were banished from the country and he was exiled.

Aldrovandus, Ulisse (1522 [or 1527]–1605). Aldrovandus was a native of Bologna and popular professor of natural history at the University of Bologna. He published his encyclopedic *Ornithologia* in three volumes, in 1599, 1600, and 1602. The text includes detailed descriptions and information on their behavior, incubation and development, food, how to capture and preserve them, edibility, medicinal uses, and use as emblems and symbols in mythology. It was the most exhaustive compilation of ornithological knowledge to date. Although his illustrations were prepared by hired artists, the results are mediocre with few exceptions, one of which is that of the Northern Cardinal (*Cardinalis cardinalis*), a live bird received from the keeper of the gardens in Pisa. The gardener's description, quoted by Aldrovandus, of the cardinal fighting its image in a mirror would ring true to every homeowner who has ever had a cardinal nest nearby. In addition to the cardinal, Aldrovandus describes the Wild Turkey (*Meleagris gallopavo*), Mourning Dove (*Zenaida macroura*), and Tufted Titmouse (*Baeolophus bicolor*).

Barrère, Pierre (1690–1755). A contemporary of Edwards, Barrère was educated as a doctor and practiced briefly before traveling to Cayenne in French Guiana in 1722. In 1727 he returned to France, where he was appointed chair of the botany department at the university in Perpignan and served as the doctor at the military hospital. His *Ornithologiae Specimen novum* is a classification system of birds based on the shape of their feet.

Belknap, Jeremy (1744–1798). Belknap's three-volume *History of New Hampshire* was written over a period of eight to ten years. The first volume was published in 1784 in Philadelphia and the latter two in Boston in 1791 and 1792, respectively. One hundred and sixteen species of birds are covered in chapter 10. Most receive some sort of Latinized name that is not recognized today. In fact, a number of bird names that were popular at the close of the eighteenth century are difficult to link to modern species, for example spring bird, winter sparrow, and quindar.[1] Other delightful names are not so difficult—for example, Wooly Woodpecker (today's Downy Woodpecker), Bird Hawk (Northern Shrike), Lord and Lady (Harlequin Duck), and Tee-arr

(Least Tern). The species accounts, such as they are, show the author's scant familiarity with birds, but include some interesting anecdotes: "In winter, turkeys frequent the seashore for the sake of picking small fishes, and marine insects which the tide leaves on the flats."[2] Belknap mentions Governor Wentworth's introduction of pheasants to New Hampshire in the eighteenth century, which was apparently the first introduction of a bird into North America, and reiterates the mistaken belief that swallows hibernate. Nonetheless, he offers one bit of advice that Wilson took to heart: "In the description of an American state it would be unpardonable not to take notice of its natural productions."[3]

Belon, Pierre (1517–1564). Little is known of Belon's birth or childhood. As a teenager he received patronage from the chancellor of France and the cardinals of Lorraine and Tournon. While in the service of the cardinals he was sent to Switzerland on a diplomatic mission, accused of heresy, and rescued by a gentleman with whom he shared a love for the poetry of Pierre de Ronsard. The same gentleman paid for his release from servitude. (And you thought ornithology was a dull business.) Belon returned to Paris where he earned a medical degree at the Collège de France in Paris. He also studied botany at the Protestant university in Wittemburg. From 1546 to 1549 he traveled in the Middle East and upon his return wrote a popular travelogue that included informal descriptions of birds (among many other animals), as well as portrayals of the area's architecture, landscapes, and geography. He then journeyed to England where he met the Venetian ambassador who interested him in fishes; Belon would subsequently publish a book on the natural history of fish. In 1555 he published his *Histoire de la Nature des Oyseaux* in which he described and illustrated in color two hundred species of birds, including a handful of North American species described from preserved specimens. He is one of the earliest authors to include a colored illustration with almost every species description. One of the most innovative aspects of the *Histoire* is his comparison of the functions of the bones of the human and avian skeletons. He placed the skeleton illustrations on facing pages and clearly explained the modifications of the bird's skeleton for flight.

In addition to his ornithological interests, Belon was eager to establish a research garden in which he could study the naturalization of exotic plants. He interested Henry II, king of France, in the project, and was promised an annual pension, which never materialized. In 1564 his patron, the French admiral Gaspard de Coligny, converted to Protestantism and became a leader of the French Huguenots. Belon, a devout Catholic, returned heartbroken to Paris, where he was accused of plagiarism. While walking home one night in April 1564 he was attacked and killed by an assassin.

Boddaert, Pieter (1730–1796). Little is known about Boddaert beyond his publication of an identification key that provides a scientific name for each species illustrated in the *Planches enluminées* (see below). Only fifty copies of his key were published, and only one has survived.[4] Despite debate over the validity of names in a book that is filled with typographical and taxonomic mistakes, Boddaert is credited with having named a substantial number of avian species.[5] Nonetheless, the book was not widely circulated after its initial publication and Wilson, who never cites Boddaert, appears to have been unaware of his work.

Buffon, Georges-Louis Leclerc Comte de (1707–1788). Buffon's *Histoire naturelle des oiseaux* had several editions, the last of which was published in 1804. It included fifty-four volumes of descriptions of species from Europe, the French colonies, and other places that French explorers had visited. Buffon realized that birds' colors and color patterns were better seen than described, so he created a series of companion volumes to his written descriptions. Known as the *Planches enluminées,* they featured engravings that are precisely rendered and hand-colored, allowing accurate identification of the species, which is important since none of the original specimens have survived and a number of the species are now extinct. Buffon was little interested in the taxonomic work of Linnaeus or Brisson and focused his energy on describing what was known about the names and biology of each species. He was not a field biologist, however, and his accounts—based on the published descriptions, field notes, and recollections of others—were salted liberally with his own often Eurocentric interpretations. Buffon's pronouncements incensed Jefferson, who argued strongly against Buffon's view that American plants and animals were degenerate forms of European species.[6] Wilson, an ardent supporter of Jefferson, also reacted strongly against Buffon's assertions.

Catesby, Mark (1682–1749). As a boy Catesby became interested in botany through the influence of his maternal grandfather and possibly John Ray, a botanist instrumental in the development of modern scientific names. Catesby made two trips to North America, from 1712 to 1719 and again from 1722 to 1726. During these trips he lived in Virginia, South Carolina, and the Bahamas. He also traveled as far inland as the Appalachians of Virginia, and sailed to Jamaica and perhaps within sight of Bermuda. In his two-volume *Natural History of Carolina, Florida, and the Bahama Islands,* written after his return to England, Catesby illustrates 109 species of birds and a substantial number of other animals, as well as 171 plants.[7] His descriptions of birds follow closely those of John Lawson. Publication of all five parts of volume 1 was completed in 1732, and the final parts of volume 2 were published in 1747. Catesby's work precedes Linnaeus's publication of the modern system of scientific names, but seventy-eight bird species—seventy-five North American and three Bahamian—were named by Linnaeus and later taxonomists based on Catesby's illustrations and descriptions.

Charlevoix, Pierre François Xavier (1682–1761). Charlevoix was one of the many Jesuit missionaries who lived and traveled in the wilderness of North America during the seventeenth and eighteenth centuries. From 1705 to 1709 he was a professor of languages and philosophy in Quebec, and on a return trip to North America in 1720 he wrote *Journal of a Voyage to North America,* a collection of thirty letters to the duchess of Lesdiguierres published in 1761. His "voyage" took him west along the St. Lawrence riverway and the Great Lakes to Michilimacinac (today's Mackinac Island, Michigan), then south on the Mississippi to New Orleans. He sailed next across the Gulf of Mexico and around Florida to the Bahamas, where he was shipwrecked. He also visited Santo Domingo before returning to France. The birds he saw are described in his letters, and although some descriptions are unclear, others, such as those of the hummingbird and cardinal, are good. In any case, his letters provide a snapshot of the abundance and diversity of birds in the interior of North America prior to its settlement.

Clavigero, Francisco Saverio (1731–1793). Clavigero was born in Veracruz, where he was educated as a Jesuit priest. Like Charlevoix, he wrote a history that included some excellent observations of Mexican birds, many of which also occur in North America. When the Jesuits were expelled from Mexico in 1767 he sailed for Bologna where he wrote his *History of Mexico,* which was published in Italian in 1780–1781 (an English translation by Charles Cullen was published in 1787). He disputes Francisco Hernandez's estimate of two hundred Mexican bird species as far too low, but the ancient Mexican names he used make his identifications hard to verify. Many of his observations are insightful; for instance he accurately describes hummingbirds' state of torpor during the night. But he also claims that hummingbirds and swallows hibernate from October to April, a commonly held belief in his day.

Du Pratz, Le Page (1700s). We know very little about Du Pratz's early years—only that he arrived in Louisiana in midlife and in the middle of the eighteenth century and that he was employed for fifteen years as an overseer of public plantations, which required that he travel through large parts of the territory belonging to the Natchez Indians. His account of these years was published in France in 1758 and condensed into a single English volume titled simply *History of Louisiana.* Du Pratz's observations on birds are contained in the twelve pages of chapter 7. Some of his observations are excellent—for example, the barbs on the tongue of woodpeckers and the fact that frigatebirds return to the coast at night to roost—but many of his identifications, descriptions, and illustrations are inaccurate.

Edwards, George (1694–1773). As a boy Edwards was tutored in the classics by clergy who inspired in him a love of learning. Later his father arranged an apprenticeship for him with a London merchant. Edwards's years in London would include the watershed event in his life, but as so often happens with parental plans, the event was very far from his father's plan for a prosperous business career. When Dr. Nicholas, a relative of Edwards's employer, died, his entire collection of natural history materials and artworks were moved into the flat occupied by the young apprentice. Edwards spent all his leisure time reading and studying the collection, developing a passion for natural history and art. From 1716 to 1720 he traveled in Europe and studied. He was so successful in his studies of birds and art that several wealthy men offered to support him. On his return from a trip to the art centers of Brussels, Antwerp, and Utrecht in 1731, he made the acquaintance of Sir Hans Sloane, president of the Royal College of Physicians, who offered him the position of librarian (the proper title is bedell or beadle) of the learned society. The position carried with it a large, free apartment and twelve pounds a year. More important to Edwards than a place to live and a salary, however, was the perquisite of access to both the library of some eight thousand volumes and Sloane's large natural history collection, which was probably the largest in the world at that time. Sloane was also supporting Mark Catesby's preparation of the second volume of the *Natural History of Carolina, Florida, and the Bahama Islands,* with the consequence that Catesby and Edwards became close friends. Catesby, who engraved all his own plates, taught Edwards engraving and instructed him in coloring. Consequently Edwards, like Catesby, engraved and colored all his own plates for the four volumes of *Natural History of Uncommon Birds,* published in 1743, 1747, 1750, and 1751, and the three volumes of *Gleanings of Natural History,* published in 1758, 1760, and 1764. The seven volumes contain 572 plates, of which ninety represent American birds, many of them newly identified. Written descriptions accompany the illustrations. Most species were described from specimens, although Edwards was able to observe some live birds in the collections of Peter Collinson, John Fothergill, and others. In 1771 Edwards published a new edition of Catesby's work to which he added the Linnaean scientific names for the plants and animals. Shortly thereafter Edwards retired to the village of Paistow, where he died of cancer on 23 July 1773.

Fernandez. Since there is no record of such an ornithologist, we suggest that Wilson's reference may be to Francisco Hernández.

Forster, Johann Reinhold (1729–1798, Forster's Tern [*Stena forsteri*]). Born near Danzig, Germany, Forster was educated for the ministry in Berlin and at Halle University and served twelve years as the minister to a country parish. Much like British parsons of the time (for example, Gilbert White), Forster occupied his free time with intensive study of natural history. In 1765 he left his parish to inspect Russian colonies east of the Volga for the Empress Catherine, but immigrated to England after relations with the Russian government soured. He taught natural history and languages at Warrington Academy for three years, but struggled financially. Thomas Pennant, who lived nearby, heard of the naturalist's financial difficulties, made small contributions to the family's economy, and involved Forster in the study of specimens of birds arriving from North America. In 1771 Forster published his *Catalogue of the Animals of North America . . . to Which Are Added Short Directions for Collecting, Preserving, and Transporting All Kinds of Natural History Curiousities.* In 1772 he published "An Account of the Birds Sent from Hudson's Bay" in *Philosophical Transactions of the Royal Society.* There he described, for the first time, the Great Gray Owl (*Strix nebulosa*), Boreal Chickadee (*Poecile hudsonicus*), Blackpoll Warbler (*Setophaga striata*), and White-crowned Sparrow (*Zonotrichia leucophrys*). In the same paper, Forster described a variant of the Common Tern (*Sterna hirundo*). Later that year, he and his son, Johann Georg, were chosen by Captain Cook as naturalists on his second Pacific voyage. Upon Forster's return three years later, father and son published a number of accounts of the voyage. These and translation work helped to repay the family's debts. Eventually Forster returned to Halle University, where he became professor of natural history, a post that he held for nineteen years until his death in 1798.

Frisch, Johann Leonhard (1666–1743). The Frisch cited by Wilson may be Johann Leonhard Frisch, a museum curator in Berlin who recommended that specimens be preserved by placing "each bird in a separate wooden or glass case. But one must take good care that these cases shut extremely tightly."[8] Imagine the joys of measuring many such specimens for a study on geographic variation.

Gmelin, Johann Friedrich (1748–1804). Born and educated in Tübingen, Gmelin received his medical degree at the age of twenty-one. He then took time off to make a "scientific tour" of England, Holland, and Austria before returning to take up the position of professor of medicine at Tübingen. He was now twenty-four. The following year he became professor of philosophy and medicine at the University of Göttingen and at the age of thirty became professor of chemistry, botany, and mineralogy, also at Göttingen, where he remained until his death. Although Gmelin was greatly interested in natural history, his contribution to ornithology is limited to the not inconsiderable task of editing the thirteenth edition of Linnaeus's *Systema Naturae,* in which he included species, many of them North American, described by John Latham and Thomas Pennant. Latham and Pennant did not accept the Linnaean system of scientific names and did not use such names in their species descriptions, but Gmelin, when adding the new species to the thirteenth edition, provided his own scientific names according to Linnaean rules. Soon after Latham published his *General Synopsis* in 1781–1785, he realized that he would not receive credit for naming the species he had just described because he had failed to give them scientific names according to Linnaean rules. He quickly set about to correct his mistake and in 1790 published his *Index Ornithologicus,* where he supplied Latin genus and species names for all his newly described species. Sadly, he was too late. In 1789, just one year earlier, Gmelin had published the section on birds for the thirteenth edition of *Systema Naturae*. Gmelin thus received credit for first naming and describing those species initially described in Latham's *General Synopsis.*

Hearne, Samuel (1745–1792). Hearne was born in London and was schooled as a young boy "without notable success."[9] At age eleven he joined the navy, where he must have mastered at least the skills of sailing, navigation, and command, because the next record of him is an account of his service in 1766, at age twenty-one, as mate on the Hudson's Bay Company whaling sloop *Charlotte*. In 1769, tired of the routine of life at sea, he applied for something more creative and was chosen by the company to conduct its first major arctic exploration of the lands west of Hudson Bay. On 7 December 1770 he left on his third walking trip to the west to find the fabled "Far Off Metal River." Upon reaching the mouth of the Copper Mine River on 17–18 July 1771, he found a four-pound piece of copper, which can be seen in the British Museum. He then returned to his headquarters at Prince of Wales Fort on the Churchill River, arriving on 30 June 1772 after his 1,300-mile trek.

Hearne's adventures were not over. In August 1782 a French force of three ships, 290 soldiers, and seventy-four large guns commanded by Jean-François de Galaup Comte de La Pérouse forced Hearne and thirty-eight civilians to surrender Prince of Wales Fort. Hearne's journal was returned to him by La Pérouse who found it during a search of the fort. La Pérouse also granted Hearne's request for one of the company's trading sloops, which Hearne and thirty-two of the civilians used to sail on a long and dangerous journey directly east to the Orkney Islands. Hearne's book *A Journey from Prince of Wale's Fort in Hudson's Bay to the Northern Ocean,* published posthumously in 1795, is rich in topographical, geological, climatological, zoological, and botanical information—indeed the level of detail resembles that of a modern book. Furthermore, his observations are not limited to the field. Hearne dissected animals and reported on their anatomy, and he kept many birds and mammals in the fort for observation. Hearne was perhaps the best of the Hudson Bay explorers. "Most importantly," according to Joseph R. Jehl, Jr. at the Smithsonian's National Museum of Natural History, he had "the mark of a good scientist—he admitted what he did not know or could not learn from the native peoples (e.g., where do Horned Waveys or Snow Birds nest), and he took pains to point these out."[10]

Hernández, Francisco (1514–1578). In 1570 Hernández, personal physician to King Philip II of Spain, was sent to the Spanish colonies to study their resources and natural history. He returned in 1577, but he died in 1578 and his voluminous notes were locked away to protect the knowledge of Spain's colonial riches from being known. Some of his writings were published in 1615 by Francisco Ximénez. Several later editions were edited and published in Mexico and Rome. In 1651 Hernández's *Historiae animalium et mineralium novae Hispaniae* was published in a collection of writings by several authors. The *Historiae* contains brief accounts of two hundred species, some of the accounts are illustrated, and nineteen of the species occur in North America.

Kalm, Peter (1716–1779). Kalm's early education in Sweden was for the ministry, but he was deflected into natural history, especially botany, by Bishop Brovallius, who was himself an ardent naturalist. Kalm was a close associate of Linnaeus, but his primary interest was practical botany. Kalm rose to prominence as a young man when he journeyed to Russia and the Ukraine in 1744. In 1747 he was appointed professor of economics at the University of Åbo. That same year, he was selected by a consortium of academic and business interests to visit the English colonies to collect plants and bring back to Sweden those that would be useful to the country's agriculture. He secured a leave of absence from the university and sailed for the colonies, where he remained until 1751. Most of his time was spent in Ontario, New York, Pennsylvania, and New Jersey. Several times he stayed with the Bartram family. He admired John Bartram greatly and was probably instrumental in his election to the Swedish Academy of Sciences in 1777. He was less impressed with the botanical knowledge and skills of William Bartram, although he admired him as well. Upon his return to Sweden, Kalm took up his teaching duties. He also produced, edited, and published his journal, *En Resa til Norra America,* between 1753 and 1761. Although a botanist, Kalm's daily notes include observations of birds beginning with those seen from the ship bringing him to North America. He provides a sensibly grouped list of birds in which woodpeckers are described in a separate list—the first such annotated list of American birds.[11] He also wrote in detail about "Maise thieves"—redwings, grackles, and Bobolinks—and argued that the use of bounties was counterproductive. His ornithological observations are not extensive, but are detailed. Kalm is remembered today much more for his descriptions of some seven hundred American plants, sixty that were new to science, and for founding the Botanic Garden in Åbo. His American trip is celebrated today in the scientific name of the Mountain Laurel *(Kalmia latifolia).*

Klein, Jacob Theodore (1685–1759). In 1750 Klein proposed grouping birds into eight families based on the number of toes:

Two toes, no hind toe—ostrich
Three toes, no hind toe—rheas, cassowaries, bustards, some Limicolae
Four toes, two forward, two back—zygodactylous birds
Four toes, simple, hind toe present—raptors, passerines, game birds
Four toes, webbed, hind toe present—
  plain-billed—ducks, geese
  cone-billed—gulls (part), loons, etc.
  anomalous-billed—avocets, skimmers
Four toes, all connected—pelecaniforms
Three toes, webbed—auks, penguins, tube-noses, gulls (part)
Four toes, lobed—coots, grebes, phalaropes, etc.[12]

Lawson, John (d. 1711). Details about Lawson's early life have been lost, but we do know that in 1700 he arrived in Charles Town (now Charleston), South Carolina. He had been appointed to survey the interior of Carolina, a task that he accomplished during a fifty-nine-day journey that took him inland through the piedmont of modern South and North Carolina to present-day High Point, North Carolina, and then east to the coast near the present town of Washington. His extensive notes on the plants and animals he saw formed the basis of his *A New Voyage to Carolina* published in London in 1709. He returned to the Carolinas in 1710, where he was captured and executed by the Tuscaroras. During his lifetime, Lawson described more than one hundred species of birds in terms that closely resemble those of Catesby.

Leems. There is no record of such a person making ornithological contributions, even though Wilson mentions his name.

Linnaeus, Carolus (1707–1778). Linnaeus began his education at Lund University, but transferred to Uppsala University after one year. In 1730 he was invited to tutor the sons of a Professor Rudbeck and to live in his house. This gave Linnaeus access to an outstanding library and to the botanic gardens established by Rudbeck's father, who was also a professor at Uppsala University. The young Linnaeus's intelligence and interest in both plants and birds impressed his professors and Rudbeck invited him to assist him in his botanical lectures. Indeed, when Rudbeck left temporarily, Linnaeus assumed all responsibility for the lectures, which soon became exceptionally popular.

When Rudbeck returned, Linnaeus decided to leave on a collecting trip to Lapland. Six months later he returned with significant quantities of botanical, zoological, and mineral specimens; a new sense of purpose; and a monumental goal—a new classification system for plants and animals. As a student Linnaeus had found that the plants and animals in his texts were described using a variety of Latin phrases. If a description was hard to follow or the specimen seemed to deviate from the printed description, a worker or scholar would simply write a new one or add to the existing description with little thought to how important the difference might be. The situation was becoming chaotic as exploration of the world was rapidly expanding the European view of the abundance and diversity of life and ever more descriptions were being added to the biological literature. Linnaeus would bring order to the chaos by creating a system in which each species received a succinct description and two names: a genus name that it shared with similar looking organisms, and a species name that was unique within the genus. Groups of organisms that looked similar to other groups were then placed in the same family and similar families in the same order. This was not an evolutionary system. Darwin was born fifty-one years after publication of the tenth edition of the *Systema Naturae,* which is the internationally recognized beginning of scientific names, and his paper on the origin of species by means of natural selection was not read before the Linnaean Society of London until a full hundred years after this publication of the 10th edition. Linnaeus provided an orderly system for assigning names to groups of organisms that looked similar. His was a system that allowed one to determine if a species had been described previously or was new.

After his return from Lapland, Linnaeus made a number of short trips over the next couple of years, but in 1735, still without a degree and unable to publish his work in Sweden, he left for Holland where he planned to take a medical degree at the University of Harderwijk and publish his scientific work. In Holland he met the botanist Jan Frederick Gronovius, who was so impressed with Linnaeus's binomial system for naming plants and animals that he and a friend, Isaac Lawson, a rich Scot, had it published at their own expense. Thereafter Linnaeus received offers to underwrite collecting trips to Africa and America. In the end his ties to his parents, to his beloved Sara, and to his professors would convince him to return to Sweden—but not before visiting England, where he met a number of famous botanists, among them Mark Catesby, who had recently returned from his second trip to America; Peter Collinson, who financed Catesby's second trip, corresponded with John and William Bartram, and bought their specimens and paintings; and Sir Hans Sloane. By the time he went to England, Linnaeus's system of names and his ideas concerning classification of plants had been widely circulated and read. Many in England were skeptical. Peter Collinson wrote in a letter to John Bartram, "The *Systema Naturae* is a curious performance for a young man but his coining of a new set of names for plants tends but to embarrass and perplex the study of botany . . . Very few like it."[13] Perhaps the reluctance evident in Collinson's comment explains William Bartram's failure to employ Linnaean classification in his list of American birds published in 1791. Moreover, this opinion must have been widely and persistently held, because John Latham failed to employ the Linnaean system in 1781–1783 and again forty years after that. Indeed the simmering resentment was still evident in 1885 when the British ornithologist Alfred Newton wrote, "He [Linnaeus] for the most part followed Ray, and where he departed from his model he seldom improved upon it."[14] Nonetheless, Linnaeus's charm must have won over many, because he left England with many plants from the Chelsea gardens.

Linnaeus returned to Sweden in 1738 despite offers from Holland, France, and Spain; established a medical practice in Stockholm; and two years later, married Sara. In 1740, Dr. Rudbeck died, and one year later Linnaeus was appointed professor of medicine and botany at Uppsala University. He continued in that position until his death in 1778.

Linnaeus's success in establishing a universal system of species names owes much to the fundamental logic and simplicity of his scheme, as well as to his exceptional charm and his voluminous correspondence. Many explorers risked their lives to collect plants and animals that were sent to Linnaeus to be named. Furthermore, Linnaeus used his exceptional skills of persuasion to convince his contemporaries to use his system. Wilson was the first American to use the Linnaean system and his monumental *American Ornithology* was a major factor in its acceptance among members of the Academy of Natural Sciences of Philadelphia and the American Philosophical Society, for example Thomas Say, the Father of Entomology.

Lorenzi. Although Wilson mentions him, we have been unable to locate any ornithologist by this name.

Marcgraf, Georg (1610–1644). Marcgraf's parents were eager for their son to learn of the world through travel, and to that end they promoted his education, which occurred at no fewer than nine different German universities. There is no record of how long Marcgraf spent at each university, but in later life he was considered a prominent scholar in theology, Latin, Greek, and natural philosophy. His accomplishments in the field of natural philosophy brought him to the attention of Dr. Johannes de Laet, managing director of the Dutch Indies Company, who invited Marcgraf to be astronomer for the company. Marcgraf accepted the offer and left Holland for Brazil on 1 January 1638. Count Johann Moritz von Nassau-Siegen, who commanded the Dutch possessions in the New World, had an observatory built for Marcgraf, but also wanted him to explore the natural history of northern Brazil. For this purpose he placed him under the command of Willem Piso, physician to the count and his soldiers. Moritz then ordered Piso and Marcgraf to explore the region around the modern states of Pernambuco, Paraiba, and Rio Grande do Norte. Marcgraf made three or more collecting trips during which he took detailed notes, but few of his journals have survived the ravages of jealousy and time. Piso's knowledge was limited to medicine whereas Marcgraf's broad knowledge of natural history, geography, and astronomy brought him much closer to Moritz. Piso resented his assistant's greater knowledge and his close relationship with Moritz. Relations between the two men deteriorated and Marcgraf, fearing that Piso would take credit for his discoveries, began writing his notes in cipher. Had he lived to decipher his own notes all might have been well despite the unpleasantness, but he did not. He died of fever in 1644. De Laet deciphered and edited Marcgraf's notes and published them with full credit to Marcgraf in 1648. A few months later de Laet was dead. Piso rewrote the entire section on natural history with all reference to Marcgraf deleted, along with many of his observations. He published this work in 1658 as *De Indiae utriusquere naturali et medica,* which featured poorly executed woodcuts based on paintings by an artist who had accompanied Marcgraf on his explorations. In 1786, however, the original paintings were rediscovered in the library of the Elector of Brandenburg, and those paintings, along with Marcgraf's descriptions as edited by de Laet, became the foundation for many species descriptions by Ray, Brisson, Buffon, Linnaeus, and Gmelin.

Marten, Humphrey (1729–1790). Hired as a "writer" by the Hudson's Bay Company in 1750, Marten worked for the company in today's Hudson Bay, mostly at York Factory, until returning to England in 1786. In 1769, while he was stationed at Fort Albany, the company asked Marten to send natural history specimens and information to the Royal Society of London. He protested that he was not trained in zoology, but warmed to his task and by 1771 was enthusiastic. He sent two shipments containing at least twenty-seven species of birds to London. The first shipment contained the first-ever specimen of the Eskimo Curlew (*Numenius borealis*), which was named and described by J. R. Forster.[15] In the spring of 1771 he put up nest boxes inside and outside Fort Albany, the first person in Canada to do so.[16] Shortly thereafter, his boxes were occupied by Tree Swallows. Marten kept notes on the swallows' arrival and departure dates for several years and did not subscribe to the myth, common at the time, that swallows hibernated in the mud at the bottom of marshes. Marten was succeeded at Fort Albany in 1774 by Thomas Hutchins. Marten was mentioned in company records in 1787 for the last time, and in 1792 Samuel Hearne referred to him as the "late Humphrey Marten."[17] Much later, Samuel Hearne, who had worked with Marten, protested that many of the natural history specimens sent from Fort Albany and credited to Hutchins had in fact been collected by Marten. There is also evidence showing that Hutchins plagiarized observations made by another colleague at Fort Albany, Andrew Graham. If Hutchins indeed stole credit for specimens sent back to the Royal Society of London, it suggests that Marten, despite his misgivings about his own ability, was an important ornithological pioneer in northern North America.

McKenzie, Alexander (1764–1820). McKenzie traveled overland from Montreal to the Canadian Pacific, reaching the ocean at Bentinck Arm in the Deane Channel. His notes on birds are few, but he observed Snow Geese migrating north along the Fraser River on 18 June 1793, an unusually late date for this migration. His observations are published in *Voyages from Montreal* (1801).

Molina, Juan Ignacio (1740–1829). Born in Guaraculén, Chile, in 1740, Molina was educated in Talca and entered the Jesuit College in Concepción in 1756. After graduation he stayed on as librarian of the college until 1768, when the Jesuits were expelled. He then settled in Bologna where he became professor of natural sciences at the university. His *Saggio sulla Storia Naturale del Chili,* published in 1782, was the first treatment of South American birds since publication of Marcgraf's *Historia naturalis Brasiliae* in 1648. It includes only thirty-three species of birds, however, some of which cannot be identified from the vague or incomplete descriptions.

Müller, Philipp Ludwig Statius (1725–1776). Müller was a professor of natural history at the University of Erlangen. He had no experience with birds outside of Europe nor did he have access to a large collection that could help him verify descriptions or improve on those that he read or received. Nonetheless he was among the first to adopt the Linnaean system of names and he compiled a large number of descriptions that he published in 1773 in his *Des Ritters Carl von Linné vollständiges Natursystem.* This work, which was widely read in Germany, led to the adoption of the Linnaean system by most German scientists.

Pennant, Thomas (1726–1798). Thomas Pennant was born on 14 June 1726. The family was an old one that counted Richard Plantagenet, duke of York, among its ancestors. The name Pennant is of Welsh origin, with "pen" meaning head and "nant" meaning a wooded dell or dingle. Indeed, Downing, the family estate in Flintshire in northeast Wales, was next to a wooded dell. It was there that the boy grew up surrounded by trees, streams, and animals. On his twelfth birthday his uncle gave him a copy of Sir Francis Willughby's *Ornithology*. His study of this and other volumes awakened a love of natural history that was to become the central interest of his life. In 1744, having completed his studies in Wrexham, he matriculated at Queen's College, Oxford, where he planned to study law. At the time he was described as "full of corporal spirits, too lively and impetuous" although also "a very honest, good-natured man."[18] At age twenty, he set out on horseback through the mining country to Cornwall, keeping a journal in which he recorded his daily observations. He would later publish his account as the first of eleven travel books, all of which went through many editions. Samuel Johnson remarked of Pennant's books that he is "the best traveler I ever read, he observes more things than any one else does."[19] In 1755 he began his correspondence with Linnaeus, who sponsored him for election to the Royal Society of Uppsala in 1757. Pennant's interest in and knowledge of natural history led to his publication of the richly illustrated *British Birds*, whose four volumes were published between 1761 and 1766. Pennant also published several scientific papers during these years. Yet these were also difficult years for him. In 1763 his father died, and shortly thereafter his wife passed away as well. Nonetheless he kept working. In 1759 the former Dutch governor of Ceylon (Sri Lanka) had come to England and shown George Edwards and Pennant his notes and paintings of 103 birds done by the Sri Lankan artist P. C. de Bevere.

Edwards used some of the paintings in his *Gleanings* and Pennant embarked on a project to describe Indian birds. In 1767, while continuing his work on Indian birds, he was elected a fellow of the Royal Society. He brought out one volume, *Indian Zoology* (1769), where he described all the birds using their scientific names, making the book the first one in English to apply Linnaean systematics throughout. In 1773 he published *Genera of Birds* based on notes he had prepared for his friend Robert Ramsay, who was to teach a course in ornithology. Most of the book is devoted to presenting accounts of ninety-five genera of birds, apparently with the intent of including illustrations to accompany the accounts. (The illustrations were included in the second edition, published in 1781.) The book is innovative in two ways: the title page features a topographic diagram of a bird with the feather areas marked off and labeled, and the preface includes a definition of ornithology, a definition of a bird, and a description of its external parts, along with a few pages of information on courtship, nidification, and eggs. Pennant's *Genera of Birds* thus seems to be the first ornithological textbook.

In 1780 Captain Cook and his companions returned from their third voyage into the northern Pacific and the Arctic Ocean. The new abundance of material from northern Asia and North America turned Pennant's attention from preparation of an American zoology to that of an *Arctic Zoology,* which was published in two volumes in 1784 and 1785. The second volume was devoted to birds, and a supplement published in 1787 included ninety-four species obtained from Samuel Hearne, Andrew Graham, and Thomas Hutchins of the Hudson's Bay Company. Pennant lists a total of 427 birds in the second volume and a later supplement, but some are subspecies, some are Asian birds mistakenly assigned to North America, and some are inaccurate descriptions of species previously described. Inexplicably, Pennant describes all the American species under vernacular names, even though he had correctly used scientific names in his *Indian Zoology* published sixteen years earlier. Despite these faults, he was the first to describe many species from northern North America. In addition, his descriptions were used by Gmelin to generate scientific names and descriptions for the thirteenth edition of the *Systema Naturae,* and his *Artic Zoology* informed Wilson's *American Ornithology,* particularly his discussions of distribution.

Ray, John (1627–1705). John Ray met Francis Willughby at Cambridge sometime after Willughby arrived in 1653. At the time Ray was lecturing on Greek and mathematics and Willughby was attracted by Ray's intellect and the personality evident in his lectures. The two men conceived a plan to describe all of nature, with Ray describing the plants and Willughby, the animals. To this end they went to the west coast of England in 1662 and carried out an intensive study of the breeding habits of seabirds. In 1663 they traveled to the continent with two other friends and visited the large colonies of herons, spoonbills, and cormorants in Holland; journeyed along the Rhine; and made their way to Vienna. They crossed over the Semmering to Venice where they arrived on 6 October 1663. They continued through Italy the following spring and then split up, with Willughby and one friend going to Spain by way of Rome and southern France, arriving back in England toward the end of 1664. Meanwhile Ray and the other student went south to explore Sicily, then headed back north to Rome where they spent the winter. In 1665 they traveled through northern Italy then on to Lucerne, Lake Geneva, and Montpellier, where they intended to spend the winter. Louis XIV ordered all Englishmen to leave France within three months, however, so in February they journeyed to Paris, then on to Calais, arriving back in England on 6 April 1666. This may seem like a pleasant, if long and leisurely, journey, but it was in fact an exciting scientific exploration: at the time there were many species to be found that were unknown to science, some of which were for sale in markets at the stalls of birdcatchers. (Imagine buying a species new to science at the supermarket!) By purchasing these birds they were able to describe them externally and dissect them as a way of investigating their internal structures, stomach contents, and internal and external parasites. They also bought pictures of birds wherever possible and drew their own to supplement their detailed notes. Back in England, Ray and Willughby pooled their notes and began to organize them; they also took the time to visit collections of live birds brought back from the Americas. Willughby proposed that the two men visit America, but in 1670 his plans were interrupted when he became ill. Over the next two years his health declined, but he continued to work on his notes and discuss ideas with Ray. In 1672 Willughby died, but his many notes and extensive discussions with Ray enabled the latter to complete their book, *Ornithologiae libri tres,* which was published in 1676 and attributed to Willughby. In 1678 Ray published an English translation, and much later he edited and updated the book to include more recent material from the travels of Sir Hans Sloane in Jamaica (1687–1689) and Paul Herman in Ceylon (Sri Lanka, 1672–1680). This edition was published in 1713, after Ray's death, with the title *Synopsis methodica avium et piscium.*

Sloane, Sir Hans (1660–1753). Like so many naturalists of the seventeenth century, Hans Sloane studied medicine, but he was unusual. He actually practiced medicine. Before settling permanently in London, Sloane traveled to Jamaica and the West Indies, where during fifteen months in 1687–1688 he collected eight hundred new species of plants. Eventually he wrote *A Voyage to the Islands Madera, Barbados, Nieves, S. Christophers, and Jamaica* based on his experiences in the West Indies. Volume 1 was published in 1707, and volume 2 in 1725. Soon after his return, Sloane began adding to his natural history collection and building a library. He had joined the Royal Society in 1685, before he left on his trip, and in 1693 he was elected secretary of the society as well as editor of *Philosophical Transactions.* He served in this post for twenty years, then was elected vice president and later president, succeeding Sir Isaac Newton. In 1716 he was the first physician to receive a hereditary title, that of baronet. He also served as president of the Royal College of Physicians. It was in the latter role that he appointed George Edwards to the position of librarian. The association with Edwards blossomed into a close friendship, which gave Edwards access to Sloane's large library and the opportunity to catalog Sloane's equally impressive collection of birds. That catalog is the basis for Edwards's four-volume *Natural History of Uncommon Birds.* Sloane also supported Mark Catesby during the preparation of his second volume of *Natural History of Carolina, Florida, and the Bahama Islands.* Alexander Light, the first of the Hudson's Bay Company bird collectors, sent his specimens to Sloane, who also bought at least one consignment of plants from John Bartram by way of Peter Collinson. When he died, Sloane left his library of some fifty thousand volumes and his large natural history collection to England, which used them to found the British Museum.

Steller, Georg Wilhelm (1709–1746; Steller's Jay [*Cyanocitta stelleri*], Steller's Eider [*Polysticta stelleri*]). A summary of Steller's short life is full of improbabilities. He was a German Jewish naturalist who sailed with a Danish captain on an expedition financed by Empress Anna of Russia to explore the coast and islands of Alaska. Steller attended the University of Wittenburg where he intended to study theology, but he was diverted to the study of botany and natural history and earned a medical degree. Unable to find a professorship in Germany, he heard of the need for scientists in St. Petersburg, signed on to a Russian ship as a doctor for soldiers injured in the siege of Danzig, and after being shipwrecked, arrived in St. Petersburg in November 1734. For two years he was the personal physician to the archbishop and spent most of his time botanizing and learning Russian. When the government decided to send additional men to join Bering's expedition, Steller was appointed. He left St. Petersburg, on the edge of western Russia, in January 1738 and arrived in Okhotsk, in far eastern Russia on the Pacific, in mid-August 1740. During the summers he explored and collected, and during the winters he completed his notes and sent them with his collections of plants and animals back to St. Petersburg. He met Bering and in the summer of 1741 sailed with him to Alaska. Steller was able to explore on Kayak Island, east of Prince William Sound, Alaska, and it was there that he discovered Steller's Jay, so named in his honor by J. F. Gmelin. He was ordered back to the ship, which was wrecked on the way back to Siberia. The crew spent the winter on one of the Aleutian Islands, and there—in a hut made of driftwood and sailcloth and partially buried under a dune for protection, a hut that he shared with seven men—Steller wrote his account of the plants, animals, and geography of eastern Russia and Alaska, *De Bestiis Marinis*. In the spring Steller and those who had survived fashioned a small ship from the remains of the wreck and returned to Okhotsk. Steller spent the summer of 1743 exploring the Kamchatka Peninsula, but in 1744 the survivors of Bering's second expedition were summoned back to St. Petersburg. Steller sent his collections and notes ahead, but died of a fever on the way back. He was buried on a high bluff overlooking the River Tura. In addition to having two birds named after him, Steller's Sea Lion and the extinct Steller's Sea Cow bear his name.

Willughby, Francis (1635–1672). Willughby, the son of a wealthy English family, met John Ray soon after he enrolled at Cambridge University in 1653. They became close friends and remained so for the rest of Willughby's short life. They shared an interest in birds, which they pursued during a study of seabirds on the west coast of England in 1662 and later during a tour of the continent. Between 1666 and Willughby's death the two men organized the notes they had taken while visiting ornithological collections and observing birds throughout Europe. After Willughby's death, Ray compiled his friend's notes and completed the three-volume *Ornithologiae libri tres,* which was published under Willughby's name in 1678. The work is vitally important in two respects. First, birds were classified by form and structure, not by function as had been the case since Aristotle. As part of the new approach, Willughby and Ray provided a strictly dichotomous classification system that provided most groups with short, diagnostic descriptions. Furthermore, many groups were similar to modern assemblages of related birds. Since publication of *Ornithologiae* classification has continued to be based on form and structure, and, since the late twentieth century, has included the structure of genes.

Second, this was the last attempt to write an exhaustive treatise on ornithology, although even in *Ornithologiae,* coverage of the different fields of ornithology was uneven. For example, the microscopic hooks and ridges that enable barbules to link adjacent barbs and thereby form the smooth yet flexible surface of a feather were first described by Robert Hooke in his *Micrographia* published in 1665, but are not mentioned by Willughby despite his descriptions of feathers that were so detailed that Ray commented, "He seems to us somewhat too painstaking in the description of the colors of individual feathers."[20] So it was that in 1678, a full 130 years before Wilson's *Ornithology,* classification of birds shifted from function to structure.

There has been considerable debate over the years about who actually wrote *Ornithologiae tres libre.* Certainly Ray organized Willughby's notes and probably devised the dichotomous classification scheme that is at the innovative heart of *Ornithologiae.* But Willughby's scientific curiosity about avian diversity, behavior, and relationships and his enthusiasm for the minutae of avian structure were clearly the stimuli that drove them both to attempt and Ray to complete their masterwork. Who did what and how much? Perhaps Ray himself provided the best answer: when he was asked why he and Willughby did not describe the mythical birds found in previous texts, he answered, "To what purpose is it eternally to wrangle about things which certainly to determine is either absolutely impossible or next door to it?"[21]

These are, to paraphrase Sir Isaac Newton's words, the giants on whose shoulders Alexander Wilson built *American Ornithology*.

APPENDIX B

Wilson's Contemporaries and Correspondents

Ornithologists

Seven other ornithologists published while Wilson was researching and writing *American Ornithology,* though his knowledge of their work was limited due to the Napoleonic wars and the British blockade, which interrupted trade and communication between continental Europe, especially France, and the United States. Here we include information about these contemporaries, as well as, in parentheses, their dates of birth and death and any North American birds that bear their names.

Bewick, Thomas (1753–1828; Bewick's Wren [*Thryomanes bewickii*]). Because of his clear descriptions and outstanding illustrations, Bewick did more than any other person to popularize ornithology in England in the nineteenth century. His *History of British Birds,* published in two volumes in 1797 and 1804, was owned by Wilson, who before his early death had planned a similarly compact edition of American birds.

Brisson, Mathurin Jacques (1723–1806). Brisson curated the ornithological collection of René Antoine Ferchauld de Réaumur (1683–1757), who at the time of his death had assembled the largest array of bird specimens in Europe. Brisson's *Ornithologie,* published in 1760, cataloged Réaumur's collection along with several smaller collections. Its descriptions of the birds were based on specimens, and whenever possible the nest and eggs. Because it was a museum catalog, the *Ornithologie* contained no natural history information; but Brisson provided clear, detailed descriptions of each species that contrasted starkly with those of Linnaeus, which were at best cryptic and at worst, incomprehensible. Furthermore, Brisson's detailed observation of many species and often multiple specimens led him to define his genera more narrowly and sharply than Linnaeus. The result was that Brisson divided birds into 115 genera in 26 orders, whereas Linnaeus had classified birds into 63 genera grouped into 6 orders. When Réaumur died on 18 October 1757 he left his entire collection to the Académie des Sciences, but it was not to be. The Comte de Buffon, who hated Réaumur, convinced King Louis XV to have the collection added to the Cabinet du Roi, where it was under Buffon's control. Brisson was effectively shut out of ornithology and spent the rest of his life as a physicist.

Brünnich, Morten Thrane (1737–1827). Brünnich, considered the founder of Danish zoology, was the son of a portrait painter. He earned his bachelor's degree at the age of twenty and was studying theology and oriental languages when he read Linnaeus and almost immediately turned his attention to natural history. He wrote the entomology section of Erik Pontoppidan's *Dansk Atlas,* which led to his appointment as curator of the large natural history collections of Counselor Thott and Judge Christian Fleischer. While in this position he turned his attention to ornithology and published *A History of the Eider Duck* (1763), followed one year later by his *Ornithologia Borealis* and his *Entomologia.* He thus wrote not only a major contribution to a book but also three books of his own in the six years after he had switched fields, and all these works were published before he was twenty-seven. He was then appointed lecturer in natural history and economy at the University of Copenhagen.

Brünnich's *Ornithologia Borealis* contains the first description of the Common Loon (Brünnich's Great Northern Diver [*Gavia immer*]) and the first recognition that the Common Murre (Guillemot in European parlance [*Uria aalge*]) and Thick-billed Murre (Brünnich's Guillemot [*Uria lomvia*]) were separate species. Despite correctly separating the two species, he then gave the scientific name *Uria Troile* to what is now known as Brünnich's Guillemot (a perhaps understandable mistake given that it was the name used in England for the Common Guillemot). Confusion reigned until 1818 when Sabine proposed *Uria Brünnichii* for the species. That suggestion was short-lived and the name *Uria lomvia,* originally proposed by Linnaeus, was adopted because it was the first name proposed—even though Linnaeus did not seem to realize that there were two species of guillemots. Sabine's intent to honor Brünnich was realized, however, when the official British name was declared to be Brünnich's Guillemot. The significance of this story is that Wilson, who got into a similar problem with his name for Lewis's Woodpecker and other species, was neither alone in encountering such problems with names nor an uneducated chump who was ignorant of scientific names and rules of precedence.

After excelling in ornithology, Brünnich published a similar list of fish, founded the museum of natural history at the University of Copenhagen, and turned his attention for the next almost sixty years to mineralogy.

Latham, John (1740–1837). John Latham is the longest lived of the ornithologists cited by Wilson, who was born when Latham was a young man and died seven years before Latham began publishing his *General History of Birds* and a full twenty-four years before Latham's own death. John Latham began life on 27 June 1740 in Eltham, Kent. He attended Merchant Taylor's School and followed his father into a medical career. He corresponded with a number of prominent naturalists, among them Thomas Pennant, the only person whose work Wilson cited more frequently than Latham's. He, like Pennant, greatly admired the pioneering ornithological work of Francis Willughby and John Ray and, probably for that reason, resisted the Linnaean system for naming species. In his first great work on birds, the *General Synopsis of Birds,* published in three volumes between 1781 and 1783, he did not use Latin binomials for the species he described. Despite realizing his mistake and rushing to prepare his *Index Ornithologicus,* his worst fears were realized in 1788 when J. F. Gmelin published the thirteenth edition of Linnaeus's *Systema Naturae* with all of Latham's birds named according to Linnaean guidelines. Nonetheless, Latham's detailed descriptions of species led Wilson to frequently cite his work while virtually ignoring the thirteenth edition of the *Systema Naturae* edited by Gmelin. After Wilson's death, Latham wrote and published his *General History of Birds* in ten volumes from 1821 to 1828. In addition to achieving remarkably complete coverage of the birds then known around the world, the book is notable for Latham's failure to supply scientific names to species, many from Australia, that he described for the first time. You have to wonder what Latham thought when he learned that Gmelin, a chemist by profession, would be remembered for naming species first described by him. Yet Latham still did not employ Latin genus or species names when forty years later he again first described new species of birds. Did he really, really dislike the Linnaean system? If so, why?

Pallas, Peter Simon (1741–1811). Perhaps Pallas's greatest contribution to ornithology, particularly American ornithology, was his synthesis of the divergent views of Buffon and Linnaeus. He was born in Berlin in 1741, the son of a professor of medicine and chief surgeon at the Charité Hospital. Initially taught at home by tutors chosen by his father, the young Pallas showed such promise as a scholar that from 1754 on he attended lectures at the Collegium Medicochirurgicum, where his father taught. He was greatly interested in animals and in 1756 wrote a tentative classification of birds. At age sixteen he went to the universities of Halle and Göttingen, having finished the anatomy course in Berlin. Four years later he went to Leiden to defend a thesis on internal parasites in which he showed that Linnaeus had misclassified these particular worms. He received his medical degree at the age of nineteen.

After receiving his degree, Pallas toured England for nine months until his father called him home to join the army and practice medicine at the Austrian-Prussian front. Fortunately for ornithology, peace broke out at that moment and Pallas returned to Holland where he earned a living for the next four years writing scientific papers and helping rich collectors classify their collections. He not only recognized the significance of the Linnaean system, but also used it consistently in his work. In addition, he took advantage of opportunities to dissect species from the stadtholder's zoo. Pallas wished to expand the characteristics on which species and generic decisions were based and to do this through a detailed study of their external and internal structures. He published his findings in a series of monographs called *Spicilegia zoologica.*

Although Pallas was now widely known among naturalists and wealthy collectors, he still had no official position. The year was 1767, and he was twenty-six years old. His father's patience had worn thin, and he demanded that his son return to Berlin and set himself up in private practice. Young Pallas wavered. He had never disobeyed his father's wishes. (No one seems to know what his mother wished. In fact she makes no appearance in accounts of Peter Simon Pallas's life, although her death is not mentioned either, and he is said to have had a happy childhood, so we assume that she was alive.) Once again fate intervened. Not peace this time, but a letter. Not from another rich collector, but from Catherine the Great, empress of Russia, requesting his presence in St. Petersburg. Pallas considered the two requests—from his father and Catherine the Great. He chose St. Petersburg. On 3 June 1769, Venus would be visible crossing the disk of the sun. To celebrate this celestial event, which would enable astronomers to measure the distance from the earth to the sun, the empress had called together all members of the St. Petersburg Academy of Sciences, to which Pallas had recently been appointed. Catherine announced that the transit would be marked by sending not only astronomers, but also naturalists, throughout the length and breadth of the empire in order to gather scientific information. They were to leave in the summer of 1768. Despite the many preparations that were necessary, Pallas found time to study the collections of Georg Wilhelm Steller from Siberia and the Bering Strait and to publish his results as another monograph in his series *Spicilegia zoologica.* Pallas left St. Petersburg in June of 1768 and spent the next six years traveling as far east as northern Mongolia and the Amur River. On 30 July 1774 he arrived back in St. Petersburg to find himself internationally celebrated for the accounts of his travels, which he had sent back to Catherine, per her instructions, and which she had had edited and published in 1771 and 1773. In these two volumes, Pallas had described and named thirteen new species, which was especially impressive because he did not have access to the existing literature during his travels. He now found that Catherine wanted him to coordinate publication of the vast wealth of information that the scientists had brought back to St. Petersburg. He acceded to her wishes, even as he also began work on the *Zoographia Rosso-Asiatica,* in which he sought to include not only species descriptions for all the animals of the vast Russian empire, but also information about these animals' morphology, variation, distribution, and habits. This was a huge undertaking, but Pallas had his own voluminous notes and those of other contemporary naturalists who had participated in Catherine's exploration; the notes of Messerschmidt, who had been commissioned by Peter the Great to explore Siberia from 1720 to 1727; and the observations and materials collected by Steller. These notes constituted a formidable information resource. At Pallas's urging, Catherine sent an expedition to the far east coast of Russia in 1786 and they returned with more than two hundred specimens of 120 species from as far east as Prince William Sound, Alaska, which was part of Russia and would continue to be so for almost eighty more years. Pallas completed one more scientific expedition, in 1793–1794, then retired to an estate in the Crimea provided him by the empress, where he began to write his *Zoographia.* In addition to their considerable taxonomic significance, the books are important because they synthesized the views of Linnaeus and Buffon. Pallas adopted Linnaeus's systematics and his use of Latin genus and species names, but he adopted Buffon's conception of species as dynamic entities subject to climatic and geographic variation. Preparation of the plates for *Zoographia* was repeatedly delayed until finally, in 1810, while his manuscript was at the printer, Pallas and his daughter moved to Berlin to supervise the preparation of the plates. He died after receiving the final proofs for the first volume and the initial proofs of the second volume. The three volumes were printed in 1811–1813, but were not released by the St. Petersburg Academy of Sciences until 1827.[1] Pallas was cited only once by Wilson, and his books were published long after Wilson's death, but the few parallels in their lives are instructive in that both men explored large regions and wrestled with the divergent views of Linnaeus and Buffon, both of which Wilson cites repeatedly. (For all authors cited by Wilson, see Table A.2.)

Turton, William (1762–1835). Thomas Say (1787–1834) lent Wilson a copy of Turton's *Linneaus,* which was published in 1806.

Vieillot, Louis Jean Pierre (1748–1831). Vieillot was unusual among ornithologists of the late eighteenth and early nineteenth centuries in his single-minded focus on ornithology, the sheer quantity and diversity of his publications, and the tragedy that beset his private and scientific lives. Louis Jean Pierre Vieillot was born on 10 May 1748 in the little Norman town of Yvetot. His childhood and educational background are unknown. As a young man he held a minor clerical position in Paris and pursued his interest in ornithology on the side. Unhappy with the progress of his career and the confines of European ornithology, he traveled to Santo Domingo in the West Indies, where he remained for some time before returning and offering his notes on birds to Buffon. Buffon, however, having finished his book on birds, declined the offer. On the eve of the French Revolution Vieillot emigrated with his family to Santo Domingo. As in France, Vieillot continued in business and devoted his spare time to collecting and describing West Indian birds. Although he had escaped the bloodshed of the French Revolution, he soon found himself threatened by the bloody rebellion of black slaves against the white landowners. In 1792, Vieillot packed up his family and fled French territory for the newly formed United States. He remained there for several years, traveling extensively in the southern states. His interest in tropical birds persuaded him to return to Santo Domingo in the mid-1790s, where he collected and described birds of the West Indies until 1798. On August 23 of that year he, his wife, and their three daughters boarded the ship *Adrastes* bound for Bordeaux. Louis was the only member of the family to arrive on September 28: one by one his wife and daughters succumbed to yellow fever. One can scarcely imagine the ordeal of caring for them on board ship and then standing at the rail as each body, wrapped in sailcloth and packed with weights, was committed to the sea. Landing alone and destitute in Bordeaux, he traveled to Paris where he settled. Charles Dumont de Sainte-Croix (1758–1830), a wealthy amateur ornithologist, offered him a clerical job in his law office, and as was his habit, Vieillot devoted all his extra time to ornithology.

There followed many years of intense study at the Paris Museum and many publications, some paid for by Vieillot himself, before he retired to Rouen. There he lived in poverty until 1830, when he received a small pension to supplement his meager income. He died, blind and alone, in 1831.

Vieillot's ornithological accomplishments were extraordinary. Early in his career Vieillot identified and catalogued the plates of African birds by François Le Vaillant. After his return to Paris, Vieillot resumed his collaboration with Jean Baptiste Audebert on the elegant and costly *L'Histoire Naturelle des Colibris, Oiseaux Mouches, Jacamares et Proméropes.* They also began planning a collaborative work on American birds, though their plans were disrupted by the unexpected death of Audebert in 1800. Vieillot completed part of the planned work, which was published as *Histoire Naturelle des Oiseaux de l'Amérique Septentrionale.* While writing *l'Amérique Septentrionale,* Vieillot gave up his opposition to Linnaean rules of classification. His rigorous use of genus and species names throughout this work is the basis of an interesting story about bird names, their authors, and international agreements.

L'Amérique Septentrionale is in two volumes and three different formats that were published separately in 1807 (as well as 1808, 1809, and perhaps 1810). The date of publication is usually given on the title page, as was customary, but French publishers were in the habit of publishing the title page and table of contents of the entire work with the first section of the book, thereby antedating the publication of subsequent sections. In this case the first section was published on 1 December 1807, but twenty-one more sections were published before Vieillot's descriptions of American birds were completed. Charles W. Richmond thought it extremely unlikely that all twenty-one were published in 1808; he felt it was far more probable that the remaining twenty-one sections were published over the next two or three years.[2] Needless to say this uncertainty over dates of publication has caused considerable confusion for those trying to determine whose description and species name should be given priority. For instance, Vieillot published his descriptions of American birds in 1807–1809, while Wilson published the first volume of his *American Ornithology* in 1808. Both men described the House Wren; Vieillot called it *Troglodytes aedon,* while Wilson named it *Troglodytes domesticus* and attributed its discovery to William Bartram. Wilson gave the bird a genus and species name, but the species name has since been changed. Vieillot, by contrast, gave the House Wren its currently accepted genus and species names.[3] So whose description was published first and, therefore, who should be credited with discovering the species? Wilson's *American Ornithology,* volume 1, which contains his description of the House Wren, along with his names in English and Latin, was published in September 1808. Vieillot's description of the House Wren, along with its English and Latin names, was published in livraison 18 of the 21 livraisons that comprise his *Histoire naturelle des oiseaux de l'Amérique Septentrionale.* But when was livraison 18 published? The cover page of livraison 18 carries the date 1807, but we know now that Vieillot's description and names for the House Wren appeared in May 1809, roughly nine months after Wilson's description in *American Ornithology.* So why is Vieillot's name given priority? To put it simply, because the French publishers were successful in establishing priority by quickly publishing the first section and giving all subsequent sections that same earliest date. Not until 1991 did we learn that Vieillot was actually the second person to describe the House Wren.[4] By then the name *Troglodytes aedon* had been in use for 182 years and the International Commission for Zoological Nomenclature voted to keep it rather than create confusion.[5] A footnote acknowledging Wilson's priority, however, has been added to the species account in the 1998 *Check-list of North American Birds*, and we feel totally justified in counting the House Wren as one of the species first described by Wilson, while also acknowledging Vieillot's provision of the name.

Vieillot deserves this small recognition; he received very little during his lifetime despite his prodigious ornithological output. In addition to the two massive works already mentioned, Vieillot published *Histoire naturelle des plus beaux oiseaux chanteurs de la zone torride* in 1805 and *Analyse d'une nouvelle ornithologie élémentaire* in 1816. The latter was a simplified classification in which he sought to portray natural relationships by introducing many new generic and family names. Although this departure from the names provided by Linnaeus was initially resented, many zoologists, including Georges Cuvier (1773–1838), later adopted Vieillot's classification and many of his generic names remain in use to this day. He also contributed extensively to the *Nouveau dictionnaire d'histoire naturelle* (1816–1819), and in 1820 began work on *Tableau encyclopédique et méthodique des Trois Règnes de la Nature, Ornithologie*, which Bonnaterre had begun in 1790 but left incomplete upon his death in 1804. In 1821, Vieillot added to his list of projects the *Faune Français,* which was left unfinished in 1828. His last completed work was with P. L. Oudart: *La Galerie des Oiseaux du Cabinet d'histoire naturelle du jardin du roi,* published 1820–1826.

How could Vieillot, who published on European, African, American, and Caribbean birds, be so little recognized in his own time and so little known today? Vieillot had the misfortune to enter French ornithology at the height of the Comte de Buffon's fame, when French ornithology and Buffon were synonymous. By the time Vieillot returned to France in 1798, Buffon was gone, but Georges Cuvier dominated French ornithology and would continue to do so until after Vieillot's death. In addition, despite his many descriptions of American birds, Vieillot's work on American birds was overshadowed by Wilson's. In this case the British were largely to blame. In the years 1808–1812 the British navy was essentially blockading both France and the United States. It is doubtful that Vieillot's work was known to anyone in America; Wilson never cites it and he seems to have considered the literature available to him pretty thoroughly (see Table A.2).

Correspondents

In addition to the literature available to him in William Bartram's library, the Library Company of Philadelphia, the American Philosophical Society, and toward the very end of his life the Academy of Natural Sciences of Philadelphia, Wilson also established a network of correspondents who provided him with information. This was no accident, but part of his plan to learn all he could about the distribution and movements of North American birds. He refers to the network as one of the few successes of his first trip into New England peddling subscriptions for his books. That Wilson's plan met with considerable success is evident in the following letter written to William Bartram:

> Philadelphia, 12 February 1812
>
> I lately recd. Some letters from Natchez on the Mississippi giving me an account of some new Birds which I am promised particularly a Singular Hawk. I was also told by the same communications that Mr. Bradbury, an English Botanist who was sent out by the Linnaean Society of London has returned from the interior of Louisiana where he has been for several years, and has brought a superb collection; as he Draws well and has found a number of Birds and has promised to my friends in Natchez to let me copy any of them, I am in hopes that on Mr. Bradbury's arrival here on his way to London I shall find something new. He comes by way of New Orleans. Mr. Abbot of Savannah lately sent me the male and female Ground Dove in excellent order—a beautiful specimen.[6]

Given the importance that Wilson attached to establishing and maintaining his network of correspondents, we provide the following notes on those with whom he is known to have corresponded concerning birds.

Abbot, John (1751–1840). Abbot's father was a lawyer and art collector who encouraged his son's interest in drawing and natural history by providing private drawing lessons and books, among them George Edwards's books on birds and Mark Catesby's *Natural History of Carolina, Florida, and the Bahama Islands.* Abbot began collecting natural history specimens and experimented with keeping insects in captivity. He started work as a law clerk in his father's office, but in 1773 sold his collection and moved to Virginia, where he settled near Jamestown. He found the collecting disappointing and after two years moved to Augusta, Georgia, where he lived for a few years before moving and settling near Savannah where he would remain for the rest of his life. During the Revolutionary War he served under Lieutenant-Colonel James McIntosh and at the end of the war, in recognition of his service, he received a tract of 575 acres. He married Penelope Warren and a son,

John, was born in 1779. He lived comfortably as a planter for some twenty years, but then appears to have given it up and turned to collecting specimens and sketching animals and plants for European collectors.

Wilson met Abbot in late February or early March 1809 when he visited Savannah. On 5 March, he wrote to William Bartram:

> There is a Mr. Abbot here, who has resided in Georgia thirty-three years, drawing insects and birds. I have been on several excursions with him. He is a very good observer, and paints well. He has published, in London, one large folio volume of the Lepidopterous insects of Georgia.[7] It is a very splendid work.[8]

During this visit, Abbot showed Wilson two new species of rails he had not seen and to which he referred in a letter to Bartram dated 11 October 1809. Much later, on 23 January 1812, Wilson wrote to Abbot thanking him for specimens of a "Small Crow, female solitary Flycatcher, and the male and female Ground dove—all in good order."[9] He indicated that the "small" Crow and "solitary Flycatcher" had already been described, but that the Ground Doves had not and asked for any information regarding them that Abbot could provide. Wilson also sent his regards to mutual friends and expressed his concern about Abbot's rheumatism. Both the tone and content of the letter suggest that Abbot and Wilson were on close personal terms. The letter also makes clear that Abbot sent both specimens and information in response to Wilson's requests and that Wilson was not shy about sending such requests. We know that Abbot was Wilson's most frequently cited correspondent, an indication that Wilson was not reluctant to give credit to his correspondents.

Bancroft. Bancroft was a friend of Thomas Jefferson to whom Jefferson entrusted a copy of his *Notes on the State of Virginia.* Nothing else is known about him beyond a passing reference by Jefferson and a single reference by Wilson.

Bartram, William (1739–1823). Of the one hundred extant letters Wilson wrote after meeting Bartram in 1802, forty were addressed to William Bartram and all relate to a greater or lesser degree to Wilson's ornithological work. The relationship was close, and important to Wilson on many different levels (see Chapter 2 for a full description).

Churchman, Mordecai (mid-1700s–early 1800s). Little is known about Mordecai Churchman except that he was a Quaker who worked in Easton, Pennsylvania, as a bank cashier and supplied Wilson with information on Chimney Swifts, which he and Wilson refer to as Chimney Swallows. Wilson spent part of 20 May 1811 with Churchman learning about Chimney Swifts according to his account of the bird in volume 5 of *American Ornithology*. Churchman also sent additional information as acknowledged in the following humorous letter of appreciation from Wilson (who thanks two other residents of Easton as well):

> To Mr. Mordecai Churchman
> Easton, Pennsylvania.
> Philadelphia, 4 November 1811
> Dear Sir,
> I was much delighted with yours of June 23rd detailing particulars of the Chimney Swallow . . .
> My best respects & thanks to Mr. Arnott and also to Mr. Abraham Hart who so obligingly explored the Chimney for me—I am sorry to send him on such a black errand, but he could bring a good deal of light out of darkness by obliging me once more with a peep up, or down the Chimney. Truth they say sometimes lies at the bottom of a well—It may also lurk at the bottom of a Chimney—Excuse my freedom and believe me with great Sincerity
> My dear Sir
> Your much obliged friend
> Alex. Wilson[10]

Dunbar, William (1749–1810). Although only a single letter to William Dunbar exists, Wilson wrote to thank him for his hospitality at his plantation, the Forest, near Natchez, Mississippi. The visit was an important one for Wilson, who had just come by himself from Nashville, Tennessee, across the Natchez Trace. The warm welcome by Dunbar, his wife, and their family; the rest; the excellent food; and the conversation with an educated fellow Scot interested in science and natural history were all wonderfully restorative. In addition Dunbar introduced Wilson to neighboring planters and gave him a letter of introduction to planters between Natchez and New Orleans as well as associates in New Orleans. As a result, Wilson garnered not only twelve subscriptions in Natchez, but sixty-four more between Natchez and New Orleans.[11]

Dunbar, the younger son of Sir Archibald Dunbar of Scotland, was born in 1749, studied mathematics and astronomy, and immigrated to the colonies in 1771. After a brief period as an Indian trader, he became a partner of John Ross, a wealthy Philadelphia merchant. They established a plantation near Baton Rouge, but lost it during the Revolutionary War. After the war Dunbar built the Forest, a large, well-managed plantation near Natchez. He surveyed the Spanish-American border with Andrew Ellicott, explored the Mississippi delta, discovered Hot Springs, Arkansas, and explored the Red River at the request of Thomas Jefferson, who was a close friend. Dunbar was also one of America's scientific pioneers, having built an astronomical observatory and published papers on the plants and animals of the Mississippi region. At the time of Wilson's visit he was confined to bed with a fatal illness, but was alert and insistent on Wilson staying at the Forest and exploring the surrounding lands, which he did gladly. In the evenings Dunbar and Wilson undoubtedly talked about birds, plants, and natural history.

William Duncan (1778–?). William was the eldest son of Alexander Wilson's older sister, Mary. He accompanied his uncle to the United States in 1794. In 1798 he and Wilson bought land near Ovid, New York, and Duncan and his sister, Elizabeth, who had recently arrived, settled there to build a farm. In 1802 he was joined by his mother and other brothers and sisters. In 1805 Duncan left the farm near Ovid in the care of Elizabeth, his younger brother

Andrew, and his mother and went to study with his uncle in Philadelphia. Later that year he took the teaching job at Milestown where Wilson had formerly taught. With both Wilson and Duncan teaching, they were able to pay off the debt incurred for the farm and help the family develop the land. Several of Wilson's letters to his nephew, especially those written after their joint trip to Niagara Falls in 1804, contain references to birds and several requests that he make notes and send these when possible. In 1810 Duncan was working for his uncle as one of the colorists for volume 2 and may have continued as a colorist for volumes 3 and 4.

Gardiner, John L. (1770–1816). Gardiner was the owner and sole resident of Gardiner's Island, a three-thousand-acre hunk of land located east of Long Island. His hobby was the study of eagles and to a lesser extent Osprey, and Wilson quoted information from his "intelligent and obliging friend" in accounts of both birds.[12]

Jefferson, Thomas (1743–1826). In his *Notes on the State of Virginia,* Jefferson published the first state bird list, which was based largely on Mark Catesby's descriptions, but to which he added thirty-three species (Table 4.1). His list, which Wilson undoubtedly knew about, also matched Catesby's Latin names to their more recent Linnaean genus species designations, listed their common names, and provided the reference of Buffon's *Histoire Naturelle des Oiseaux.* But Jefferson's importance to Alexander Wilson extends well beyond the list. When Wilson as a young man wrote about the unfair labor practices of the mill owners and the rights of workers, he deeply respected Jefferson, whose vision of individual rights he shared. After Wilson arrived in the United States he avoided overt political activity, with one exception: Jefferson's inauguration in 1801. Wilson celebrated the day by accepting an invitation to speak in Milestown where he was a teacher. The text of his speech was published in several prominent newspapers.

After Wilson moved to Gray's Ferry and met William Bartram, he learned that Bartram and Jefferson were correspondents. Bartram encouraged him to write to Jefferson, which he did after his return from Niagara Falls, where he had shot and later painted two birds that he had never seen. He sent the painting to Jefferson with a cover letter from William Bartram. The president responded with the following letter:

To Alexander Wilson.

Monticello, 7 April 1805

I received here yesterday your favor of Mar. 18 with the elegant drawings of the new birds you found on your tour to Niagara, for which I pray you accept my thanks. The jay is quite unknown to me. From my observations, while in Europe, on the birds & quadrupeds of that quarter, I am of opinion there is not in our continent a single bird or quadruped which is not sufficiently unlike all the members of it's family there to be considered as specifically different. On this general observation I conclude with confidence that your Jay is not a European bird.

The first bird on the same sheet I judge to be a Muscicapa from it's bill, as well as from the following circumstance. Two or three days before my arrival here a neighbor killed a bird, unknown to him & never before seen here as far as he could learn. It was brought to me soon after I arrived; but in the dusk of the evening, & so putrid that it could not be approached but with disgust. But I retain a sufficiently exact idea of it's form & colours to be satisfied it is the same with yours. The only difference I find in yours is that the white on the back is not so pure, and that the one I saw had a little of a crest. Your figure, compared with the white-bellied Gobemouche 8. Buff. 342 P. enlum 566. shews a near relations. Buffon's is dark on the back.

As you are curious in birds there is one well worthy of your attention, to be found or rather heard in every part of America, & yet scarcely ever to be seen. It is in all the forests, from spring to fall, and never but on the tops of the tallest trees from which it perpetually serenades us with some of the sweetest notes, & as clear as those of the nightingale. I have followed it miles without ever but once getting a good view of it. It is of the size & make of the Mockingbird, lightly thrush-coloured on the back, & greyish-white on the breast

& belly. Mr Randolph, my son in law, was in possession of one which had been shot by a neighbor. He pronounces this also a muscicapa, and I think it much resembling the Moucherolle de la Martinique 8. Buffon 374. Pl. enlum. 568. As it abounds in all the neighborhood of Philadelphia, you may perhaps by patience & perseverance (of which much will be requisite) get a sight, if not a possession of it. I have for 20. years interested the young sportsmen of my neighborhood to shoot me one, but as yet without success. Accept my salutations & assurances of respect.

Th. Jefferson[13]

Wilson was pleased with the response and wrote to Bartram shortly after receiving Jefferson's response:

To William Bartram.

18 April 1805

My dear Sir,

. . .

Mr Jefferson speaks of a very strange bird. Please let we know what it is. I shall be on the look-out, & he must be a sly fellow if he escape me. I shall watch his motions and the sound of his serenade pretty closely, to be able to transmit to our worthy President a faithful sketch of a bird which he has been so long curious to possess.[14]

Wilson also alerted his nephew, William Duncan, then living on the family farm in Ovid, New York, between lakes Seneca and Cayuga.

To William Duncan

Gray's Ferry, 8 May 1805

. . . I informed you in my last of sending Mr. Jefferson drawings of the Falls, and some birds, which I found on the Mohawk, and which it seems have never been taken notice of by any naturalist. He returned me a very kind and agreeable letter, . . . describing to me a bird, which he is very desirous of possessing, having interested the young sportsmen of his neighbourhood, he says, these twenty years, to shoot him one, without success. It is of the size and make of the Mocking-bird, lightly thrush-coloured on the back, and grayish-white on the breast; is never heard but from the tops of the tallest trees, whence it continually serenades us with some of the sweetest notes, and as clear as those of the nightingale. Mr. Bartram can give no account of this bird, except it be the Wood Robin, which I don't think it is; for Mr. Jefferson says, '*it is scarcely ever to be seen*'; and 'I have followed it for miles without ever, but once, getting a good view of it.' I have been on the look-out ever since, but in vain. If you can hear of such a bird, let me know. I wish you also to look for the new bird which I discovered.[15]

On May 31 Wilson wrote to his nephew about William's pending trip to Gray's Ferry and remarked: "I wrote you respecting the letter I had from the President. I have never been able to get a sight of the bird he mentions. I hope you will not neglect to bring your gun with you, and look out as you come along."[16] Finally, in the early autumn Wilson again wrote Jefferson, having arrived at an answer to his question.

Kingsessing, 30 September 1805

I had the honour last spring of presenting your Excellency with drawings of two Birds which I suppos'd to be both non descripts until the receipt of your very condescending Letter to me of Ap. 7th. referring to 8 Buffon 342. Pl. enlum. 566. which I find to contain a Bird of the same Species with one of those sent but unnoticed by me before. Allow me Sir as an atonement for this mistake once more to beg your acceptance of another Sheet of Drawings being my poor efforts to represent faithfully 4 of our most capital Songsters among which is (I believe) the Bird so particularly and accurately described in your Excellency's Letter to me. This being the only Bird I can find among all our Songsters corresponding in every respect with the description there given. The clearness and plaintive Sweetness of its notes—its shy solitary disposition—continually serenading us from the tops of the tallest trees—its colour size and resemblance to

> the Moucherolle de la Martinique of Buffon, as observed by your Excellency, designate this, (and my friend Mr. Bartram is of the same opinion) to be the Bird so justly esteemed by your Excellency.[17]

This delightful exchange, with its dated use of titles and terms of deference—for example "Excellency" when referring to the president—resembles in other respects those between modern birders communicating by telephone or email. It is the first example of Wilson's later use of correspondents to help develop his data on the habits, species diversity, abundance, and distribution of birds. Jefferson continued to correspond with Wilson, but he did far more. In January 1809, with a copy of volume 1 of *American Ornithology* under his arm, Wilson knocked on the door of the White House, at the time simply referred to as the President's House. As described more fully in Chapter 2, the consequent conversation between the two men led to Jefferson's purchasing his own subscription and encouraging others to do the same. Wilson's southern trip was consequently such a success that the press run of two hundred sets had to be increased to four hundred sets.

Lawson, Alexander (1773–1846). Born in Ravenstruthers, Scotland, Alexander Lawson emigrated to the United States in 1794, the same year as Alexander Wilson. We do not know just when they met, but sometime before March 1804 Wilson wrote the following letter, which clearly indicates a friendship of some months, probably years:

> To Mr. Lawson, Engraver.
> Philada.
>
> Gray's Ferry, 12 March 1804
>
> Dear Sir,
> I dare say you begin to think me very ungenerous and unfriendly in not seeing you for so long a time. I will simply state the cause, and I know you will excuse me . . . 5 days . . . are occupied in . . . pedagoguing matters and the other 2 are sacrificed to that itch for drawing, which I caught from your honourable self.[18]

Clearly the relationship had existed long enough for Lawson to have suggested that Wilson take up drawing, which can be regarded as the first important step toward development of the *American Ornithology,* also long enough for Wilson to apologize for his absence. Wilson would address many of his travel letters to Lawson, who was by then engraving plates for the *American Ornithology.* Later Lawson was commissioned by Charles Lucien Bonaparte for his revision and expansion of Wilson's *American Ornithology.* Lawson died in 1846 in Philadelphia.

Lewis, Meriwether (1774–1809). Lewis spent the first ten years of his life as a neighbor of Thomas Jefferson. In 1784 his family moved to Georgia and the boy spent his adolescence hunting and studying Latin, mathematics, and science. His father died when he was eighteen and he took over management of the family estate, but two years later, during the Whiskey Rebellion, he joined the local militia, then enlisted in the regular army. In 1801 Jefferson was inaugurated president and invited Lewis to become his private secretary. In January 1803, following months of discussion about a land route to the Pacific, Jefferson asked Congress for permission to send an expedition in search of such a route. In June, Lewis asked William Clark to share command of a force of about forty-five men, who assembled toward the end of the year at a site eighteen miles north of St. Louis. When Wilson and Lewis met soon after Lewis's return in 1806 from the western expedition, Lewis gave Wilson specimens then unknown to science that in Wilson's hands would become Lewis's Woodpecker, Clark's Nutcracker, and the Western Tanager (*Piranga ludoviciana*). In addition Lewis provided Wilson with notes from the expedition, perhaps most importantly the diary of John Ordway, and put Wilson in touch with Ordway who provided Wilson with much additional information.

Lewis began writing an account of the expedition to the Pacific, but had to leave to take up his duties as governor of the Louisiana Territory. It was while returning alone to Washington along the Natchez Trace that Lewis died at Grinder's Stand, Tennessee, under mysterious circumstances that have never been resolved. On his own solitary trip along the Natchez Trace,

Wilson stopped at Grinder's Stand to investigate Lewis's death. Although he spoke with Mr. and Mrs. Grinder, he was not able to determine whether Lewis had been murdered or had taken his own life. Before leaving, Wilson paid Grinder to fence in and tend Lewis's grave, and the two men signed a written agreement to that effect. After his return to Philadelphia, Wilson wrote a report of Lewis's death as described by the Grinders and their servants that was published in *Port Folio.*

Michaux, François André (1770–1855). François Michaux and William Bartram were both sons of famous botanists. Michaux first came to America in 1785 with his father, who had been sent by the French government to study trees throughout eastern North America—from Hudson's Bay, south through the Appalachians and eastern states, and on into Spanish Florida. Michaux remained in or near New York and later Charleston, South Carolina, until 1790 when he returned to France and became an ardent supporter of the French Revolution. In 1801 the government sent him to the United States to sell the nurseries his father had established. He spent the next two years traveling through Ohio, Kentucky, Tennessee, and the Carolinas before returning to France in 1803. Back in France he completed work on his *Voyage à l'ouest de monts Alléghanys,* which was published in Paris in 1804.

Michaux returned to the United States in 1806. During this trip he was the guest of William Bartram on more than one occasion and it was at Bartram's house that he and Alexander Wilson met. After his return to Paris in 1809 he wrote *Histoire des Arbres forestiers de l'Amérique Septentrionale,* which was published in four volumes from 1810 to 1813. Wilson greatly admired the plates in the latter work and persuaded a friend who read French to read portions of the text to him and William Bartram, both of whom were "highly satisfied" with its information.[19] In addition to describing his work on and plans for *American Ornithology,* Wilson corresponded with Michaux about a possible translation of Michaux's volumes on American trees. Wilson was unable to arrange with Samuel Bradford for publication of an English edition, although one was published in 1817 with the title *The North American Sylva.*

Mitchell, Samuel L. (1764–1831). Samuel Mitchell was born in Hempstead, Long Island. His father was a remarkably intelligent and industrious Quaker farmer, but it was his uncle, Dr. Samuel Latham, who helped the boy obtain a sound education and with whom he initially studied medicine. In 1783 Mitchell enrolled at the University of Edinburgh and in 1786 he received a medical degree. He spent a year traveling around England before returning to the United States, where he studied law and participated in negotiations with the Oneidas and Onondagas for the western parts of New York State. In 1790 he was elected to the New York State legislature, and in 1792 he was appointed professor of chemistry, natural history, and philosophy at Columbia College. His mineralogical survey of the Hudson Valley helped establish his reputation as an early pioneer in geology. In 1801 he was elected to the U.S. House of Representatives and in 1804 he was appointed to the Senate. He returned to the House in 1809 and continued to serve until 1813. In 1808 he was given the position of professor of natural history at the College of Physicians and Surgeons of the City of New York.

Over his lifetime, Mitchell published in many European and American journals, including *Transactions of the American Philosophical Society*, of which he and Wilson were both members. Mitchell was on friendly terms with Meriwether Lewis, as was Wilson, and he published his only surviving letter from Wilson, which was ostensibly a recommendation of Major Robert Carr, but also devotes more than half the space to a discussion of several species of hawks.[20] In this letter, Wilson refers to previous and future correspondence as well, indicating that he and Mitchell were well acquainted. Wilson also mentions the imminent publication of volume 5 of *American Ornithology* and the excellence of its plates, about which he positively glows. This is a letter between friends.

Ord, George (1781–1866). In 1810 Alexander Wilson lived in Philadelphia, in a rented room in the house of William Jones. A few blocks away on South Front Street lived George Ord, a wealthy young man who was tall and lean, with strong but good-looking features, the son of a wealthy Swedish woman and a British sea captain who had founded a very successful chandlery business in the city. Little is known of Ord's boyhood or education, except that he was well educated and fond of words. (In fact, before he met Wilson he had amassed a large collection of obsolete words, which he offered to Noah Webster. When Webster declined, Ord was furious and never forgave Webster, although his word collection was accepted by the editor of the British revision of Dr. Johnson's dictionary.) Despite his prickly nature, Ord was soft spoken and could be kindly.

After his father's death, Ord became head of the family chandlery business, although it was his mother who actually ran the company: Ord was a hunter and sportsman who devoted his time to hunting, and like many of Philadelphia's hunters he visited Duck Island on the Delaware, Little Egg Harbor, and Cape May. Exactly when or where Ord and Wilson met is unclear. Ord may have responded to the appeal published in the preface of the third volume of *American Ornithology:*

> To gentlemen of leisure, resident in the country, whose taste disposes them to the pleasing and rational amusement of natural history, and who may be in possession of facts, authentic and interesting, relative to any of our birds which have not yet made their appearance in this work, the author respectfully addresses himself . . . for the direction of those who may be disposed to honour the Author with their correspondence, the following list is subjoined; containing the common popular names of the most interesting of our Land Birds, whose history we have yet to detail, and of whose manners any authentic particulars will be gladly received.

Alternatively Ord may have met Wilson at Little Egg Harbor where he was hunting and Wilson was collecting specimens. Or he may have arranged to happen across Wilson as he left his lodgings a few blocks from Ord's house. Whatever the circumstances of that first meeting, it is fairly certain that Ord had been collecting information on Wilson long enough to have formed an impression of the solitary author and naturalist, whose poetry and prose were published often in *Port Folio,* America's leading literary magazine, and whose

American Ornithology was a literary, scientific, and artistic success. Because of Wilson's literary successes, Ord assumed he was a man of means. Because Wilson was well read and cited all the important ornithological works, Ord assumed he owned a large library. Because Wilson kept meticulous records of the birds he shot, as well as of when and where and how many birds he saw, and because he wrote long and detailed travel letters that were later edited and published, Ord assumed that Wilson kept voluminous journals of his field experiences. This was a man on the verge of greatness. This was a man with whom Ord wished to associate and did. Probably not until September 1813—as Ord, co-executor of Alexander Wilson's last will and testament, sat in his friend's upstairs room in the house of William Jones—did he realize that Wilson was not wealthy and in fact owed Bradford & Inskeep $2,284.22; that Wilson's library consisted of two books, Bewick's *History of British Birds* volumes 1 and 2; and that Wilson's notes and journals were far from voluminous. Every one of Ord's assumptions concerning Alexander Wilson had been wrong. Unlike Ord, Wilson had lived for his work. He had had no other life.

We cannot know what George Ord thought as he sat alone in the room where Alexander Wilson had lived and died, but despite his surprise at what he didn't find, despite the superficial reasons that had led him to befriend Wilson, he had become Wilson's ardent admirer and genuinely interested in birds. Therein lies the final tragedy of Wilson's life: Ord set out to firmly establish Wilson's reputation as the foremost American ornithologist of all time. He oversaw the proofreading of volume 8 of *American Ornithology,* which was at the press when Wilson died; edited what few notes and plates Wilson had left for volume 9; and saw this last volume through publication. He even wrote a memorial to Wilson for *Port Folio,* though it contained important omissions, for example, Wilson's date of birth, which he could have gotten easily from William Duncan or Wilson's sister. It also contained outright errors, for example his assertion that Wilson's southern trip had produced no subscriptions when in fact he had signed 250 subscribers including Thomas Jefferson, James Madison, and James Monroe. The editors of *Port Folio* transformed Ord's pretentious, sharp-edged memoir into a shortened, graceful memorial, but Ord objected strongly to their editorial liberties. The editors then offered to publish a note to accompany their memorial indicating that the original had been altered, but Ord was not so easily appeased. He insisted that they publish his original memoir. Thus *Port Folio* carried two accounts of Wilson's life and an editorial note, while volume 9 of *American Ornithology* carried a version of Ord's memoir of Wilson that essentially matched that which had been submitted to *Port Folio*. The dispute and Ord's intransigence were a portent of what was to come.

In 1824–1825 Ord oversaw the publication of an expanded *American Ornithology* in an octavo format, which was smaller than the original and so could be marketed more widely. In April 1824, Audubon arrived in Philadelphia and was introduced to Charles Lucien Bonaparte, nephew of Napoleon, a young man deeply interested in birds who planned to revise and update Wilson's *American Ornithology*. Bonaparte looked at Audubon's portfolio, was deeply impressed, and invited Audubon to attend a meeting of the Academy of Natural Sciences that very evening as his guest. At the meeting, Bonaparte introduced Audubon to George Ord, then vice president of the Academy, who looked through Audubon's portfolio and objected to his mingling of plants and birds.[21] Others at the meeting were impressed. We can never know if Ord immediately saw Audubon as a threat to Wilson's reputation as an ornithologist and scientist (and indirectly his own reputation), but we do know that he blocked Audubon's election to the academy.

Ord was certainly in a position to influence Audubon's reception in the literary and scientific capital of the United States. He had been elected to the Academy of Natural Sciences in 1815 and served as its collections curator until 1817 shortly after his election as vice-president (1816–1834). He had served on the Academy's publications committee (1817–1821; 1832–1833) and was elected president in 1851, a position he held until 1858. He was also active in the American Philosophical Society following his election in 1817. Not only did he proofread volume 8 of Wilson's *American Ornithology* and write volume 9 from Wilson's notes; not only did he edit and publish a revised and expanded quarto edition in 1824–1825 and an enlarged biography of Wilson in 1828; not only did he write the ornithological sections of William Guthrie's *Zoology*, which was the first systematic work on American zoology; but he also published several original articles based on his own observations of birds.[22]

George Ord published his last ornithological article in 1836, twenty-three years after Wilson's death and nine years after Audubon began publication of his elephant folio.[23] The entire article, which is based on hours of observation at the nests of birds parasitized by cowbirds, reflects the subtle influence of Alexander Wilson. From his observations of living birds Ord draws a number of logical conclusions that refute the ideas of previous authors. The article is important to the development of field ornithology in its emphasis on Wilson's core belief in the need to observe the living bird in its own habitat, if one is to understand its natural history. Yet even in this, his most important study of birds, written so many years after Wilson's death and Audubon's triumph, Ord cannot resist attacking Audubon. Beginning with a rude reception on an April evening in the Philadelphia Academy of Natural Sciences, Ord's intransigence and Audubon's ego led to increasingly bitter accusations of plagiarism on both sides until, tragically, Wilson's accomplishment was lost in the haze of charge and countercharge and even Ord's own ornithological contributions were overlooked.

Peale, Charles Wilson (1741–1827). As a young boy, the famous painter Charles Wilson Peale obtained a rudimentary education in Kent County, Maryland, where his father was the schoolmaster. At the age of thirteen he was apprenticed to a saddler and after completing his indenture established his own saddlery. He was bankrupted during the Stamp Act crisis for his support of the Sons of Freedom and so turned to his interest and talent in portraiture to earn a living. He attracted commissions and several wealthy patrons who encouraged him to go to England to study, which he did under Benjamin West from 1767 to 1769. Peale returned to Maryland and then moved to Philadelphia in 1776, where he painted portraits of the delegates to the Continental Congress. He served in Washington's army and saw action at Trenton and Princeton, New Jersey, and in the evacuation of Philadelphia. During this period he painted portraits of many of his fellow officers. He was elected to the General Assembly of Philadelphia in 1779 as a representative of Philadelphia. After the Revolution he opened a public gallery to display his portrait collection, which because he had long been an amateur naturalist included natural curiosities. In 1794 his growing collection was moved to rooms at the American Philosophical Society and in 1802 to rooms in Independence Hall. In 1805, Peale helped found the Philadelphia Academy of Fine Arts. Still later the collection was incorporated as the Philadelphia Museum and after 1810 was managed by his sons. Peale's museum was the repository of 279 of Wilson's specimens, the very specimens from which he painted the plates for his *American Ornithology*. Equally important, Peale worked closely with Wilson to design paints that would achieve the exact color and transparency that he insisted on for the coloring of his plates. The latter collaboration reflected not only Wilson's attention to detail, but also his dream that everything about *American Ornithology* should be American and so contribute to shaping American culture.

Peale, Rubens (1784–1865). Rubens Peale managed the Peale Museum, founded by his father, until its dissolution in 1846. He also founded similar museums in Baltimore and New York.

Notes

1. *Themes in Wilson's Life and Writings*

1 W. Jardine, "Illustrative Notes, and Life of Wilson," in Alexander Wilson, *American Ornithology,* with a continuation by Charles Lucian Bonaparte, vol. 1 (London: Whittaker, Teacher, and Arnot; Edinburgh: Stirling and Kenny, 1832), p. xxxvi.

2 A. Wilson, *American Ornithology,* vol. 1 (Philadelphia: Bradford & Inskeep, 1808), p. 3.

3 Ibid., pp. iii–iv. Emphasis in original.

4 Ibid., pp. 1–2.

5 Ibid., p. 1. Emphasis in original.

6 Ibid., p. 4.

7 Ibid.

8 Ibid., pp. 7–8.

2. *A Varied Life*

1 C. Hunter, *The Life and Letters of Alexander Wilson* (Philadelphia: American Philosophical Society, 1983), 26. Readers interested in holding and reading letters written by Wilson can visit the collection at Houghton Library, Harvard University.

2 W. Jardine, "Illustrative Notes and Life of Wilson," in Alexander Wilson, *American Ornithology,* ed. Charles Lucien Bonaparte, vol. 1 (London: Whittaker, Treacher, & Arnot; Edinburgh: Stirling and Kenny, 1832), p. xv.

3 J. S. Wilson, *Alexander Wilson, Poet-Naturalist: A Study of His Life with Selected Poems* (Washington, D.C.: Neale Publishing, 1906), p. 19.

4 Ibid., p. 20.

5 Jardine, "Illustrative Notes," p. xiv.

6 Ibid.

7 On the apprenticeship being against his wishes, see ibid., p. xv.

8 R. Cantwell, *Alexander Wilson: Naturalist and Pioneer* (Philadelphia: J. B. Lippincott, 1961), p. 35.

9 Jardine, "Illustrative Notes," p. xviii.

10 B. Mearns and R. Mearns, *Biographies for Birdwatchers: The Lives of those Commemorated in Western Palearctic Bird Names* (London: Academic Press, 1988), p. 418.

11 A. Wilson, *American Ornithology,* vol. 1 (Philadelphia: Bradford & Inskeep, 1808); R. Plate, *Alexander Wilson, Wanderer in the Wilderness* (New York: David McKay, 1966).

12 Cantwell, *Alexander Wilson,* pp. 36–37, Hunter, *Life and Letters,* pp. 26–27. Both authors recount the incident and offer circumstantial evidence against John Craig, but there is nothing to implicate Alexander Wilson beyond his and Meg's residence in the tower during the summer of 1782 and his frequent visits to the tower in 1783 and 1784 (after he had moved to Lochwinnoch).

13 Hunter, *Life and Letters,* p. 26. See also forty-six manuscript volumes by A. Cairn Crawfurd of Lochwinnoch on the history of Lochwinnoch—which were compiled from conversations of those who knew the Wilsons and others in Lochwinnoch during the second half of the eighteenth century—located in the archives of the Paisley Library, Paisley, Scotland.

14 "Bobbin girls" were those individuals, usually teenage girls, who kept thread wound onto the spools that were feeding into the cloth as the weavers wove. Wilson was unusual in employing a grown woman by the name of Peggy Orr. See Cantwell, *Alexander Wilson,* p. 36.

15 Jardine, "Illustrative Notes," p. xvi.

16 Ibid.

17 Ibid., p. xviii.

18 Ibid., p. xxii.

19 T. Crichton, *Biographical Sketch of the Late Alexander Wilson to a Young Friend* (Paisley, Scotland: J. Neilson, 1819).

20 A. Wilson, *Poems* (Paisley, Scotland: J. Neilson, 1790).

21 Jardine, "Illustrative Notes," p. xxviii.

22 A. Wilson, *The Poems and Literary Prose of Alexander Wilson, the American Ornithologist,* ed. A. B. Grosart (Paisley, Scotland: Alexander Gardner, 1876), vol. 2, p. x.

23 Cantwell, *Alexander Wilson,* p. 75.

24 J. S. Wilson, *Alexander Wilson, Poet-Naturalist,* p. 58.

25 Cantwell, *Alexander Wilson,* p. 268.

26 Ibid., p. 275.

27 Cantwell, *Alexander Wilson,* pp. 265–276.

28 Wilson, *Poems and Literary Prose,* vol. 2, p. xxxvii.

29 Jardine, "Illustrative Notes," pp. xxx–xxxi.

30 Cantwell, *Alexander Wilson,* p. 276.

31 Hunter, *Life and Letters,* pp. 158–160.

32 Ibid.

33 Wilson, *Poems and Literary Prose,* vol. 2, pp. 323–333.

34 Hunter, *Life and Letters,* pp. 183–185.

35 Ibid., pp. 181–183.

36 Ibid., pp. 190–192.

37 W. Faxon, "Early Editions of Wilson's *Ornithology*," *Auk* 18 (1901): 216–218.

38 Hunter, *Life and Letters,* p. 202.

39 Ibid., p. 203; T. Crichton, *Biographical Sketch of the Late Alexander Wilson to a Young Friend* (Paisley, Scotland: J. Nielson, 1819).

40 E. H. Burtt, Jr. and W. E. Davis, "Historic and Taxonomic Implications of Recently Found Artwork in Arithmetic Books of Students of Alexander Wilson," *Wilson Bulletin* 107 (1995): 193–213.

41 Hunter, *Life and Letters,* pp. 202–203.

42 Burtt and Davis, "Historic and Taxonomic Implications," pp. 193–213.

43 Hunter, *Life and Letters,* p. 204.

44 Ibid., pp. 204–205.

45 Ibid., p. 205; Crichton, *Biographical Sketch.*

46 Hunter, *Life and Letters,* p. 209.

47 Ibid., p. 74.

48 Burtt and Davis, "Historic and Taxonomic Implications," pp. 193–213.

49 Hunter, *Life and Letters,* p. 211.

50 Ibid., pp. 207–212.

51 Ibid., pp. 219–220.

52 W. B. Fetters, *Maximilian Leech (1763–1815) of Blockley and Kingsessing* (Decorah, Iowa: Anundson, 2003), p. 63.

53 T. Jefferson, *Notes on the State of Virginia,* ed. W. Peden (Chapel Hill: University of North Carolina Press, 1982); A. Feduccia, ed., *Catesby's Birds of Colonial America* (Chapel Hill: University of North Carolina Press, 1985).

54 Hunter, *Life and Letters,* pp. 232–234.

55 For the letter to Bartram, see ibid., pp. 238–239; for the letter to Duncan, see ibid., pp. 239–240; Wilson, *Poems and Literary Prose.*

56 Hunter, *Life and Letters,* pp. 244–245.

57 W. Bartram, "Description of an American Species of *Certhia*, or Creeper," *Philadelphia Medical and Physical Journal* 1 (1804): 103–106.

58 Hunter, *Life and Letters,* pp. 249–251; G. Ord, *Sketch of the Life of Alexander Wilson, Author of the "American Ornithology"* (Philadelphia: H. Hall, 1828), pp. xlvi–xlvii.

59 Hunter, *Life and Letters,* p. 254.

60 Ibid., pp. 267–268.

61 Ibid., pp. 268–272.

62 Ibid.

63 Ibid.

64 Ibid., pp. 260–261; Wilson, *Poems and Literary Prose*; Ord, *Sketch of the Life of Alexander Wilson.*

65 Hunter, *Life and Letters,* pp. 261–262.

66 Ibid., pp. 262.

67 Ibid., pp. 286–290.

68 Ibid., pp. 291–292.

69 Ibid., *Life and Letters,* p. 274.

70 A. Feduccia, ed., *Catesby's Birds of Colonial America* (Chapel Hill: University of North Carolina Press, 1985).

71 W. Bartram, *Travels and Other Writings* (New York: Library of America, Literary Classics of the United States, 1996).

72 Cantwell, *Alexander Wilson,* p. 185.

73 Hunter, *Life and Letters,* pp. 313–315.

74 Ibid., pp. 318–319.

75 Ibid.

76 Jardine, "Illustrative Notes," p. ci; Wilson, *Poems and Literary Prose,* vol. 1, pp. 207–217; ibid., pp. 358–370.

77 Cantwell, *Alexander Wilson,* p. 198.

78 Hunter, *Life and Letters,* pp. 326–339.

79 Lewis's Woodpecker was originally named *Picus torquatus* by Wilson. Unbeknownst to him, however, Boddaert had applied that name in 1783 to the Ringed Woodpecker *(Celeus torquatus)* of northeastern South America and the Amazon basin. In 1849 Gray corrected Wilson's mistake, but respected his wish to honor Meriwether Lewis by giving the bird the scientific name *Picus Lewis,* later revised to its current form *Melanerpes lewis.* The other species brought back by the Lewis and Clark expedition and named by Wilson are Clark's Nutcracker *(Nucifraga columbiana)* and the Western Tanager *(Piranga ludoviciana).* In both cases Wilson is credited with naming the species.

80 Alexander Wilson, "Particulars of the Death of Capt. Lewis," *Port Folio* 7 (1811): 24–47.

81 Hunter, *Life and Letters,* pp. 103–104.

82 The list of subscribers was published in the ninth and last volume of *American Ornithology.* Cantwell, in *Alexander Wilson,* has reprinted the list with brief biographies of the many subscribers who could be traced.

83 Hunter, *Life and Letters,* pp. 398–399.

84 Ibid., pp. 401–402.

85 Ibid., pp. 403–405; Ord, *Sketch of the Life of Alexander Wilson,* pp. civ–cv.

86 Hunter, *Life and Letters,* pp. 405–406.

87 Ibid., pp. 406.

88 Ibid., p. 112; Anon. 2006, Paisley History Museum, available online at http://www.paisley.org.uk/history/index.php.

3. Illustrating *American Ornithology*

1 A. S. Blum, *Picturing Nature: American Nineteenth-Century Zoological Illustration* (Princeton, N.J.: Princeton University Press, 1993), pp. 26–46.

2 Wilson to Bartram, 4 January 1806, in C. Hunter, ed., *The Life and Letters of Alexander Wilson* (Philadelphia: American Philosophical Society, 1983), pp. 243–244.

3 A. Wilson, *American Ornithology,* 9 vols. (Philadelphia: Bradford & Innskeep, 1808–1814), vol. 1, p. 7.

4 A. Johnson et al., eds., *Dictionary of American Biography,* 22 vols. (New York: N.p., 1928–1958), vol. 11, p. 56.

5 Wilson, *American Ornithology,* vol. 1, p. 7.

6 Alternatively, it might read "Engraved by George Murray" or "Engraved by George Warnicke," because these men each engraved a few plates.

7 If a species' scientific name is given in the commentary that accompanies its portrait, the name is not given in the text. For species that are not portrayed, the scientific name follows the first mention of the common name.

8 The common names of birds as recognized by the American Ornithologists' Union have initial capital letters (e.g., Gray Catbird). General names do not (e.g., catbird).

9 Wilson to Bartram, 22 May 1807, in Hunter, *Life and Letters,* pp. 263–264.

10 W. B. Fetters, *Maximilian Leech (1763–1815) of Blockley and Kingsessing* (Decorah, Iowa: Anundson, 2003), p. 63.

11 Throughout, if the current scientific name is the same as the original, the name of the person who described the species follows the species' name without parentheses. If the genus has changed, the original author's name is in parentheses.

12 Wilson, *American Ornithology,* vol. 1, p. 30.

13 Ibid., p. 46.

14 Ibid., p. 64.

15 Ibid., p. 75.

16 A. Wilson, "Pine Grosbeak," in George Ord, ed., *American Ornithology; or, The Natural History of the Birds of the United States by Alexander Wilson,* vol. 2 (New York: Collins & Company, 1828), p. 156.

17 Wilson, *American Ornithology,* vol. 1, p. 87.

18 Ibid., p. 132.

19 Ibid., pp. 95, 96.

20 Wilson, *American Ornithology,* vol. 2, pp. 117–118.

21 Ibid., p. 77.

22 Wilson, *American Ornithology,* vol. 3, p. 20.

23 Ibid., p. 26.

24 Ibid., p. 38.

25 Ibid., p. 32.
26 Ibid., p. 29.
27 Ibid., p. 46.
28 Ibid., p. 41.
29 Ibid., p. 59.
30 Ibid., p. 53.
31 Ibid., p. 55.
32 Ibid., p. 70.
33 Ibid., p. 72.
34 Ibid., p. 78.
35 Ibid., p. 74.
36 Ibid., p. 76.
37 Ibid., p. 90.
38 Ibid., p. 101.
39 Ibid., p. 103.
40 Ibid., p. 100.
41 Wilson, *American Ornithology,* vol. 4, p. 9.
42 Ibid., p. 13.
43 Ibid., p. 17.
44 Ibid., p. 19.
45 Ibid., p. 27.
46 Ibid., pp. 22–23.
47 Wilson, *American Ornithology,* vol. 1, p. 142.
48 Wilson, *American Ornithology,* vol. 4, p. 40.
49 Ibid., pp. 35–36.
50 Ibid., pp. 43.
51 Ibid., pp. 75–77.
52 Ibid., p. 66.
53 Ibid., p. 68.
54 Ibid., pp. 89, 93–94.
55 Wilson, *American Ornithology,* vol. 5, pp. 14–15.
56 Ibid., p. 27.
57 Ibid., p. 32.
58 Ibid., p. 31.
59 Wilson, *American Ornithology,* vol. 7, p. 27.
60 Wilson, *American Ornithology,* vol. 5, p. 34.
61 Wilson, *American Ornithology,* vol. 8, p. 41.
62 Wilson, *American Ornithology,* vol. 5, p. 46.
63 Ibid., p. 58.
64 Ibid., p. 48.

65 Ibid., p. 64.
66 Ibid., pp. 66, 69.
67 Ibid., p. 73.
68 Ibid., p. 84.
69 Ibid., p. 96.
70 Ibid., p. 98.
71 Ibid., p. 100.
72 Ibid., p. 114.
73 Ibid., p. 113.
74 Ibid., pp. 104–105.
75 Wilson, *American Ornithology,* vol. 6, p. 15.
76 Ibid., p. 27.
77 Ibid., pp. 46–47.
78 Ibid., p. 67.
79 Buffon remarks that this owl rarely constructs a nest of its own; but not unfrequently occupies that of others, particularly the Magpie.
80 Wilson, *American Ornithology,* vol. 6, p. 73.
81 Ibid., p. 70.
82 Ibid., pp. 75–76.
83 Ibid., p. 86.
84 Ibid., p. 82.
85 Wilson, *American Ornithology,* vol. 1, p. 28.
86 Wilson, *American Ornithology,* vol. 6, p. 88.
87 Wilson, *American Ornithology,* vol. 2, p. 85.
88 Wilson, *American Ornithology,* vol. 6, pp. 84–85.
89 Ibid., p. 99.
90 Ibid., p. 96.
91 Wilson, *American Ornithology,* vol. 7, pp. 22–23.
92 Ibid., p. 30.
93 Ibid., p. 28.
94 Ibid., p. 25.
95 Wilson, *American Ornithology,* vol. 3, p. 137.
96 Ibid., p. 36.
97 Ibid., p. 42.
98 Ibid., p. 32.
99 Ibid., p. 43.
100 Ibid., p. 39.
101 Although we cannot absolutely rule out the Long-billed Dowitcher (*Limnodromus scolopaceus*) as the species described by Wilson, his emphasis on its coastal distribution and his description of its whistled call suggest to us that he was in fact observing the Short-billed Dowitcher.

102 Wilson, *American Ornithology,* vol. 7, p. 32.
103 Ibid., pp. 45–46.
104 Ibid., pp. 45–47.
105 Ibid., p. 48.
106 Ibid., p. 55.
107 Ibid., p. 63.
108 Ibid., pp. 60–61.
109 Ibid., p. 41.
110 Ibid., pp. 71–72.
111 Ibid., pp. 73–74.
112 Ibid., p. 86.
113 Ibid., pp. 68–69.
114 Ibid., p. 80.
115 Ibid., p. 77.
116 Ibid., p. 86.
117 Ibid., pp. 93–96.
118 Ibid., p. 87.
119 Ibid., p. 98.
120 Ibid., p. 106.
121 Ibid., pp. 99–100.
122 Ibid., p. 104.
123 Ibid., p. 108.
124 Ibid., p. 119.
125 Ibid., p. 127.
126 Wilson, *American Ornithology,* vol. 8, p. 33.
127 Ibid., p. 36.
128 Ibid., p. 38.
129 Ibid., p. 14.
130 Ibid., p. 27.
131 Wilson, *American Ornithology,* vol. 7, pp. 114–115.
132 Ibid., pp. 123–124.
133 Wilson, *American Ornithology,* vol. 8, p. 20.
134 Ibid., p. 13.
135 Ibid., p. 15.
136 Ibid., p. 23.
137 Ibid., p. 30.
138 Ibid., p. 26.
139 Ibid., p. 35.
140 Ibid., p. 37.
141 Ibid., p. 43.

142 Ibid., p. 62.
143 Ibid., p. 65.
144 Ibid., p. 49.
145 Ibid., p. 51.
146 Ibid., p. 60.
147 Ibid., pp. 53, 56.
148 Ibid., pp. 69–70.
149 Now extinct, the last known Laborador Duck was shot in Elmira, New York, on 12 December 1878.
150 Wilson, *American Ornithology,* vol. 8, p. 96.
151 Ibid., p. 89.
152 Ibid., p. 89.
153 Ibid., p. 83.
154 Ibid., p. 86.
155 Ibid., p. 84.
156 Ibid., pp. 94–96.
157 Ibid., pp. 97–98.
158 Ibid., p. 113.
159 Ibid., p. 101.
160 Ibid., p. 104.
161 Ibid., p. 110. Red-headed Widgeon, Pochard, and Dun Bird, as well as Poker, refer to what is now called the Pochard.
162 Ibid., p. 133.
163 Ibid., p. 145.
164 Ibid., p. 143.
165 Ibid., p. 135.
166 Ibid., p. 137.
167 Ibid., p. 139.
168 Ibid., p. 141.
169 Ord, *American Ornithology,* vol. 9, pp. 72–73.
170 Ibid., p. 69.
171 Bartram, *American Ornithology,* vol. 9, p. 62.
172 Ord, *American Ornithology,* vol. 9, p. 77.
173 Ibid., pp. 75–76.
174 Abbot, *American Ornithology,* vol. 9, p. 83.
175 Ord, *American Ornithology,* vol. 9, p. 90.
176 Ibid., pp. 94, 95.
177 Ibid., p. 84.
178 Ibid., pp. 97–98.
179 Ibid., p. 105.

180 Ibid., pp. 116–117.

181 Ibid., pp. 120, 121.

182 R. H. Welker, *Birds and Men: American Birds in Science, Art, Literature, and Conservation, 1800–1900* (1955; repr. New York: Atheneum, 1966), pp. 21–23.

183 Hunter, *Life and Letters,* pp. 86–88.

184 Blum, *Picturing Nature,* p. 37.

185 Hunter, *Life and Letters,* p. 211.

186 E. Coues, "Behind the Veil," *Bulletin of the Nuttall Ornithological Club* 5 (1880): 194.

187 E. Coues, *A Key to North American Birds; Containing a Concise Account of Every Species of Living and Fossil Bird at Present Known from the Continent North of the Mexican and United States Boundary* (New York: Dodd and Mead, 1872), pp. xvi.

188 Blum, *Picturing Nature,* p. 45.

4. Pioneer Ornithologist

1 C. Hunter, ed., *The Life and Letters of Alexander Wilson* (Philadelphia: American Philosophical Society, 1983), pp. 148–151.

2 Ibid., pp. 164–166.

3 Ibid., pp. 167–170.

4 Ibid., p. 203.

5 E. Coues, "Behind the Veil," *Bulletin of the Nuttall Ornithological Club* 5 (1880): 193–204; R. Cantwell, *Alexander Wilson: Naturalist and Pioneer* (Philadelphia: J. B. Lippincott, 1961).

6 On Wilson the poet and Wilson's address at Jefferson's inauguration, see A. Wilson, *The Poems and Literary Prose of Alexander Wilson,* ed. A. B. Grosart, 2 vols. (Paisley, Scotland: Alexander Gardner, 1876). On Wilson's earning money as a surveyor, see Cantwell, *Alexander Wilson;* and R. Plate, *Alexander Wilson: Wanderer in the Wilderness* (New York: David McKay Company, 1966).

7 Hunter, *Life and Letters,* p. 203.

8 Ibid., p. 202.

9 Ibid., p. 205.

10 Personal communication with J. T. Fry, curator, Bartram's Garden, Philadelphia, based on John Bachman's reminiscences of his youth in Philadelphia.

11 G. Edwards, *Natural History of Uncommon Birds,* 4 vols. (London: College of Physicians, 1743–1751); G. L. Comte de Buffon, 1750. *Histoire naturelle . . . avec la description du cabinet du roi,* 44 vols. (1750; English trans. London: W. Smellie, 1781).

12 W. Turton, *A General System of Nature: Through the Three Grand Kingdoms of Animals, Vegetables, and Minerals* . . . (London: Lackington, Allen, and Co., 1806).

13 E. H. Burtt, Jr. and A. P. Peterson, "Alexander Wilson and the Founding of North American Ornithology," in W. E. Davis, Jr. and J. A. Jackson, eds. *Contributions to the History of North American Ornithology: Memoirs of the Nuttall Ornithological Club* (Cambridge, Mass.: Nuttall Ornithological Club, 1995), pp. 359–386.

14 A. Wilson, *American Ornithology,* vol. 1 (Philadelphia: Bradford & Inskeep, 1808).

15 D. Barrington, "Experiments and Observations on the Singing of Birds, by the Hon. Daines Barrington, Vice Pres. R. S. in a Letter to Mathew Maty, M. D. Sec. R. S." *Philosophical Transactions* 63 (1773): 249–291.

16 Wilson, *American Ornithology,* vol. 1.

17 Ibid.

18 Ibid.

19 *Pl. enl. 623* refers to the 623rd plate of the *Planches enluminées* (1764–1783), 973 colored plates painted by F. N. Martinet under the supervision of the younger Daubenton and published as a supplement to Buffon's *Histoire.*

20 Wilson, *American Ornithology,* vol. 1.

21 Ibid.

22 Ibid.

23 Ibid.

24 Hunter, *Life and Letters,* p. 262. Emphasis in original.

25 Wilson, *American Ornithology,* vol. 1, pp. 3–4.

26 Ibid., pp. 2–4.

27 Ibid.

28 Ibid., p. 1.

29 Hunter, *Life and Letters,* pp. 210–211.

30 American Ornithologists' Union, *Checklist of North American Birds,* 7th ed. (Washington, D.C.: American Ornithologists' Union, 1998).

31 Ibid.

32 For the Pine Creeper of Catesby, see A. Feduccia, *Catesby's Birds of Colonial America* (Chapel Hill: University of North Carolina Press, 1985) p. 111; for the Field Sparrow, see W. Bartram, *Travels and Other Writings,* ed. T. P. Slaughter (New York: Library of America, Literary Classics of the United States, 1996), p. 243; and for the Marsh Wren, see J. Latham, *General Synopsis of Birds* (London: Leigh & Sotheby, 1781).

33 American Ornithologists' Union, *Checklist of North American Birds,* pp. 480–481.

34 Bartram, *Travels and Other Writings,* p. 243.

35 L. P. Vieillot, *Histoire naturelle des oiseaux de l'Amerique Septentrionale . . .,* vol. 1 (1807–1808) and vol. 2 (1808–1809). (Paris: Chez Desfray, 1807–1809).

36 C. W. Richmond, "On the Date of Lacépède's 'Tableaux,'" *Auk* 16 (1899): 323–325.

37 American Ornithologists' Union, *Check-list of North American Birds,* 6th ed. p. 532 (Washington, D.C.: American Ornithologist Union, 1983).

38 M. R. Browning and B. L. Monroe, Jr., "Clarifications and Corrections of the Dates of Issue of Some Publications Containing Descriptions of North American Birds," *Archives of Natural History* 18 (1991): 381–405.

39 American Ornithologists' Union, *Checklist of North American Birds,* pp. 481; R. C. Banks and M. R. Browning, "Comments on the Status of Revived Old Names for Some North American Birds," *Auk* 112 (1995): 633–648.

40 On Vieillot, see P. H. Oehser, "Louis Jean Pierre Vieillot (1748–1831)," *Auk* 65 (1948): 568–576.

41 G. R. Gray, *The Genera of Birds,* vol. 3 (London: Taylor & Francis, 1849).

42 Burtt and Peterson, "Alexander Wilson and the Founding of North American Ornithology."

43 Based on data in ibid.

44 Ibid.

45 E. Stresemann, *Ornithology from Aristotle to the Present* (Cambridge: Harvard University Press, 1975).

46 Burtt and Peterson, "Alexander Wilson and the Founding of North American Ornithology."

47 E. Stringham, *Alexander Wilson: A Founder of Scientific Ornithology* (Kerrville, Tex.: Stringham; 1958); Stresemann, *Ornithology from Aristotle to the Present.*

48 Wilson, *American Ornithology,* vol. 4.

49 Wilson, *American Ornithology,* vol. 1.

50 J. F. Gmelin, *Caroli a Linne, Systema naturae per regna tria naturae,* 13th ed., vol. 1 (Lyon, France: J. B. Dellamolliere, 1788).

51 Burtt and Peterson, "Alexander Wilson and the Founding of North American Ornithology."

52 Wilson, *American Ornithology,* vol. 2.

53 Wilson, *American Ornithology,* vol. 4.

54 Wilson, *American Ornithology,* vol. 2.

55 Wilson, *American Ornithology,* vol. 4.

56 Ibid.

57 Ibid.

58 Huntor, *Lifo and Lottors,* p. 85.

59 Wilson, *American Ornithology,* vol. 2.

60 Ibid.

61 Ibid.

62 W. Faxon, "Relics of Peale's Museum," *Bulletin of the Museum of Comparative Zoology* 59 (1915): 119–148.

63 Wilson, *American Ornithology,* vol. 3.

64 Ibid.

65 Ibid.

66 Wilson, *American Ornithology,* vol. 1.

67 Wilson, *American Ornithology,* vol. 5.

68 Wilson, *American Ornithology,* vol. 4.

69 Ibid.

70 On Catesby, see Feduccia, *Catesby's Birds of Colonial America.* On Lawson, see J. Lawson, *A New Voyage to Carolina . . .,* ed. Hugh T. Lefler (Chapel Hill: University of North Carolina Press, 1967).

71 Wilson, *American Ornithology,* vol. 1.

72 Ibid.

73 Ibid.

5. Wilson's Legacy

1 E. G. Allen, "The History of American Ornithology before Audubon," *Transactions of the American Philosophical Society* 41 (1951): 387–591.

2 A. Feduccia, *Catesby's Birds of Colonial America* (Chapel Hill: University of North Carolina Press, 1985), p. 8.

3 Ibid., p. 10.

4 R. Cantwell, *Alexander Wilson: Naturalist and Pioneer* (Philadelphia: J. B. Lippincott, 1961), p. 238.

5 G. Ord, *Sketch of the Life of Alexander Wilson, Author of the "American Ornithology"* (Philadelphia: Harrison Hall, 1828), p. 98. Emphasis ours.

6 Cantwell, *Alexander Wilson,* p. 200; R. Rhodes, *John James Audubon: The Making of an American* (New York: Alfred A. Knopf, 2004), pp. 67–68.

7 C. Hunter, *The Life and Letters of Alexander Wilson* (Philadelphia: American Philosophical Society, 1983), p. 371.

8 Cantwell, *Alexander Wilson,* p. 201.

9 Allen, "History of American Ornithology," pp. 387–591.

10 Rhodes, *John James Audubon*, pp. 71–72.

11 Cantwell, *Alexander Wilson,* p. 251.

12 N. Pick, *The Rarest of the Rare: Stories behind the Treasures at the Harvard Museum of Natural Zoology* (New York: HarperCollins, 2004), pp. 144–145.

13 Cantwell, *Alexander Wilson,* p. 298.

14 Ibid., p. 236.

15 Ibid.

16 B. Mearns and R. Mearns, *Audubon to Xántus: The Lives of Those Commemorated in North American Bird Names* (London: Academic, 1992), pp. 95–101.

17 Ibid., pp. 116–121.

18 Mearns and Mearns, *Audubon to Xántus,* pp. 43–55.

19 J. Kastner, *A World of Watchers* (New York: Alfred A. Knopf, 1986), p. 18.

20 Ibid., p. 17–29.

21 R. T. Peterson, *Peterson Field Guide to Birds of North America* (Boston: Houghton Mifflin, 2008); D. P. Sibley, *The Sibley Field Guide to Birds of Eastern North America* (New York: Alfred A. Knopf, 2003).

Appendix A. On the Shoulders of Giants

1 John T. Zimmer—as cited in E. G. Allen, "The History of American Ornithology before Audubon," *Transactions of the American Philosophical Society* 41 (1951): 529—suggests that this is an onomatopoetic rendition of the squawk of a heron, although which species is unknown.

2 Allen, "History of American Ornithology," p. 529.

3 Ibid.

4 M. Walters, *A Concise History of Ornithology* (New Haven: Yale University Press, 2003).

5 On the debate over these names' validity, see G. M. Mathews and T. Iredale, "On the 'Table des Planches Enlum.' of Boddaert," *Australian Avian Record* 3 (1915): 31–51.

6 T. Jefferson, *Notes on the State of Virginia,* ed. W. Peden (Chapel Hill: University of North Carolina Press, 1982).

7 A. Feduccia, *Catesby's Birds of Colonial America* (Chapel Hill: University of North Carolina Press, 1985), p. 5.

8 C. Mylius, "Johann Leonhard Frischens Vorstellung der Vögel," *Hamburgisches Magazin* 4 (1749): 412, as cited in E. Stresemann, *Ornithology from Aristotle to the Present* (Cambridge: Harvard University Press, 1975).

9 R. Glover, as cited in S. Houston, T. Ball, and M. Houston, *Eighteenth-Century Naturalists of Hudson Bay* (Montreal: McGill-Queen's University Press, 2003), p. 80.

10 J. R. Jehl, as cited in ibid., p. 90.

11 Allen, "History of American Ornithology," p. 508.

12 J. T. Klein, *Historiae avium prodromus* (Lübeck, Ger.: Schmidt, 1750) pp. 83–84.

13 Ibid., p. 489.

14 A. Newton, "Ornithology," in *Encyclopaedia Britannica,* 9th ed., vol. 18 (Edinburgh: Encyclopaedia Britannica, 1885), pp. 2–50.

15 J. R. Forster, "An Account of the Birds Sent from Hudson's Bay," *Philosophical Transactions of the Royal Society* 62 (1772): 370–381.

16 Houston, Ball, and Houston, *Eighteenth-Century Naturalists of Hudson Bay,* p. 56.

17 Ibid., p. 57.

18 Ibid., p. 24.

19 Ibid.

20 Stresemann, *Ornithology from Aristotle to the Present*, p. 45.

21 J. Ray, as cited in S. J. Gould, *The Hedgehog, the Fox, and the Magister's Pox: Mending the Gap between Science and the Humanities* (Cambridge, MA: Harvard University Press, 2011), p. 47.

Appendix B. Wilson's Contemporaries and Correspondents

1 E. Stresemann, "Date of Publication of Pallas Zoographia Rosso-Asiatica" *Ibis* 93 (1951): 316–318.

2 C. W. Richmond, "On the Date of Lacépède's 'Tableaux,'" *Auk* 16 (1899): 327fn.

3 American Ornithologists' Union, *Checklist of North American Birds,* 7th ed. (Washington, D.C.: American Ornithologists' Union, 1998), pp. 480 481.

4 M. R. Browning and B. L. Monroe, Jr., "Clarifications and Corrections of the Dates of Issue of Some Publications Containing Descriptions of North American Birds," *Archives of Natural History* 18 (1991): 381–405.

5 R. C. Banks and M. R. Browning, "Comments on the Status of Revived Old Names for Some North American Birds," *Auk* 112 (1995): 633–648; "*Bombycilla cedrorum* Vieillot, [1808] and *Troglodytes aedon* Vieillot, [1809] (Aves, Passeriformes): Proposed Conservation of the Specific Names," *Bulletin of Zoological Nomenclature* 53 (1996): 187–190.

6 C. Hunter, ed., *The Life and Letters of Alexander Wilson* (Philadelphia: American Philosophical Society, 1983), pp. 398–399.

7 The book referred to by Wilson is J. E. Smith, ed., *The Natural History of the Rarer Lepidopterous Insects of Georgia: With 104 Plates Based on Abbot's Illustrations* (London: T. Bensley, 1797).

8 Hunter, *Life and Letters,* pp. 308–309.

9 Ibid., p. 397.

10 Ibid., pp. 394–395.

11 Ibid., pp. 101–103; R. Cantwell, *Alexander Wilson: Naturalist and Pioneer* (Philadelphia: J. B. Lippincott, 1961), pp. 277–305.

12 A. Wilson, *American Ornithology,* vol. 2 (Philadelphia: Bradford & Inskeep, 1810).

13 Hunter, *Life and Letters of Alexander Wilson,* pp. 232–233.

14 Ibid., pp. 238–239.

15 Ibid., pp. 239–240.

16 Ibid., pp. 241–243.

17 Ibid., pp. 244–245.

18 Ibid., pp. 206–207.

19 Ibid., p. 400.

20 Ibid., p. 341.

21 R. Rhodes, *John James Audubon: The Making of an American* (New York: Knopf, 2004), pp. 123–138.

22 G. Ord, "An Account of an American Species of the Genus Tantalus or Ibis," *Journal of the Academy of Natural Sciences of Philadelphia* 1 (1817): 53–57; G. Ord, "Observations on Two Species of the Genus Gracula, Latham," *Journal of the Academy of Natural Sciences of Philadelphia* 1 (1817): 253–261; G. Ord, "An Account of the Florida Jay, Bartram," *Journal of the Academy of Natural Sciences of Philadelphia* 1 (1817): 345–347; G. Ord, "Some Observations on the Moulting of Birds," *Transactions of the American Philosophical Society* 3 (1830): 292–300; G. Ord, "Observations on the Cow Bunting of the United States," *Magazine of Natural History* 9 (1836): 57–71.

23 Ord, "Observations on the Cow Bunting of the United States."

Selected Bibliography

Abbot, J. B. "An Artist Looks at Wilson." *Frontiers* 23 (1943): 11–14.

Allen, E. G. "The History of American Ornithology before Audubon." *Transactions of the American Philosophical Society* 41 (1951): 387–591.

Allen, J. L. "Alexander Wilson." *Natural History* 19 (1919): 396–403.

Anon. "Memento of Wilson, the Naturalist." *Port Folio* 18 (1824): 487–488.

Bair, J. "The Emergence of the Modern Bird Guide in America." *Snowy Egret* 42 (1979): 37–48.

Bartram, W. "Description of an American Species of *Certhia,* or Creeper." *Philadelphia Medical and Physical Journal* 1 (1805): 103–106.

———. *Travels and Other Writings.* New York: The Library of America, Literary Classics of the United States, 1996.

Blum, A. S. "Picturing Nature: American Nineteenth-Century Zoological Illustration." Princeton, N.J.: Princeton University Press, 1993.

Bridson, G. D. R. "From Xylography to Holography: Five Centuries of Natural History Illustration." *Archives of Natural History* 16 (1989): 121–141.

Browning, M. R., and B. L. Monroe, Jr. "Clarifications and Corrections of the Dates of Issue of Some Publications Containing Descriptions of North American Birds." *Archives of Natural History* 18 (1991): 381–405.

Burns, F. L. "Alexander Wilson. I: The Audubon Controversy." *Wilson Bulletin* 20 (1908): 3–18.

———. "Alexander Wilson. II: The Mystery of the Small-Headed Flycatcher." *Wilson Bulletin* 21 (1908): 63–79.

———. "Alexander Wilson. III: The Unsuccessful Lover." *Wilson Bulletin* 21 (1908): 130–145.

———. "Alexander Wilson. IV: The Making of the *American Ornithology.*" *Wilson Bulletin* 20 (1908): 165–185.

———. "Alexander Wilson. V: The Completion of the American *Ornithology.*" *Wilson Bulletin* 21 (1909): 16–35.

———. "Alexander Wilson. VI: His Nomenclature." *Wilson Bulletin* 21 (1909): 132–151.

———. "Alexander Wilson. VII: Biographies, Portraits, and a Bibliography of the Various Editions of His Works." *Wilson Bulletin* 21 (1909): 165–186.

———. "Alexander Wilson. VIII: His Early Life and Writings." *Wilson Bulletin* 22 (1910): 79–96.

———. "Alexander Wilson in Bird Census Work." *Wilson Bulletin* 19 (1907): 100–102.

———. "The Mechanical Execution of Wilson's *American Ornithology.*" *Wilson Bulletin* 41 (1929): 19–23.

———. "Miss Lawson's Recollections of Ornithologists." *Auk* 34 (1917): 275–282.

Burtt, E. H., Jr. and W. E. Davis, Jr. "Historic and Taxonomic Implications of Recently Found Artwork in Arithmetic Books of Students of Alexander Wilson." *Wilson Bulletin* 107 (1995): 193–213.

Burtt, E. H., Jr. and A. P. Peterson. "Alexander Wilson and the Founding of North American Ornithology." In W. E. Davis, Jr. and Jerome A. Jackson, eds., *Contributions to the History of North American Ornithology: Memoirs of the Nuttall Ornithological Club,* no. 12. Cambridge, Mass.: Nuttall Ornithological Club, 1995.

Cantwell, R. *Alexander Wilson: Naturalist and Pioneer.* Philadelphia: J. B. Lippincott, 1961.

Christy, B. H. "Alexander Lawson's Bird Engravings." *Auk* 43 (1926): 47–61.

———. "The Earliest List of Pennsylvanian Birds." *Oriole* 11 (1946): 30–33.

Coues, E. "Behind the Veil." *Bulletin of the Nuttall Ornithological Club* 5 (1880): 193–204.

———. "Private Letters of Wilson, Ord and Bonaparte." *Penn Monthly* (June 1879): 443–455.

Davis, R. A. "Printing Alexander Wilson's Copper Plates." *Bulletin* (Field Museum of Natural History) 44 (1973): 12–15.

Davis, W. E., Jr. and M. K. LeCroy. "A Watercolor and Notes by Alexander Wilson." *Archives of Natural History* 21 (1994): 237–242.

Eckhardt, G. H. "Alexander Wilson Painted Birds Long before Audubon Quit Storekeeping." *New York Sun,* Oct. 29, 1943.

Egerton, F. N. "Notes Chiefly on the Plates of Wilson's *American Ornithology.*" *Journal for the Society of the Bibliography of Natural History* 4 (1963): 125–137.

Elman, R. *America's Pioneering Naturalists.* Chap. 4: "Alexander Wilson." Tulsa, Okla.: Winchester Press, 1982.

Faxon, W. "Early Editions of Wilson's *Ornithology.*" *Auk* 18 (1901): 216–218.

Feduccia, A. *Catesby's Birds of Colonial America.* Chapel Hill: University of North Carolina Press, 1985.

Fetters, W. B. *Maximilian Leech (1763–1815) of Blockley and Kingsessing.* Decorah, Iowa: Anundson Publishing, 2003.

———. *William Fetters (1794–1857) and Mary B. Leech: Ancestors and Descendents.* Decorah, Iowa: Anundson Publishing, 1991.

Hammond, N. *Modern Wildlife Painting.* New Haven, Conn.: Yale University Press, 1998.

Houston, S., T. Ball, and M. Houston. *Eighteenth Century Naturalists of Hudson Bay.* Montreal: McGill-Queen's University Press, 2003.

Huddleston, E. L. "Topographical Poetry in the Early National Period." *American Literature* 38 (1966): 303–322.

Hunter, C., ed. *The Life and Letters of Alexander Wilson.* Philadelphia: American Philosophical Society, 1983.

Huth, H. *Nature and the American.* Chap. 2: "Scientists, Philosophers, and Travelers." Berkeley: University of California Press, 1957.

Jefferson, T. *Notes on the State of Virginia.* Ed. W. Peden. Chapel Hill: University of North Carolina Press, 1982.

Lawson, J. *A New Voyage to Carolina.* Ed. H. T. Lefler. 1709; repr. Chapel Hill: University of North Carolina Press, 1967.

Mathews, G. M., and Iredale, T. "On the 'Table des Planches Enlum' of Boddaert." *Australian Avian Record* 3 (1915): 31–51.

McAtee, W. L. "Georgian Records in John Latham's 'General History of Birds.'" *Oriole* 11 (1946): 1–11.

McNamara, K. R. "The Feathered Scribe: The Discourses of *American Ornithology* before 1800." *College of William and Mary Quarterly* 47 (1990): 210–234.

McWilliam, J. M., and C. A. Wood. "The James Craw Portrait of Alexander Wilson." *Auk* 58 (1941): 236–238.

Nicholson, C. P. "Alexander Wilson's Travels in Tennessee." *Migrant* 57 (1986): 1–7.

Oehser, P. H. "Louis Jean Pierre Viellot (1748–1831)." *Auk* 65 (1948): 568–576.

Olsen, H. A. "Alexander Wilson: The Frustrated Poet Who Pioneered in Ornithology." *Snowy Egret* 27 (1963): 6–34.

Ord, G. "Life of Alexander Wilson." *Port Folio* 4, 3d ser. (1814): 343–362.

———. *Sketch of the Life of Alexander Wilson, Author of the "American Ornithology."* Philadelphia: Harrison Hall, 1828.

Pasquier, R. F., and J. Farrand, Jr. *Masterpieces of Bird Art: 700 Years of Ornithological Illustration.* New York: Abbeville Press, 1991.

Peabody, W. B. O. "Life of Alexander Wilson." In J. Sparks, ed., *The Library of American Biography.* New York: Harper and Bros., 1840.

Peck, R. M. 1983. "Books from the Bartram Library." Pp. 46–50 in A. Wheeler, ed., *Contributions to the History of North American Natural History.* Edinburgh: Society for the Bibliography of Natural History, 1983.

Plate, R. *Alexander Wilson: Wanderer in the Wilderness.* New York: David McKay Company, 1966.

Rosen, J. "William Bartram's Sketches: The Field and the Image." *Tipularia* 7 (1992): 8–11.

Skipworth, P. *The Great Bird Illustrators and Their Art, 1730–1930.* New York: A & W Publishers, 1979.

Slatkin, C. A. "Alexander Wilson." Pp. 554–556 in J. A. Garraty and M. C. Carnes, eds., *American National Biography.* New York: Oxford University Press, 1999.

Smith, G. V. "Scenes from the Life of Alexander Wilson: The First Volume of the *American Ornithology.*" *Oologist* 10 (1893): 269–271.

———. "Scenes from the Life of Alexander Wilson: His Early American Career." *Oologist* 10 (1893): 147–148.

———. "Scenes from the Life of Alexander Wilson: His Last Days." *Oologist* 11 (1894): 174–177.

———. "Scenes from the Life of Alexander Wilson: His Second Southern Tour." *Oologist* 11 (1893): 89–92.

———. "Scenes from the Life of Alexander Wilson: His Second Southern Tour—Continued." *Oologist* 11 (1894): 142–145.

———. "Scenes from the Life of Alexander Wilson: His Southern Subscription Tour." *Oologist* 10 (1893): 321–325.

———. "Scenes from the Life of Alexander Wilson: Journey to the Falls of Niagara." *Oologist* 10 (1893): 221–223.

———. "Scenes from the Life of Alexander Wilson: The Planning of His *American Ornithology*." *Oologist* 10 (1893): 206–208.

———. "Scenes from the Life of Alexander Wilson: The Progress of His Undertaking." *Oologist* 10 (1893): 245–247.

———. "Scenes from the Life of Alexander Wilson: Wilson and Bartram the Naturalist." *Oologist* 10 (1893): 175–176.

———. "Scenes from the Life of Alexander Wilson: Wilson in Scotland." *Oologist* 10 (1893): 110–112.

Stevens, O. A. "The First Descriptions of North American Birds." *Wilson Bulletin* 48 (1936): 203–215.

Stone, W. "Some Unpublished Letters of Alexander Wilson and John Abbot." *Auk* 23 (1906): 362–368.

———. "Wilsoniana." *Cassinia* 17 (1913): 1–5.

Stringham, E. *Alexander Wilson: A Founder of Scientific Ornithology*. Kerrville, Tex.: Stringham, 1958.

Stringham, E., and F. N. Egerton. "The American Ornithologist in Carolina." *Chat* 24 (1960): 63–65.

Welker, R. H. *Birds and Men: American Birds in Science, Art, Literature, and Conservation, 1800–1900*. 1955; repr. New York: Atheneum, 1966.

Wilson, A. *American Ornithology*. 9 vols. Philadelphia: Bradford & Inskeep, 1808–1814.

———. "Particulars of the Death of Capt. Lewis." *Port Folio* 7 (1811): 24–47.

———. *The Poems and Literary Prose*. Ed. A. B. Grosart. 2 vols. Paisley, Scotland: Alexander Gardner, 1876.

Acknowledgments

In developing and writing this book, we have received encouragement and support from many sources. Our interest in Alexander Wilson began in the summer of 1967 when EHB read Cantwell's biography, a present from his college roommate, Edward O. Minot, now an ornithologist in New Zealand. John W. Collins III provided further encouragement when he asked us to determine if a small painting and doodles in student exercise books given to Harvard's Guttman Library were indeed the work of Alexander Wilson as claimed by the donor. Neither the painting nor the doodles in the student work books were signed or dated. Intrigued, our study of Wilson's art had begun. When WED later conceived the idea of a book on the history of American ornithology and asked EHB to write a chapter on Wilson's contributions, we realized that a book devoted to Wilson's role in the founding of American ornithology could illuminate the scientific and artistic gifts that he brought to the field.

The majority of Wilson's surviving letters and draft artwork for his *American Ornithology* are archived at Harvard. Our thanks to Douglas Siegel-Causey and Scott Edwards, who graciously provided office space and access to the bird collection and library at the Museum of Comparative Zoology during the two semesters that EHB spent there studying the collections of Wilson's specimens and papers. Constance Rinaldo encouraged us and ensured that we had access to Harvard's collections and Dana Fisher was an invaluable guide to Wilson's art in the archives of the Ernst Mayr library. Thanks also to those at the Houghton Library who provided unlimited access to their collection of Wilson's correspondence.

In 2005 EHB visited Philadelphia at the suggestion of Robert M. Peck, who showed him Wilson's drawings in the archives of the Philadelphia Academy of Natural Sciences. Thanks to him for being our point person in Philadelphia and for alerting us to resources at the Academy, the American Philosophical Society, the Library Company of Philadelphia, Independence

National Historical Park, and Bartram's Garden—and to Cathy Buckwalter and Megan Gibes for scanning Wilson's drawings for us. Joel T. Fry, curator at Bartram's Garden, graciously added to our knowledge of the relationship between Bartram and Wilson and provided photographs of Bartram House and of the study used by Wilson on his visits to the Bartram family. We are indebted as well to Andrea Ashby, Karie Diethorn, and Doris Devine Fanelli of the Independence National Historical Park for their help in obtaining a copy of the portrait of William Bartram for us. Charles Greifstein made available the magnificent portrait of Alexander Wilson by Rembrandt Peale that is housed at the American Philosophical Society Library. And Mary K. LeCroy at the American Museum of Natural History arranged for us to have a copy of Wilson's drawing of a Great Egret with his extensive notes. We are grateful for these wonderful additions to our illustrations.

Professor Pat Monaghan at the University of Glasgow welcomed EHB to Scotland and provided advice on where information on Wilson might be found. Bernie Zonfrillo, another Scot with knowledge of Wilson, shared that knowledge freely and even sent copies of some old texts that helped a twentieth-century American understand eighteenth-century Scotland. Nicola MacIntyre, archivist at the Paisley Public Library and Museum, was an invaluable resource person. She found material on Wilson deep in the archives that neither we nor earlier students of Wilson's life had found, and generously arranged for these to be photographed for the book.

We were fortunate indeed to have learned of Thomas Blanton, a private collector of historical material on natural history, which, of course, included material about Wilson. Most importantly he had a collection of proofs of plates gathered by the daughter of Robert Carr, Wilson's printer. We have reproduced several of these to illustrate the process of making colored plates in the early nineteenth century and to show the care that Wilson and the colorists took to accurately portray the colors and patterns of American birds.

Fate was kind in leading us to William B. Fetters and Walt Webb, descendents of Mary B. Leech who studied under Wilson at the Gray's Ferry school and who later became one of Wilson's colorists. Pictures in their family files of Mary Leech and the schoolhouse were important additions we could never have anticipated, and they graciously allowed us to use them in this book.

We discussed our interest in and ideas concerning Wilson with many colleagues, friends, and students over the years, but Alan Peterson, Dennis Williams, Sean Williams, and Christopher Wilson deserve special mention for their insightful contributions to our thinking. We also thank those individuals who read all or part of the manuscript in its early versions: John Kricher, Jack Stenger, and Peter Stettenheim. Many thanks also to John Kricher and an anonymous referee who reviewed the manuscript for Harvard University Press. Their comments stimulated our thoughts and guided our final revision. We also thank Michael Fisher, Tim Jones, Lauren Esdaile, and many others at Harvard University Press whose skills have helped to produce a wonderful tribute to Alexander Wilson.

Pam "Thelma" Burtt and Betsy Davis have lived with us and Alexander Wilson for the last ten years, and we are grateful for their patience and good humor throughout. They have laughed at the anecdotes, even on the tenth telling; eased our frustrations; and expressed joy at our insights, progress, and pleasure. Without their support we could not have finished this book.

Index

Page numbers in italics refer to figures.

Abbot, John, 53, 262, 276, 277, 286, 319, 336, 395–396
Academy of Natural Sciences (Philadelphia), 322, 338, 346, 347, 375, 394, 409, 410
Albin, Eleazar, 355
Aldrovandus, Ulisse, 362
Allan, John, epitaph, 17
American Ornithologists' Union, 305, 306, 347, 350
American Ornithology (Wilson), ix, 1, 42–52, 59, 60, 63, 279; Jefferson and, 48, 50–51, 403, 408; ninth volume (Ord), 57, 273, 331, 338, 408, 409, 416n82; prospectus for, 42–44, 47–48; quality of, 4, 8, 10, 11–12, 42–43, 406, 411; reactions to, 1, 10, 45, 349, 350, 375; revision of (Bonaparte), 203, 314, 338–339, 347, 348, 404, 409; as science, 12, 44, 50, 52, 60, 285, 299, 324, 325, 352, 375. *See also* color; colorists; engravers; engraving and plates; printing *American Ornithology*
American Ornithology, finances of, 42, 49, 52, 57, 59, 308–309, 333; subscriptions, 3–4, 26, 42, 44, 47–51, 52, 54, 55, 56, 57, 58, 75, 82, 289, 309, 333, 334–335, 336–337, 394, 398, 403, 408, 416n82
American Philosophical Society (Philadelphia), 38, 60, 282, 285, 338, 346, 355, 375, 394, 409, 411
Anhinga, *260,* 261–262, *263,* 265
Audubon, John James, ix, 57, 128, 279, 285, 308, 314, 331, 333–346, 347, 348, 349, 350, 352, 409–410; inspired by and plagiarizing Wilson, 337–344, 346; meeting Wilson, 55, 333–336, 342–343; non-subscriber to *American Ornithology,* 334–335; and Ord, 338, 339, 344, 346, 347, 409, 410
Auklet, Cassin's, 349
Bachman, John, 285
Baird, Spencer, 314, 349–350
Bakewell, Benjamin, 334
Bakewell, Thomas, 343
Barrère, Pierre, 362
Barrington, Daines, 295
Bartram, Anne (Nancy), 41, 75–76, 80, 276, 284
Bartram, John (Sr.), 29, 30, 37, 281, 356, 371, 374, 381
Bartram, John (Jr.), 30
Bartram, William, 30, 35, 284, 285, 355, 374, 394, 397, 405; bird description, as model for Wilson, 40–41; as illustrator, 80, 85, 87, 112, 130, 220, 276, 277, 319, 351; list of bird names, 10, 31, 83–84, 257, 289, 290, 305, 332, 374, 393, 402; as mentor to Wilson, 7, 31–35, 36–38, 39, 44–47, 50, 52–54, 56, 57, 59–60, 65, 194, 276, 282–283, 285, 286, 301–302, 304, 326, 399–400; as naturalist, 6, 7, 30–31, 35, 47, 50, 195, 283, 287, 310, 352, 371
Bechstein, Johann, 194, 306–307
Belknap, Jeremy, 362–363
Bell, John, 342, 349
Belon, Pierre, 363–364
Bent, Arthur Cleveland, 317–318
Bewick, Thomas, 243, 286–287, 289, 290, 385, 408
Bittern: American, 213, 222, *223,* 224–225; Least, 213, 222, *223,* 225, *225*
Blackbird: Brewer's, 348; Red-winged, 126, *127,* 128, 129, 340; Rusty, 105, *106,* 107
Blue Bird, Eastern, 72, *73*
Bobolink, 316, 371
Boddaert, Pieter, 83, 103, 364, 416n79
Bonaparte, Charles Lucien, 203, 257, 314, 338–339, 341, 343, 346–348, 350, 404, 409
Bradford, Samuel, 42, 43, 44, 47, 48, 57, 58, 59, 81, 308, 405

Bradford & Inskeep, 42, 43, 52, 408
Brant, 246, 246, 247–248, 250, *251*, 252, 278–279
Brewer, Thomas, 348, 350
Brisbane and Brannan, 48
Brisson, Mathurin Jacques, 297, 356, 364, 376, 385–386
Brodie, David, 17–18, 20
Browning, M. Ralph, 306
Brünnich, Morten Thrane, 386–387
Bufflehead. *See* Duck, Bufflehead
Buffon, Georges-Louis Leclerc Comte de, 3, 4, 286, 295, 296, 299, 314, 364–365, 376, 386, 388, 390, 391, 394, 419n78, 423n19; *Histoire Naturelle*, 39, 205, 208, 297, 315, 356, 364, 399, 400, 401, 402, 403. *See also* "degenerate" American species
Bunting: Indigo, 326; Painted, 110, *111*, 112–113
Burns, Robert, 2, 5, 8, 18, 21
Buzzards, 158

Canvasback. *See* Duck, Canvasback
Cardinal, 366; Northern, 362
Carr, Major Robert, 54, 57, 290, 406, 434
Cassin, John, 349
Catbird, 327; Gray, *74*, 75
Catesby, Mark, 6, 11, 41, 50, 83, 121, 278, 286, 287, 289, 315, 316, 327, 332, 340, 355, 356, 360, 365, 367, 374, 381, 395; as illustrator, 63, 87, 276, 277, 332, 352, 365, 367; Jefferson as follower of, 10, 39, 287, 399; list of bird names, 10, 39, 287, 288, 305, 314, 332, 360, 372; *Natural History of Carolina, Florida, and the Bahama Islands*, 276, 332, 356, 359, 365, 367, 381, 395
Charlevoix, Pierre François Xavier, 365–366
Chat, Yellow-breasted, 300, 326
Checklist of North American Birds, 305–306, 347, 350
Chickadee: Black-capped, 76, *77*; Boreal, 368
Chuck-will's-widow, *176*, 177–178
Churchman, Mordecai, 397
citizen science, 4, 319–320, 349, 407
Clark, William, 103, 130, 286, 404, 416n79
Clavigero, Francisco Saverio, 366
Collinson, Peter, 367, 374, 381
color, in drawings. *See* wash; watercolor
color, in specimens, 318, 319
coloring instructions, from Wilson, 75–77, 82, 206, 233, 318
colorists, 38, 52, 56, 59, 74, 75, 76, 78, 80, 81, 82, 125, 132, 189, 220, 399; Wilson as, 59, 82, 309
Cook, Captain, voyages, 286, 368, 379
Cooper, William, 347
Coot, American, *253*, 254, *255*, 257. *See also* Sora
copper plates. *See* engravings and plates
copying, 33, 35, 65, 355, 372, 395; Audubon, of Wilson, 339–340, 341–342, 410; plagiarism, 340, 364, 377
Coues, Elliot, 278, 279, 350
Cowbird, Brown-headed, 318–319, 352
Crane: Sandhill, 55, 334; Whooping, 218, *219*, 220
Creeper: Brown, 40; Kauai, 349
Crichton, Thomas, 19
Crows, 91, 278; American, 172; Common, 139, 270; Fish, 136, 137, *138*, 139, 172, 312
Cuckoos, 298–299; Black-billed, *119*, 120, 121, 312; Yellow-billed, 118, *119*, 120–121, 348
Curlew, 376; Long-billed, 218, *219*, 221–222
Cuvier, Georges, 310, 333, 356, 394
Cyclopaedia (Rhees), 41, 42, 44, 47, 48, 57

Darwin, Charles, 290, 313, 373
Davis, Jefferson, 349
"degenerate" American species, 3, 4, 50, 295–297, 310, 364–365
Dickcissel, *73*
Dove: Common Ground-, 159–160, *159*, 396; Mourning, 362
Dovekie, 260, *261*, 261, 262–263, *264*, 265
Dowitcher: Long-billed, 419n101; Short-billed, 189, 190, *190*, 191, *191*, 192, 27, 419n101
Ducks, 135, 172, 271, 310; American Black, 72, *246*, 247, 250, *253*, 254; American Wigeon, 235, *236*, 239; Bufflehead, 229, *230*, 232; Canvasback, ix, 67, *69*, *70*, 239, *241*, 242, 244, *245*, 245, 312; Common Goldeneye, 229, *230*, 230–231; Greater Scaup, 235, *236*, 236, 239; Green-winged Teal, 241, *241*, 242,

243–244; Harlequin, *246*, 247, 249, *252*, 252, 254, 362; Labrador, 235, *236*, 237, *240*, 240, 250; Long-tailed, 67, *71*, 240, *241*, 242; Mallard, 241, *241*, 243; Northern Shoveler, 229, *230*, 231; Redhead, *70*, 241, *241*, 242, 244; Ring-necked, 229, *230*, 231, 232, *233*, 233; Wood, 231, 240, *241*, 243. *See also* Mergansers; Scoters
Dunbar, William, 56, 215, 322, 323, 336, 398
Duncan, Isabel, 27
Duncan, Meg, 16
Duncan, William, 15, 18, 24, 26, 27, 39, 40, 57, 82, 398–399, 401, 408
Dunlin, 178, *179*, 181–182, *182*
Dunlop, William, 31
Du Pratz, Le Page, 366

Eagles, 158, 351, 399; Bald, 134–136, *134*, *136*, 287
economic ornithology, 48, 327–330
Edwards, George, 41, 63, 112, 220, 256, 276, 277, 278, 286, 319, 357, 366–367, 378–379, 381, 395
Egret: Great, 208–212, *209*, *211*, 212; Snowy, 213
Eider, Steller's, 382
engravers, 52, 65, 67, 72, 287, 339, 340, 367; Wilson as, 27, 64, 65, 67; Lawson as, 48, 56, 65, 67, 279, 284, 339, 347, 404
engravings and plates, 9, 41, 63, 80, 165, 229, 278, 364; Wilson's drawings for, 65–66, 67, 82, 91, 94, 97, 120, 160, 165, 172, 198, 214, 220, 229, 268, 272, 273, 278, 279; techniques, 63, 65, 67, 69, 70, 72, 75, 114, 258 (*see also* iron oxide)
evolution and alternatives, 5, 290–291, 295, 299, 313, 317, 373

Fairman, Gideon, 339
Falcon, Peregrine, 268, *269*, 271
field guides, modern, 53, 78, 123, 237, 242, 250, 348, 351, 352
Finch, Cassin's, 349
fixity of species. *See* evolution and alternatives
Flickers, 36–37, 299; Northern, *73*, *86*, 87, 297–298, 316–317; Yellow-shafted, 339–340
Flycatchers, 40, 118, 126, 335; Acadian, 97, *98*, 99, 100; "Small-headed," 308, 340–341; White-eyed, 327
Forster, Georg, 286, 319, 324, 368
Forster, Johann Reinhold, 357, 368, 376
Franklin, Benjamin, 285, 355
Frigatebirds, 366
Frisch, Johann Leonard, 368–369

Gallinule, Purple, 254, *255*, 256, 258, *259*
Gannet, Northern, 15
Gardiner, John L., 399
Gilmour, Robert, 21
Gmelin, Johann Friedrich, 83, 290, 297, 300, 308, 310, 313, 324, 369, 376, 382, 388. *See also Systema Naturae* (13th ed.)
Goatsucker, 148
Godwit, Marbled, 178, *179*, 180
Goldeneye, Common. *See* Duck, Common Goldeneye
Goldfinch: American, *64*, 65, *66*; Lawrence's, 350
Goose, 135, 271; Canada, 229, *230*, 232–233; Snow, 235, *236*, 237–238
Grackles, 371; Common, 105, *106*, 107; Great-tailed, 339, 347; Purple, 326
Gray, G. R., 103, 307, 311, 416n79
Grosbeaks, 112; Blue, 110, *111*, 112, 113; Pine, *76*, *90*, 91, 92
Grosart, Rev. Alexander, 21, 25
Grouse, Ruffed, 163–164, *164*, 341–342, *342*, *343*
Guillemot, 386–387
Gulls, 135; Bonaparte's, 346; Herring, 258, 273, *274*; Laughing, 260, *261*, 261, 262, *265*, 266, 273
Guthrie, William, 409–410

Harrier, Northern, 165, *166*
Havell, Richard, 340
Hawks, 158, 278, 406; Broad-winged, 307, 312, 341; Cooper's, 347; Long-winged, 172; Red-shouldered, 170, *171*, 172; Red-tailed, ix, 167, *168*, *169*, 170, 175, 277, 287; Rough-legged, 170, *171*, 172, *174*, *175*, 175, 278; Sharp-shinned, 159, 289, 307, 312
Hearne, Samuel, 369–370, 377, 379

Henry, William, 23
Hernández, Francisco, 366, 370–371
Heron: Great Blue, 213, 222, *223*, 224; Green, 206–208, *207*, 212; Little Blue, 212, 220; Tricolored, 213, 218, *219*, 220
Hill, Peter, 21
Humboldt, Alexander von, 285
Hummingbird, 366

Ibis: Scarlet, 140, *141*, 142–143, 227; White, *226*, 227, *228*, 228
ink, and engraving plates, 70, 75, 78, 80
ink drawings, ix, 68, 89, 130, 134, 152, 206, 215, 236, 267, 273, 286
iron oxide, 67, 69, 72, 114, 148, 149, 250–252, 258, 259

Jackdaw, 105
Jardine, William, 25, 260, 341
Jay, 39, 91; Blue, *64*, 65, *66*, 80, 102, 130, 318; Steller's, 382
Jefferson, Thomas, 1, 3, 41, 44, 60, 308, 310, 333, 398, 399–403, 404; admired by Wilson, 1, 3, 4, 27, 28, 282, 399, 422n6; and *American Ornithology*, 48, 50–51, 403, 408; and Catesby's bird names, 10, 39, 287, 288, 332, 399; correspondence with Wilson, 28, 39–40, 41, 48, 302 399–401, 402–403; and Linnaean taxonomy, 10, 39, 289, 332–333, 399; meeting Wilson, 50, 403; and nature, 3, 7, 295, 310, 364–365; *Notes on the State of Virginia*, 39, 287, 288, 310, 396, 399
Jones, Lynds, 351
Jones, William, 61, 62, 331, 406, 408

Kalm, Peter, 371–372
Kennedy, James, 19, 20
Kestrel, American, 53–54, 72, 97, *98*
Killdeer, 196, *197*, 198, 199
Kingbird, Cassin's, 349
Kingfisher: Belted, *108*, 109; Common, 109
Kinglet, Ruby-crowned, 303–304
Kite: Mississippi, 312, 322–324, 340; Swallow-tailed, 165, *166*, 167
Klein, Jacob Theodore, 372
Knot, Red. *See* Red Knot
Kuhl, Heinrich, 203, 347

Lapwing, 199
Larks, 172; Horned, 75, *76*, *90*, 91, 92
Latham, John, 121, 277, 296, 300, 308, 310, 313, 324, 387–388; bird classifications, 45–46, 301, 305, 307, 314–315, 369, 388; classification system, 10, 290, 291–294, 300, 302, 313, 369, 374, 387–388
Latham, Samuel, 406
Lawson, Alexander, 48, 49, 55, 56, 57, 64, 65, 67, 69, 72, 165, 214, 220, 279, 282, 284, 285, 327, 335, 336, 339, 340, 347, 403–404
Lawson, John, 327, 332, 365, 372
Leech, Isaac, 29, 38, 39, 80
Leech, Mary, 29, 80, 81
Leech family, 29, 38, 39, 80
Leslie, Charles Robert, 74, 81–82
Leslie, Eliza, 82
Lewis, Meriwether, 103, 130, 286, 307, 404–405, 406, 416n79; death of, 55–56, 307, 404–405
Lewis and Clark expedition, 103, 130, 286, 311, 416n79
Library Company of Philadelphia, 38, 285, 355, 394
Linnaean Society of London, 373, 395
Linnaean taxonomy. *See Systema Naturae*
Linnaeus, Carolus, 30, 83, 314, 317, 332, 360nb, 371, 373–375, 378; and fixity of species, 290–291, 295, 299, 317
Loon, Common, 261, *261*, 265, *266*, *267*, 386

MacGillivray, William, 339
Madison, James, 310, 408
Magpies, 91; Black-billed, 130, *131*, 132, 277
Manakins, 300
Marcgraf, Georg, 375–376, 377
Marten, Humphrey, 376–377
Martin, Purple, 143, *144*, 145, 327
McDowall, William, 19
McKenzie, Alexander, 377
McMurtrie, Henry, 339
Meadowlark, Eastern, 100, *101*, 102
Mease, James, 338
Merganser: Common, *234*, 234–235; Hooded,

235, *236,* 238; Red-breasted, 235, *236,* 236, 238
Michaux, François André, 405
misidentifications, 155, 295, 296–297, 299, 308, 314–317, 324; due to incomplete information, 50, 99, 260, 340–341, 346; due to migration, 53, 316, 324, 346–347; due to sex, age, and seasonal variation, 53–54, 99, 152, 175, 183, 186, 210, 262, 287, 288, 289, 297, 314, 316, 329
Mitchell, James, 23, 24, 25
Mitchell, Samuel L., 406
Mitchell, William, 23, 25
Mockingbird, Northern, 295
Molina, Juan Ignacio, 377
Monroe, Burt, 306
Monroe, James, 408
Müller, Philipp Ludwig Statius, 377–378
Murre: Common, 386; Thick-billed, 386
Museum of Comparative Zoology, Harvard University, 218, 322, 341

Natural History of Carolina, Florida, and the Bahama Islands (Catesby), 276, 332, 356, 359, 365, 367, 381, 395
Neilson, John, 19, 20, 21, 24
Nighthawk, Common, 146, 147
Night-Heron: Black-crowned, 212; Yellow-crowned, 213, 222, *223,* 224
Nightingale, European, 295–296
Nutcracker, Clark's, 103, *104,* 105, 311, 404
Nuthatch, 102, 296–297, 301; Common, 189; Red-breasted, 297, 327–328; White-breasted, 296–297, 327–328
Nuttall, Thomas, 348

Ord, George, 57, 59, 60, 69, 177, 245, 258, 260, 285, 338, 340, 341, 344–346, 347, 406–410; and Audubon, 338, 339, 344, 346, 347, 409, 410; and Wilson's posthumous reputation, 57, 339, 344, 346, 408–409; and last volumes of *American Ornithology,* 57, 258, 260, 273, 285, 287, 307, 331, 338, 409; writings on birds, 256–258, 262, 265, 268, 270–271, 287, 409–410
Ordway, John, 404
Orioles, 314, 315, 316; Audubon's, 333; Baltimore, *64,* 65, *66,* 170, *171,* 173, 314, 315, 318, 326, 352; Orchard, 67, *68, 88,* 89, 314, 315, 318, 325, 326, 327–328, 351
Orr, Charles, 28, 281
Orr, John, 17
Osprey, 136, 137, *138,* 399
Ovenbird, *74,* 75
Owls, 92, 102, 278; Barn, 165; Great Gray, 369; Great Horned, 167; Long-eared, 165, *166,* 167; Northern Saw-whet, 130, *131,* 132, 278. *See also* Screech-owl
Oystercatcher, American, 218, *219,* 220, 221, 237

Paine, Thomas, 1, 24
Pallas, Peter Simon, 300, 324, 357, 388 390
Parakeets, 54–55; Carolina, 114–115, *115, 116,* 117; Poll (pet), 54–55, 56, 115
Parula, Northern, *119,* 120, 121, 317
Peale, Anna, 82
Peale, Charles Wilson, 4, 95, 285, 309–310, 319, 321, 322, 335, 357, 410–411
Peale, Rembrandt, 339, 434
Peale, Rubens, 411
Peale, Titian, 338, 347
Peale family, 337
Peale Museum (Philadelphia), 4, 215–216, 310, 311–312, 319, 322, 335, 411
pen and ink. *See* ink drawings
pencil sketches, x, 63, 67, 68, 69, 72, 87, 89, 93, 95, 123, 130, 134, 152, 160, 161, 169, 189, 206, 215, 226, 233, 236, 250, 252, 253, 254, 258, 263, 265–267, 273, 275, 343
Pennant, Thomas, 45, 277, 290, 296, 300, 301, 307, 310, 314, 315, 324, 357, 368, 369, 378–379, 387
Peterson, Roger Tory, 250, 351
Petrel: Leach's, 346; Wilson's Storm-, ix, 202, *202,* 203, 204–205, 346–347
Phalaropes, 278; Red, 254, *255,* 255, 258; Wilson's, ix, 254, *255,* 255, 256, 258, *260,* 260
Pheasants, 363
Philadelphia Athenaeum, 338
Phoebe: Eastern, 326; Say's, 347
Pigeons, 334; Martha (last Passenger), 157;

Pigeons (*continued*)
Passenger, xi, 55, 156, *156*, 157, 158, 277, 325–326
Plover, 192–193; American Golden-, 185, 196, *197*, 198; Black-bellied, 182, *184*, 185, 196, *197;* Piping, 140; Semipalmated, 137, *138*, 139–140; Small Ringed, 172; Wilson's, ix, 254, *255*, 257, 346
Poorwill, Common, 348
Porphyrion, 256
printers, 52, 54; Carr, 54, 290, 434; Neilson, 19, 24
printing, of *American Ornithology*, 1, 48, 54, 57–58, 63, 65, 67, 71, 74, 75; materials, 4, 42. *See also* engraving, techniques

Rail, Clapper, 214–215, *214*, 279. *See also* Sora
Raven, 135; Common, 268, *269*, 270–271, 272, 273
Ray, John, 365, 374, 376, 380–381, 383–384, 387
"Red-bird," 92
Red Knot, 182, 183–184, *184*, 186, 187, *187*, 199
Redpoll, Common, 128, *129*
Redstart, American, 313
Redwings, 371
Rhees, Abraham, *Cyclopaedia*, 41, 42, 44, 47, 48, 57
Rider, Alexander, 80
Rively, William, 29
Robins, 327
Rozier, Ferdinand, 335, 337, 342–343

Sanderling, 183, *184*, 186, 199, *200*, 201, 287
Sandpipers, 172; Baird's, 349; Curlew, 181; Least, 137, *138*, 139; Solitary, 312; Spotted, 193–196, *194;* Upland, 193–195, *194*, 332
Say, Thomas, 290, 338, 347, 375, 391
Scaup, Greater. *See* Duck, Greater Scaup
Scoter: Black, *246*, 247, 248–249; Surf, 229, *230*, 231; White-winged, *246*, 247, 249
Screech-Owl, Eastern, 67, *71*, 100, *101*, 102, 151, 152, *153*, 289
Sharp, William, 24, 25
Shearwater, Audubon's, 333
Shoveler, Northern. *See* Duck, Northern Shoveler
Shrike, Northern, *90*, 91–92, *93*, 94, 277, 362
Sibley, David Allen, 351
Siskin, Pine, 311
Skimmer, Black, 199, *200*, 201, 202, *202*, 204
Sloane, Sir Hans, 286, 319, 357, 367, 374, 381
Snipe, 139; Wilson's, ix
Sora, 160, *161*, *162*, 163
Sparrow: American Tree, 311; Baird's, 349; Brewer's, 348; Cassin's, 349–350; Chipping, 327; Field, *34*, 305, 311, 352; Fox, *108*, 109, 110, 351; Grasshopper, *111*, 112, 114, 312; Savannah, *108*, 109, 110; Seaside, 132–133, *133*, 351; Song, 311, 327; Swamp, 327; White-crowned, 368
specimens, 76, 78, 175, 217, 295, 319, 348, 349, 364, 376–377, 381, 385, 386, 390; as basis of study, 4, 10, 44, 54, 91, 255, 258, 276, 286, 290–291, 303, 317, 329, 340, 363, 367, 368, 369, 373, 385; given to Wilson, 56, 103, 177, 215, 263, 271, 286, 307, 322, 336, 396, 404; method of preparation, 10, 319, 320–321, 336, 368–369; Wilson's, 40, 51, 52, 59, 289, 310, 335, 407, 411. *See also* type specimens
Spoonbill: Roseate, 56, 215, *216*, *217*, 218, 322, 336; White, 218
Steller, Georg Wilhelm, 382, 389, 390
Stephens, 308
Stilt, Black-necked, 191, *191*, 192–193
Storm-Petrel, Wilson's. *See* Petrel, Wilson's Storm-
subscriptions, as means of financing publication, 26, 333; for Wilson's poetry book, 19, 20; for *Cyclopaedia*, 47, 48. *See also American Ornithology*, finances of, subscriptions
Sully, Thomas, *frontispiece*, 338, 339
Swallows, 363, 366, 376; Bank, 140, *141*, 143; Barn, 140, *141*, 142, 327; Tree, 142, 307, 376–377
Swan, Tundra, 258, *275*, 275
Swift, Chimney, 143, *144*, 145, 327, 397
Systema Naturae (Linnaeus), 10, 83, 286, 290, 313–314, 317, 340, 357, 361, 365, 373, 374, 376, 388; bird classifications, 39, 121, 203, 296, 297–298, 300, 302, 305
Systema Naturae (Linnaeus), followers and non-followers of, 83, 290, 309–310,

332–333, 364, 367, 369, 374, 375, 377–378, 379, 386, 387, 388, 390, 392, 393–394; Audubon, 343–344; Wilson, 11, 84, 289, 290, 300, 302, 307–308, 313, 329, 333, 375; Bartram, 31, 40, 83, 287, 289, 332, 374; Jefferson, 10, 39, 289, 332–333, 399
Systema Naturae (13th ed., Gmelin), 290, 313, 359, 361, 369, 379, 387–388

Tanager: Summer, 95, *96*, 97; Western, 103, 311, 404
Tannahill, Robert, 5
Teal, Green-winged. *See* Duck, Green-winged Teal
Tern: Black, 205–206, *206;* Common, 201, *202*, 203–204, 368; Forster's, 368; Gull-billed, *246*, 247, 248; Least, 201, *202*, 203, 362–363; Sooty, 246, *246*, 248
Thrushes, 310; Hermit, 151, *154*, 155; Tawny (*see* Veery); Wood, 40, *84*, 85, 302, 307–308
Titmouse, 102, 317; Tufted, 76, 77, 362
Towhee, Eastern, 170, *171*, 173
Turkeys, 363; Wild, 362
Turnstone, Ruddy, 183, *184*, 185, *188*, 189, *190*
Turton, William, 53, 260, 286, 290 391
Tyler, John, 51
type specimens, 295, 310, 311–312, 322, 332, 340

Veery, 151, *153*, 155, 307–308
Vieillot, Louis Jean Pierre, 94–95, 99, 260, 300, 305–307, 311, 312, 314, 324, 356, 357, 391–394
Vireo, 327; Bell's, 349; Cassin's, 350; Yellow-throated, 307
Vultures, 135, *272*, 273; Black, 268, *269*, 270, *272*, 273; Turkey, 268, *269*

Warblers, 112, 126, 289, 310, 313, 317; Bachman's, 285; Bay-breasted, 156, *156*, 157, 289, 311; Black-and-white, 100, *101*, 102; Blackburnian, 289; Blackpoll, 125–126, *125*, *129*, 287, 310, 368; Black-throated Blue, 151, 152, 289, *154*, 155, 289; Black-throated Green, 156, *156*, 157, 340–341; Canada, *116*, 117, 118; Cape May, *176*, 177, 341, 346; Cerulean, 289, 311; Connecticut, 143, *144*, 146; Grace's, 350; Hooded, *116*, 117; Kentucky, 55, 312; Magnolia, 312; Mourning, *74*, 75; Nashville, 55, 312; Parula (*see* Parula, Northern); Pine, 305, 311; Prothonotary, 110, *111*, 112, 113; Tennessee, 55, 312, 318; Wilson's, *116*, 117, 118, 312; Worm-eating, 110, *111*, 112, 113; Yellow, 118, *119*, 120, 326; Yellow-rumped, 32, 33
wash, 74, 75, 80, 93, 102, 123, 146, 220
watercolor, ix, 87, 123, 146, 152, 177, 235, 337
Webster, Noah, 344, 407
West, Benjamin, 81, 410
Whimbrel, 178, *179*, 179–180
Whip-poor-will, 178, 312; Eastern, 148, *149*, 151
White, John, 331–332
Wigeon, American. *See* Duck, American Wigeon
Willet, 178, *179*, 180–181
Willughby, Francis, 239, 318, 357, 378, 380–381, 383–384, 387
Wilson, Alexander: ambition and passions of, x, 8, 9, 10, 282–284, 344, 346; citizenship, 1, 3, 5, 35, 331; death of, 61–62, 245, 331; education, 5, 13–14, 17, 282; emigration, 3, 26, 281, 284; finances of, 58, 59, 408; interest in birds, 3, 15–16, 33, 281, 283, 285, 304, 346; marksmanship and shooting, 15, 16, 33, 63, 114, 112, 318, 334, 335, 340; as musician, 13, 15, 23, 33, 283; and nature, 1, 3, 5–8, 13, 15–16, 27, 43, 59, 304, 346; as observer, x, 1, 6, 18, 20, 21, 26, 37, 47, 48, 63, 93, 135, 137, 277, 285, 304, 324, 329; politics, 2, 3, 4, 23–26, 27; possible son of, 16; and religion, 5, 13–14, 43–44; as teacher, 3, 27, 28–29, 35–36, 38, 281, 324, 399; work as weaver, 15, 16, 17–18,19, 21, 23, 24, 282
Wilson, Alexander, as artist, 33, 35–36, 351; as colorist, 57–58, 59, 82, 309; as engraver, 27, 64, 65, 67. *See also* engravings and plates, Wilson's drawings for; ink drawings; pencil sketches
Wilson, Alexander, as ornithologist, 9, 331, 333, 344; accomplishments, numerical summaries of, 287, 289, 306, 333; on ecology and behavior of birds, 7, 9, 11, 48, 304, 313, 314, 317; and evolution, 5, 313;

Wilson, Alexander (*continued*)
as "Father of American Ornithology," ix, x, 1, 62, 329, 331, 333, 338, 348; keeping live birds, 51, 54–55, 56, 115, 124, 278, 304, 329; and live birds in nature, x, 4, 10, 11, 15, 26, 31, 44, 46–47, 51, 54–55, 56, 63, 93, 132, 148, 285, 303, 317, 318–319, 324, 329, 352, 410; quantification, 324–327, 329; scientific method, 322–324, 329, 333, 347, 348; and taxonomy, 10–11, 285, 289–290, 299, 301, 313, 317, 333, 347–348

Wilson, Alexander, as poet, 2, 8, 9, 10, 15, 17–18, 19, 20–21, 27–28, 47, 48, 62, 281, 282, 304, 329, 407; and Burns, 2, 18, 21; criminal charges against, 23–26; subscribers for poetry book, 19, 20, 26; WORKS: "The Hollander," 23, 24, 25; "Jefferson and Liberty," 27–28; "Watty and Meg," 8, 21, 24

Wilson Ornithological Society, 350–351

Witherspoon, John, 13

Witherspoon, Tom, 19, 24

Wood, William, drawings in exercise book of, 29, 32, 34

Woodpeckers, 36, 37, 87, 298–299, 366, 371; Downy, 362; Ivory-billed, 51, 78, *79*, *122*, 123–124; Lewis's, 56, 103, *104*, 105, 307, 311, 404; Nuttall's, 348; Pileated, 121, *122*, 123, 351; Red-headed, *122*, 123, 124, 328. *See also* Flickers

Wren: Bewick's, 385; "Disconsolate Wren" (poem), 281; European, 296; House, *94*, 94–95, 305–306, 309, 311, 326, 393; Marsh, 305, 311; Winter, 296

Yellowlegs, Lesser, 191, *191*, 193

Yellowthroat, Common, 287, 318–319, 328, 351–352